Design Thinking

Design Thinking is a set of strategic and creative processes and principles used in the planning and creation of products and solutions to human-centered design problems.

With design and innovation being two key driving principles, this series focuses on, but is not limited to, the following areas and topics:

- User Interface (UI) and User Experience (UX) Design

- Psychology of Design

- Human-Computer Interaction (HCI)

- Ergonomic Design

- Product Development and Management

- Virtual and Mixed Reality (VR/XR)

- User-Centered Built Environments and Smart Homes

- Accessibility, Sustainability and Environmental Design

- Learning and Instructional Design

- Strategy and best practices

This series publishes books aimed at designers, developers, storytellers and problem-solvers in industry to help them understand current developments and best practices at the cutting edge of creativity, to invent new paradigms and solutions, and challenge Creatives to push boundaries to design bigger and better than before.

More information about this series at https://link.springer.com/bookseries/15933

Instructional Design Unleashed

Unlocking Professional Learning Potential with UX, Agile and AI Methods

Ankita Jiyani Mangtani

Apress®

Instructional Design Unleashed: Unlocking Professional Learning Potential with UX, Agile and AI Methods

Ankita Jiyani Mangtani
Jaipur, Rajasthan, India

ISBN-13 (pbk): 979-8-8688-0415-1   ISBN-13 (electronic): 979-8-8688-0416-8
https://doi.org/10.1007/979-8-8688-0416-8

Managing Director, Apress Media LLC: Welmoed Spahr
Acquisitions Editor: James Robinson-Prior
Development Editor: James Markham
Editorial Assistant: Gryffin Winkler

Cover designed by eStudioCalamar

Distributed to the book trade worldwide by Springer Science+Business Media New York, 1 New York Plaza, Suite 4600, New York, NY 10004-1562, USA. Phone 1-800-SPRINGER, fax (201) 348-4505, e-mail orders-ny@springer-sbm.com, or visit www.springeronline.com. Apress Media, LLC is a California LLC and the sole member (owner) is Springer Science + Business Media Finance Inc (SSBM Finance Inc). SSBM Finance Inc is a **Delaware** corporation.

For information on translations, please e-mail booktranslations@springernature.com; for reprint, paperback, or audio rights, please e-mail bookpermissions@springernature.com.

Apress titles may be purchased in bulk for academic, corporate, or promotional use. eBook versions and licenses are also available for most titles. For more information, reference our Print and eBook Bulk Sales web page at http://www.apress.com/bulk-sales.

Any source code or other supplementary material referenced by the author in this book is available to readers on GitHub. For more detailed information, please visit https://www.apress.com/gp/services/source-code.

If disposing of this product, please recycle the paper

For my husband, who always pushes me to reach for the stars and inspires a boundless growth mindset within me. Your unwavering support and belief in my potential have been the guiding light on this journey.
This book is dedicated to you, my constant source of encouragement and inspiration.

To my parents, who planted the seed of knowledge in my mind and nurtured it with love and guidance.
Your belief in my ability to forge my own path, to explore and choose my own educational and career journey, has been an invaluable gift. Thank you for empowering me to follow my passions and dreams, knowing that your unwavering support will always be there, no matter which path I choose to tread.

Table of Contents

About the Author

Ankita Jiyani Mangtani, MTech, is a seasoned Instructional Designer with a deep-seated passion for creating engaging and impactful eLearning courses. Over the years, she has amassed a wealth of experience designing and developing educational programs catering to diverse audiences, ranging from K-12 students to professionals in corporate compliance training. Her expertise spans a variety of instructional design strategies, including gamification, microlearning, and scenario-based learning, ensuring that each course is not only informative but also highly interactive and enjoyable.

Ankita holds a Diploma in Instructional Design, which has provided her with a solid foundation in both the theoretical and practical aspects of the subject. Her academic background, combined with her hands-on experience, enables her to develop pedagogically sound and technologically advanced courses.

In addition to her extensive work in traditional instructional design, Ankita is particularly enthusiastic about integrating artificial intelligence (AI) tools within eLearning environments. She continually explores AI's potential to enhance learning experiences, personalize education, and improve learner outcomes. Her forward-thinking approach and innovative mindset have positioned her as a thought leader in eLearning.

Throughout her career, Ankita has worked with various clients, including educational institutions, nonprofit organizations, and multinational corporations. Her ability to adapt her design approach to meet the unique needs of each client has earned her a reputation for excellence and reliability.

Ankita is also a sought-after speaker and consultant, frequently sharing her insights on the latest trends and best practices in instructional design at industry conferences and workshops. Her contributions to the field have been recognized through various awards and accolades, reflecting her commitment to advancing the quality and effectiveness of eLearning.

When she is not designing courses, Ankita enjoys staying updated with the latest developments in educational technology, reading about AI advancements, and engaging with the instructional design community through online forums and social media.

Ankita continues to push the boundaries of what is possible in eLearning, driven by her unwavering dedication to making education accessible, engaging, and effective for all learners.

About the Technical Reviewer

Yvonne Dawydiak, BEd, MET, was born in a small coastal community on Northern Vancouver Island, on the lands of the Quatsino First Nation, and has continued to live, learn, and work most of her life near the Pacific Ocean on the traditional unceded lands of Coast Salish peoples. Yvonne has a Masters in Education Technology and a long-time K-12 and postsecondary educator passionate about designing effective and inclusive learning experiences, both "in place" and "virtually," that help connect students with their teachers, with one another, and with content and concepts.

As Learning Design Manager in Teacher Education at UBC, Yvonne has the privilege of collaborating with faculty and teacher candidates to design learning experiences that align with 21st-century approaches including digital technology integration for face-to-face, blended, and hybrid contexts.

Acknowledgments

As I reflect on the journey of writing this book, I am overwhelmed with gratitude toward those who have been instrumental in its creation.

First and foremost, I owe a debt of gratitude to my parents, whose unwavering support and encouragement ignited my passion for knowledge and paved the way for my continuous pursuit of learning.

To my beloved husband, your steadfast belief in me and your unwavering support have been the bedrock upon which this endeavor was built. Without your love and encouragement, this book would have remained but a dream.

I am also deeply thankful to all my colleagues, whose diverse perspectives and experiences, from my days as an assistant professor to my time as an instructional designer, have enriched my understanding of the professional world and the field of instructional design.

Special appreciation goes to Ms. Purnima Valiathan, whose work and journey inspire me greatly.

Last, but certainly not least, I extend my heartfelt thanks to the dedicated team at Apress. Your unwavering support, guidance, and encouragement throughout the writing process were instrumental in bringing this project to fruition. Without your collaboration and belief in my work, I would not have been able to share my knowledge and experiences with the instructional design community.

To each and every one of you who have contributed to this journey, thank you for being a part of this remarkable chapter in my life.

Introduction

As I embarked on my journey to explore the multifaceted world of instructional design, I found myself navigating through a vast landscape of theories, methodologies, and best practices scattered across various platforms and resources. From textbooks to online forums, from workshops to webinars, I roamed far and wide in search of a comprehensive guide that would bring together all the essential elements of instructional design in one place. Yet, much to my surprise, such a resource seemed elusive, leaving me to piece together fragments of knowledge from disparate sources.

It was this journey, this quest for knowledge, that ignited within me a sense of curiosity and a desire to delve deeper into the science and art of instructional design. As I delved further into the intricacies of the field, I began to uncover hidden gems of insight and understanding, each discovery fueling my passion for learning and discovery.

It was during this process of exploration and discovery that the idea for this book began to take shape. Fueled by a desire to share my experiences and insights with others who, like me, sought a comprehensive guide to instructional design, I set out to create a resource that would serve as a beacon of knowledge in an otherwise fragmented landscape.

This book is the culmination of that journey, a testament to the power of curiosity and the transformative potential of knowledge. Within these pages, you will find a comprehensive guide to the science and art of instructional design, drawing upon my own experiences, insights, and learnings gathered from years of exploration and study.

INTRODUCTION

Whether you are a seasoned instructional designer looking to deepen your understanding of the field or a newcomer eager to learn the fundamentals, this book is designed to be your companion on the journey to mastering the craft of instructional design. From foundational principles to advanced techniques, from theoretical frameworks to practical applications, each chapter is crafted to provide you with the knowledge and tools you need to succeed in the dynamic and ever-evolving field of instructional dcsign.

My hope is that this book will serve as a road map for those who, like me, once found themselves wandering aimlessly in search of guidance. May it be a source of inspiration, insight, and practical wisdom as you embark on your own journey of discovery in the fascinating world of instructional design.

PART I

Science of Instructional Design

The field of instructional design, like a well-composed symphony, skillfully combines a variety of components to create learning experiences that are impactful. This part explores the anatomy of instructional design, drawing comparisons to the work of a film director and contrasting its intricacies and subtleties with the artistic sensibilities of filmmaking. It examines the advantages as well as the drawbacks, explaining how careful planning can promote effective information transfer.

The incorporation of learning theory, a key component in avoiding cognitive overload, is at the center of this investigation. Instructional designers create learning experiences that maximize retention and comprehension by utilizing cognitive concepts. By using them wisely, they strike a careful balance between providing knowledge and not overwhelming students.

The discussion of learning modalities, which acknowledges the various cognitive paths that students take, clarifies this story even more. Instructional designers effectively create materials to accommodate auditory, visual, kinesthetic, and other learning preferences. Through embracing this variability, learning resources become more than just tools for distribution; they become inclusive, customized learning experiences.

As we stride into the 21st century – an era of new pedagogical approaches, characterized by innovative approaches that transcend the traditional approaches – instructional designers must incorporate modern approaches into instructional materials, including gamification, project-based learning, microlearning, and many more, to increase participation and create a dynamic, engaging learning environment.

Most importantly, inclusiveness is the compass that directs efforts in instructional design. Fair access to education is crucial at a time of growing diversity. Consequently, the last chapter of this part clarifies methods to guarantee that educational resources are globally accessible, meeting the needs of learners with varying skill levels. Inclusivity infuses every aspect of instructional design, from including motor skill-enhancing activities to providing aural descriptions for visually impaired learners.

This part essentially reveals the complex web of instructional design, which combines inclusion, science, and the arts to create transforming learning environments. Instructional designers shape the educational landscape, enabling learners to go on exciting journeys of knowledge acquisition and growth, much like a great director brings life to a movie masterpiece.

CHAPTER 1

Anatomy of Instructional Design

The field of instructional design is not a new phenomenon; it has deep roots that trace back to World War II. However, it gained significant traction and popularity with the advancement of technology in subsequent years.

In the crucible of World War II, a pressing demand arose to swiftly and efficiently train many military personnel. In response, a group of dedicated minds joined forces – psychologists, educators, and technologists – to forge a new path in instructional design. With a shared purpose, they set out to create a systematic framework that would unlock the secrets of effective training. With an unwavering focus, they developed practical exercises to bridge the gap between theory and practice, equipping trainees with the skills and knowledge needed for the battlefield. Thus, the seeds of instructional design were sown – a structured approach that would forever transform the education and training landscape. From the crucible of war emerged a framework that would shape the future, laying the groundwork for the design and delivery of instruction as we know it today.

Following the war, instructional design gained popularity in various domains. It incorporated theories of learning, cognition, and human behavior, influencing the design of instructional materials, assessments, and learning environments. The field matured as instructional designers applied systematic processes and models to create effective and engaging learning experiences.

© Ankita Jiyani Mangtani 2024
A. J. Mangtani, *Instructional Design Unleashed*, Design Thinking,
https://doi.org/10.1007/979-8-8688-0416-8_1

As eLearning platforms, online courses, and virtual learning environments mushroomed with the Internet's and digital technologies' advancement, they leveraged multimedia and interactive elements to deliver instruction online. Instructional designers began to embrace learner-centered approaches, personalized learning, and technology integration into their practices.

The winds of change persistently blow, carrying the whispers of technological progress and the ever-shifting landscape of education. Today, this dynamic field dances to the tune of emerging educational needs. It embraces innovative approaches such as mobile learning, augmented reality, and virtual reality to deliver an immersive learning experience to enhance learner engagement and interactivity. Data-driven decision-making becomes the compass guiding the way, unlocking valuable insights through learning analytics, which then can be used to develop adaptive learning systems that can tailor instruction to individual learner needs. Instructional designers now play a vital role in designing blended learning experiences that combine face-to-face instruction with online components.

The current state of the art in instructional design is characterized by a multidisciplinary approach, drawing from cognitive science, educational psychology, human-computer interaction, and data analytics. It emphasizes creating learner-centric experiences that promote active engagement, collaboration, and meaningful learning outcomes. Instructional designers continuously explore emerging technologies and new pedagogical strategies and leverage data to improve instruction's effectiveness and impact.

What Is Instructional Design?

Instructional design (ID) creates an **engaging learning experience** in **various forms** by providing easy-to-understand instructions to learners to navigate the **learning journey** while ensuring they effectively meet their **learning objectives**.

Engaging learning experience refers to an activity, event, or process that motivates the learners to actively participate, explore, and deeply interact with the content. It enables learners to acquire new knowledge, develop skills, or change their attitudes or behaviors.

Learning experiences can take **various forms**, including formal education in schools and universities, online courses, workshops, seminars, hands-on training, simulations, group discussions, problem-solving activities, scenario-based activities, and more.

Learning journey refers to the continuous and dynamic process of acquiring knowledge, skills, and experiences over time. It emphasizes the progression and growth of the learner's learning from a starting point to a desired objective.

Learning objectives are clear and specific statements describing what learners are expected to know, understand, or be able to do by the end of a learning experience.

What Difference Does Instructional Design Make?

Instructional design bridges the gap between content and learning by assessing learners' current state and needs and establishing appropriate instructional goals. By adjusting pedagogies (teaching methods and strategies), the instructional design aims to create efficient, effective, and engaging learning environments that cater to various learning styles and preferences. Gone are the days of one-way communication, where learners were "talked at" and expected to memorize information. The instructional design recognizes the need for interactive and participatory learning experiences.

Two Aspects of Instructional Design

Content and learners are two critical aspects of instructional design (Figure 1-1). It is essential to consider what needs to be taught (content) and how it should be taught (instructional methods).

Logically structured and accurate content should be thoroughly analyzed to determine key concepts, facts, procedures, and skills to be taught.

Characteristics, needs, and learning preferences should be analyzed to determine the learning strategies for the learners to design their learning journey. Other factors such as age, educational background, prior knowledge, cultural diversity, and individual learning styles should also be examined.

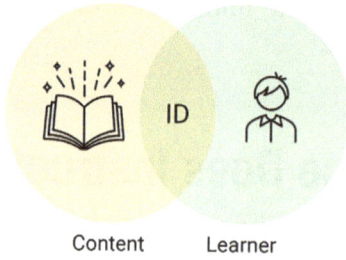

Figure 1-1. *Aspects of Instructional Design*

Three Interdependencies of Instructional Design

To ensure a successful learning experience through instructional design, it is crucial to establish a strong interdependence among three key components: Objectives, Learning Activities, and Assessments (Figure 1-2). If any of these pieces are missing, the learning journey will falter and prove ineffective. Similarly, if any of these components are not designed optimally, learners may experience discouragement, confusion, boredom, or dissatisfaction, which can have a detrimental impact on learning outcomes.

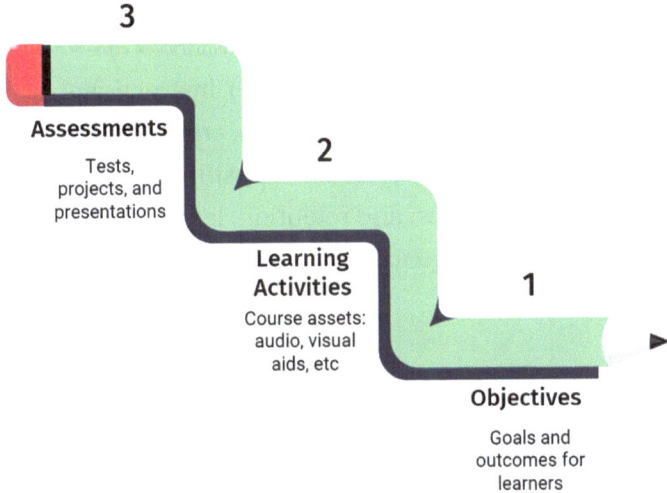

Figure 1-2. *Interdependencies of Instructional Design*

Four Shades of Instructional Design

To design a successful learning journey, you must get painted with all four shades of instructional design. The four shades are learning, creativity, technology, and business. The hue of all different shades is listed in Figure 1-3.

Figure 1-3. *Four Shades of Instructional Design*

Five I's of Instructional Design

By employing the following five I's (Figure 1-4), instructional designers can enhance learner engagement, motivation, and willingness to actively participate in the learning journey. It sets a positive foundation for effective knowledge acquisition and retention, leading to a more fruitful and impactful learning experience.

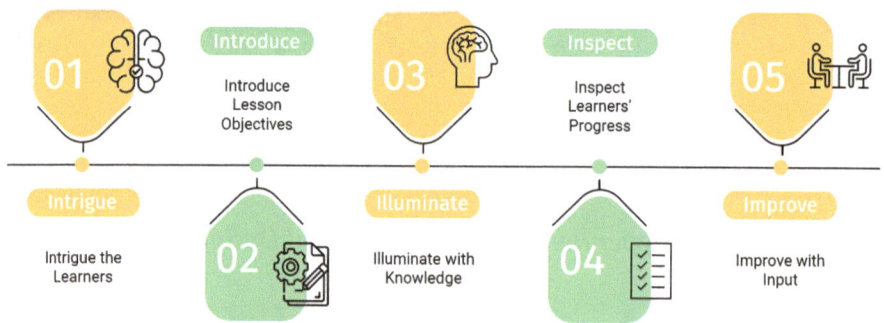

Figure 1-4. *Five I's of Instructional Design*

1. **Intrigue the Learner:** Intriguing the learners from the beginning of the learning experience is a crucial task many instructional designers need help to achieve correctly. It requires a deliberate and conscious effort to create a hook while ensuring learners are engaged and receptive. The initial theme sets the playground and expectations for the learner. Practical approaches to intrigue the learner:

   a. Present a problem that correctly captures the learner's interest and triggers their problem-solving skills.

   b. Share surprising facts or statistics to captivate the learner's attention by providing fascinating or thought-provoking information.

    c. Portraying a problem as a scenario or role play appeals the learner's active engagement and simulates their critical thinking.

    d. Presenting a compelling story that resonates with the learners emotionally can create an immediate connection and draw them into the learning process.

2. **Introduce Learning Objectives:** Informing learners about the course objectives sets a road map for learners to gauge their progress and success. Learners can continually assess their understanding and skills about the stated objectives, enabling them to monitor their learning journey. This self-assessment promotes a sense of ownership and responsibility for their learning outcomes. Additionally, clear objectives guide learners to prioritize their efforts and allocate time and resources accordingly.

   By clearly articulating the desired learning outcomes, educators empower learners to organize their thoughts and channel their learning efforts effectively. When learners clearly understand what they will be able to do or achieve upon course completion, they can align their learning goals and expectations accordingly.

3. **Illuminate with Knowledge:** In the realm of eLearning development, a principle stands as an indispensable foundation for creating compelling online learning experiences. This principle, which catalyzes instructional design, evokes a fundamental question: How can information be presented in an

interactive and accessible manner? This question is at the core of an instructional mindset that encourages creative thinking and paves the way for engaging and impactful eLearning modules. Practical approaches to present the information:

a. Storytelling

b. Branching scenarios

c. Role plays

d. Gamification

e. Interactive videos

However, it is vital to note that these techniques should always be tailored to align with the learning objectives and desired outcomes. Breaking information into manageable chunks is crucial to preventing cognitive overload and enhancing knowledge retention.

4. **Inspect Learner's Progress:** To mark the successful completion of the learning journey, learners must reflect on the information they have covered. By incorporating interactive and engaging assessment methods such as learning games, word puzzles, drag-and-drop interactions, flashcards, and scenario activities, educators can create assessments that go beyond mere evaluation and actively engage learners in the learning process. Furthermore, when aligned with the learning objectives, providing learners a second chance to attempt examinations can enhance their learning outcomes and promote a growth mindset.

5. **Improve with Input:** To foster meaningful learning experiences, well-crafted constructive feedback shapes the learner's assumptions, enhances their understanding, and drives continuous improvement. By incorporating relatable adjectives and customizing feedback to align with the course or module theme, educators can enrich learners' learning experiences and promote more profound engagement. Strategies for delivering feedback that goes beyond simple "Correct and Incorrect" labels are as follows:

   a. **Constructive Feedback:** Provide specific and detailed feedback that highlights areas of improvement and offers suggestions for further development. Focus on the process and effort put forth by the learner rather than solely on the outcome.

   b. **Descriptive Feedback:** Instead of generic labels, provide descriptive feedback explaining why a response is correct or incorrect. Offer insights into the reasoning behind the feedback, guiding learners to understand better the concepts and principles involved.

   c. **Personalized Feedback:** Tailor feedback to the individual learner, considering their strengths, weaknesses, and learning style. Acknowledge their progress and provide guidance that aligns with their unique needs.

   d. **Contextual Feedback:** Relate the feedback to real-world examples or scenarios relevant to the course or module. By connecting the feedback to practical applications, learners can better grasp the significance and impact of their learning.

e. **Timely Feedback:** Provide feedback on time, preferably soon after completing an activity or assessment. This helps learners connect the feedback with their performance and promptly make necessary adjustments or improvements.

f. **Goal-Oriented Feedback:** Align the feedback with the course or module's educational goals and desired outcomes. Clearly articulate how the feedback supports the learner's progress toward achieving these goals, reinforcing their motivation and commitment to learning.

g. **Appreciative Feedback:** Recognize and appreciate the strengths and accomplishments of the learners. Acknowledge their efforts, celebrate their achievements, and provide positive reinforcement encouraging them to continue their learning journey.

Who Is an Instructional Designer?

An instructional designer can be compared to a movie director. Like a director's filmmaking role, an instructional designer oversees the entire process of creating and delivering educational content. Table 1-1 provides a detailed overview of the role of instructional designers.

Table 1-1. *Instructional Designer*

Roles and Responsibilities	Instructional Designer	Movie Director
Vision and Concept	Envisions the desired learning outcomes and conceptualizes the instructional content and activities	Starts with a vision for the film, outlining the story, themes, and objectives
Scripting and Storyboarding	Collaborates with subject matter experts to develop a curriculum or course outline, detailing lesson's structure, content, and sequence through storyboards or instructional blueprints	Collaborates with screenwriters to develop a script that guides the movie's narrative
Design and Production	Coordinates the development of multimedia elements, instructional materials, and interactive activities to enhance the learning experience	Oversees the creative and technical aspects of filming, including set design, cinematography, sound, and special effects
Directing and Facilitating	Guides instructors or trainers on effectively presenting the content, engaging learners, and facilitating discussions	Guides actors in delivering their lines, portraying emotions, and achieving the desired performance
Editing and Polishing	Reviews and revises instructional materials, incorporates feedback, and polishes the content to ensure clarity, accuracy, and alignment with the learning objectives	Works closely with editors to refine the footage, ensure continuity, and create the final product
Audience Engagement and Impact	Focuses on engaging learners, creating interactive experiences, and designing assessments that measure learning outcomes effectively	Aims to captivate the audience, evoke emotions, and leave a lasting impression

What Are Instructional Materials?

Instructional materials are tools to facilitate active learning and assessment. They encompass various resources employed to assist learners in their learning journey. The chosen instructional materials must align with the learning objectives, ensuring that the materials effectively support the intended outcomes. A famous Chinese proverb concludes that

- Listening alone leads to forgetting

- Seeing helps in remembering

- Doing enables understanding

Instructional designers should aim to design instructional materials in a way that not just involves the sense of hearing but also the sense of sight and touch.

According to Kindler, people generally remember information in varying degrees, as given in the chart (Figure 1-5). Read the chart as "Learners tend to remember 10% of what they Read."

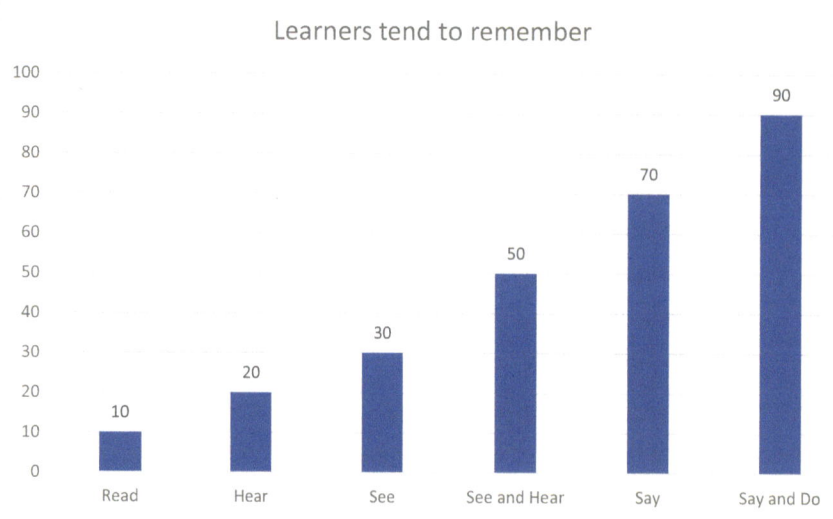

Figure 1-5. *Role of Different Senses in Learning*

Why We Need Instructional Materials

According to Rhert Heinich, instructional materials serve several purposes, which include the following:

- Capturing and maintaining the learner's attention
- Offering visual representations of processes or techniques
- Directing focus toward critical key points
- Creating a significant impact
- Enhancing the comprehension of abstract explanations
- Providing a shared experience for a large group of learners
- Stimulating a sense of reality

Types of Instructional Materials

According to Agu Okogbuo, instructional materials could be classified into four major categories, as explained in Table 1-2. Please note that the definitions provided are brief explanations to help understand the types of instructional materials.

Table 1-2. *Types of Instructional Materials*

Types of Instructional Materials	Definition	Examples	Use Cases
Printed Nonprojected Media	A traditional, concrete, and abstract form of materials	Printed materials include books, reference books, specimens, study guides, and pamphlets	• Reading comprehensions • Vocabulary development • Assessment handouts • Equipment instruction manuals • Flashcards
Visual Material	It stimulates the sense of sight	Visual materials include pictures, diagrams, maps, and charts	• Concept visualization • Data representation • Illustration of relationships • Anatomy of science education • Graphic information • Historical events and timelines • Art and design education

(*continued*)

Table 1-2. (*continued*)

Types of Instructional Materials	Definition	Examples	Use Cases
Audio Material	It stimulates the sense of hearing	Audio materials include tape recordings, cassettes, radio, and teleconferencing	• Pronunciation and accent training • Music education • Language listening skills • Audiobooks and storytelling
Audio-Visual Material	It stimulates both the sense of sight and hearing	Audio-visual materials include video recordings, motion pictures, slide films, and multimedia	• Demonstrations and simulations • Visual tours, virtual tours • Multimedia case study • Digital interactive content • Interviews and expert talk • Foreign language learning

Scope and Context of Instructional Design

Instructional design principles can be applied to any situation worldwide where individuals are expected to carry out a task. The scope of instructional design includes analyzing the current state of learners, understanding the content to be taught, and designing interventions to facilitate learning.

The content of instructional design encompasses the development of instructional materials tailored to specific industries. These materials may include training manuals, guides, handbooks, online resources, interactive modules, and simulations. They cover various topics such as policies, procedures, safety protocols, technical skills, compliance, regulations, and more. Table 1-3 provides a detailed overview of the scope of instructional design along with the context and instructional materials.

Table 1-3. *Scope and Context of Instructional Design*

Industry	Audience	Content	Instructional Material
Healthcare	• Nurses • Technicians • Staff members	• Equipment handling • Patient handling • Clinical guideline • Hygiene maintenance	• SOPs • Handbooks and manuals • Training videos • Case studies • Hands-on training • Learning apps
Education	• Learners • Teachers • Professors • Education professionals	• New technique • New subject • Updates in school/ college software	• Textbooks • Workbooks • Visual aids • Laboratory materials • Reference materials
Corporate	• Fresher employee • Experienced employee	• Onboarding • Compliance • Leadership development • Customer service • Sales techniques	• Training manuals • Online learning modules • Job aids • Presentation and slide decks • Learning apps
Manufacturing	• Employee	• Equipment operation • Safety protocols • Quality control • Production process	• SOPs • Handbooks and manuals • Training videos • Safety training guide • Quality control and inspection guide

(continued)

Table 1-3. (*continued*)

Industry	Audience	Content	Instructional Material
Technology	• Employee • Customer	• Using and troubleshooting software, hardware, and IT systems	• User manuals and guides • Knowledge base and FAQs • Weblnars and WBTs • Documentation and API references • Online forums and communities • Training programs
Retail	• Employee • Customer	• Customer service skills • Product knowledge • Point-of-sale systems • Store operations	• Training manuals and guides • POS system guides • Customer service training • Visual merchandising guidelines • Onboarding and orientation materials
Hospitality and Tourism	• Staff personnel	• Guest service • Food preparation • Housekeeping • Event management	• SOPs • Handbooks and guides • Training materials (food and beverage, front desk) • Onboarding and orientation materials

(*continued*)

Table 1-3. (*continued*)

Industry	Audience	Content	Instructional Material
Aviation	• Pilots • Cabin crew • Ground staff • Maintenance personnel	• Safety procedures • Aircraft operations • Emergency protocols	• Flight training manuals • Aviation regulations and manuals • Simulation (flight simulator) • Aircraft operation and maintenance manuals
Government	• Employee	• Law enforcement • Public administration • Emergency response • Regulatory compliance	• Onboarding and orientation materials • Job aids • Training handbook and guides • Policy and procedure manuals • Legislative materials • Brochures and pamphlets • Online modules
Nonprofit and Social Service	• Volunteer • Staff members	• Specific programs • Community engagement • Social issues	• Program handbooks • Volunteer orientation materials • Workshop and training materials • Webinars • Grant writing and fund-raising guides

Benefits and Limitations of Instructional Design

The following are some of the benefits of instructional design:

1. **Cost Savings:** Skillful instructional design can lead to notable cost savings by optimizing the return on training investment.

2. **Enhanced Learning Outcomes:** Well-designed instruction yields superior learning outcomes compared to inadequately designed training. Comprehensive analyses have demonstrated substantial positive effects of properly crafted instruction on performance.

3. **Improved Performance and Team Efficiency:** Implementing instructional design interventions has positively influenced performance and team efficiency. Studies focusing on healthcare and error management training have revealed significant positive effects, highlighting the advantages of well-designed instruction over the absence of training or poorly designed alternatives.

4. **Customized Training:** Skilled instructional designers can tailor the learning experience to individual learners' needs and preferences. Instructional designers can create more engaging and effective personalized lessons by understanding each learner's unique characteristics, learning styles, and abilities. This customization can involve adapting the pace, content, and instructional methods to cater to the diverse needs of learners.

5. **Learner-Centered Learning:** By actively engaging in the learning process, learners find acquiring and retaining knowledge effortless, thereby transforming knowledge acquisition into a practical endeavor.

6. **Result Oriented:** This approach focuses on attaining specific learning outcomes and goals. Learning experiences that are designed with clear objectives ensure learners acquire the necessary knowledge, skills, and competencies. This approach minimizes unnecessary distractions and tangents, allowing instructional designers to direct their efforts toward achieving desired results.

The following are some of the limitations of instructional design:

1. **Cost:** On the one hand, low-cost instructional material benefits learners, making educational resources more accessible. However, developing high-quality instructional material requires a significant investment in human resources, advanced technology, and equipment. This costs a lot of money. As a result, the producers of instructional material may suffer.

2. **Quality:** Instructional designers possess specialized knowledge and skills in designing effective and engaging learning experiences. When these experts craft instructional material, it is more likely to be of high quality, incorporating appropriate pedagogical strategies, clear explanations, and interactive elements to engage the learners. However, the instructional material crafted without the expertise

of instructional designers can induce the risks, such as inadequate content, ineffective instructional methods, and a lack of alignment with educational standards.

3. **Requires Adequate Training:** Using new technology in the instructional material requires facilitators to undergo sufficient training to use and operate these tools effectively. As technology evolves and new educational resources are introduced, facilitators need to familiarize themselves with the features and functionalities of these tools. Adequate training ensures that facilitators can optimize the use of instructional material, integrate it seamlessly into their teaching practices, and provide valuable guidance to learners.

4. **Lack of Resources:** Online instructional material heavily relies on the availability of appropriate resources such as electronic devices and a stable Internet connection. Learners need access to computers, laptops, tablets, or smartphones to effectively engage with online educational resources. A decent Internet connection is also necessary to access and utilize these materials without interruptions. However, accessing online instructional material may be challenging in regions or communities with limited resources or inadequate infrastructure. This digital divide can create disparities in educational opportunities and hinder the widespread adoption of online instructional resources.

5. **Requires Guidance:** While instructional material can be valuable, it may only sometimes provide comprehensive or detailed information on specific topics. In such cases, the presence of a facilitator becomes crucial to guide learners through the instructional material. Facilitators can supplement the content, clarify doubts, provide additional explanations, and facilitate discussions to ensure a deeper understanding of the subject. The guidance of a knowledgeable facilitator enhances the effectiveness of instructional material by contextualizing the content and addressing individual learning needs.

Summary

- Instructional design (ID) creates engaging learning experiences to help learners meet their objectives.

- ID bridges the gap between content and learning by adjusting pedagogies and promoting interactive learning.

- Content and learners are critical aspects of instructional design, involving analysis of key concepts and learner characteristics.

- Interdependencies among objectives, learning activities, and assessments are essential for effective instructional design.

- Instructional design involves four shades: learning, creativity, technology, and business.

- The five I's of instructional design enhance learner engagement: intrigue, introduce objectives, illuminate with knowledge, inspect progress, and improve with input.

- An instructional designer is comparable to a movie director, overseeing the entire process of creating and delivering educational content.

- The scope of instructional design includes various industries, such as healthcare, education, corporate, manufacturing, technology, retail, hospitality and tourism, aviation, government, and nonprofit sectors.

- The instructional materials developed for these industries include manuals, guides, handbooks, online resources, interactive modules, simulations, and more. They cover equipment handling, safety protocols, compliance, customer service, software usage, and social issues.

- Benefits of instructional design include cost savings, enhanced learning outcomes, improved performance and team efficiency, customized training, learner-centered learning, and a result-oriented approach.

- Limitations, including cost, quality concerns without instructional designers, the need for adequate training, lack of resources, and the necessity of guidance from facilitators.

Let's Brainstorm

These mini-scenarios will make you scratch your head and scribble on your pad.

Mini-scenario 1

You have been hired as an instructional designer for a large healthcare organization. The organization wants to improve the training program for its nursing staff, technicians, and other staff members in various departments. They expect you to analyze the current state of the learners, understand the content that needs to be taught, and design interventions to facilitate effective learning. How would you approach this task, considering the specific industry, audience, and content areas such as equipment handling, patient handling, clinical guidelines, and hygiene maintenance? What instructional materials and strategies would you recommend to enhance the training program for healthcare professionals?

Mini-scenario 2

You work in a manufacturing company where safety is a top priority. You have been assigned to a team responsible for ensuring adherence to safety protocols in the production process. What instructional materials would be essential for training employees on safety protocols and quality control in your manufacturing facility?

Mini-scenario 3

As an instructional designer, you are working to design the course for a manufacturing company. The company wants you to create instructional materials that promote problem-solving skills and decision-making abilities among employees. How would you design a simulation-based scenario where employees must troubleshoot and resolve complex production issues within a time-constrained environment?

Mini-scenario 4

You are designing employee compliance training as an instructional designer for a multinational company. You want to create a tricky scenario to ensure they grasp the importance of adhering to policies. Design a scenario where employees encounter ethical dilemmas that require them to make difficult decisions, weighing personal interests against company values. How would you challenge them to choose the ethically correct option and explain the consequences of their choices?

Mini-scenario 5

As an instructional designer, you are developing a training module for restaurant staff to provide exceptional customer service. Design a scenario where the staff encounters a demanding customer dissatisfied with their dining experience and demands special treatment, such as complimentary items or discounts. The scenario becomes trickier when challenging customer demands are unreasonable and may disrupt the restaurant operations. How would you guide the staff to handle the situation diplomatically, de-escalate tensions, and find a suitable resolution that balances customer satisfaction and the restaurant's policies?

CHAPTER 2

Science of Learning and Its Theories

As we understand the anatomy of instructional design, it is time to ascend to the next level. How frequently does an instructional designer incorporate learning science principles into their instructional designs? Are their instructions crafted carefully, or are they built on a foundation of assumptions? Let's explore the beliefs that may underpin an instructional journey.

Assumption 1: How learners learn and respond to instruction based on their understanding of cognitive processes, prior knowledge, and individual differences. These assumptions about the learner's functioning can be accurate or inaccurate.

Assumption 2: Having certain expectations and preconceived notions about what learners should do or achieve with the materials and information provided. These assumptions shape the design and delivery of instruction and may sometimes be explicitly stated, but more often, they remain unspoken and seldom discussed.

By acknowledging and discussing these assumptions, instructional designers can gain deeper insights into their instructional decisions and their influence on the learning experience. It encourages instructional designers to reflect on their beliefs, particularly those related to learners' behavior and learning processes, to ensure that they are rooted in evidence and harmony with the goals of effective instruction.

© Ankita Jiyani Mangtani 2024
A. J. Mangtani, *Instructional Design Unleashed*, Design Thinking,
https://doi.org/10.1007/979-8-8688-0416-8_2

If an incorrect assumption is laid as a foundation of instructional design, the outcome can significantly differ from learners' expectations.

Designing effective instruction requires one magical ingredient, which is a solid understanding of how the learners' mind functions. By understanding cognitive processes and mechanisms involved in learning, designers can make informed decisions about instructional strategies, content organization, and delivery methods.

Over the past five decades, significant knowledge has been acquired regarding the functioning of the mind through research conducted in cognitive science, neuroscience, and education.

What Is the Need for Learning Theories?

The importance of learning theories in the design of instructional materials can be derived from the following key points:

Verified Instructional Strategies: Learning theories are a valuable source of verified instructional strategies, tactics, and techniques. Knowing various strategies is crucial when seeking practical solutions to specific instructional problems.

Informed Strategy Selection: Learning theories provide a solid foundation for intelligent and reasoned strategy selection. Designers must possess a sufficient pool of strategies and know when and why to utilize each. This ability to match task demands with appropriate instructional strategies is essential for creating a practical learning experience.

Integration within Instructional Context:
Integrating chosen strategies within the
instructional context is essential. Learning theories
and research offer valuable insights into the
relationships among instructional components and
the overall instructional design. This knowledge
helps designers determine how specific techniques
and strategies can be best aligned with the given
context and the unique needs of the learners.

Reliable Prediction: Time and resources are often
limited when designing instructional materials.
Therefore, selecting and implementing strategies
with the highest likelihood of success becomes
paramount. Relying on research-based solid
theories for choosing a strategy ensures a more
reliable approach than relying solely on anecdotal
"instructional phenomena."

How Does Learning Occur?

Human learning occurs through a complex interplay of cognitive processes,
neural connections, and environmental factors. With its remarkable
capacity for adaptation, the human brain is at the center of this process.

Human learning occurs majorly in the following five phases:

Phase 1 – Unfamiliarity: When a learner is entirely
new to a concept, the learning process starts with
establishing initial neural connections. As they
engage with the information, neurons in the brain
communicate through synapses, forming neural
networks specific to that concept. These networks
act as the foundation for further learning.

Phase 2 – Initial Awareness: As the learner gains exposure and experience, these neural networks become more refined and interconnected. Each interaction strengthens the neural pathways through repeated activation and practice. This reinforcement enables the learner to grasp the fundamental understanding of the concept, building a solid framework for further learning.

Phase 3 – Basic Skills: As the learner progresses, they acquire basic skills related to the concept. These skills involve the application and execution of knowledge in practical contexts. The neural networks associated with these skills become more robust and efficient through practice and deliberate repetition. The learner becomes proficient in performing tasks and solving problems related to the concept, gaining a sense of mastery at a foundational level.

Phase 4 – Analyze and Experiment: The learner can analyze and experiment with different aspects of the concept. They have a solid understanding and skill set, allowing them to delve deeper into the intricacies of the subject matter. They can critically evaluate information, identify patterns, and make connections between different aspects of the concept. By exploring and experimenting, they can customize their experiences with the idea, tailoring it to their needs or interests.

Phase 5 – Mastery: With continued learning and experience, the learner becomes more adept and knowledgeable, eventually reaching a state of mastery. At this advanced phase, the neural networks associated with the concept have matured significantly. The learner has a deep understanding, exceptional skills, and a nuanced perspective. They can effortlessly apply the concept in various contexts, adapt it to new situations, and contribute to its further development through their insights and innovations.

The neural networks are constantly refined, updated, and expanded throughout this learning journey. New connections are established and trimmed weaker connections, resulting in more efficient and specialized networks. This neural plasticity is essential in the learner's transition from beginner to expert because the brain changes to accommodate and support the learner.

Table 2-1 summarizes the concept with the help of an analogy from the baking industry.

Table 2-1. *Phases of Learning*

Phases	Progression	Define	Describe
Phase 1: Unfamiliarity		"I do not know what baking is."	At this phase, the learner has no knowledge or understanding of baking. It is an entirely new concept for a learner.
Phase 2: Initial Awareness		"Now I know what baking is."	The learner has gained some knowledge about baking. The learner understands that it involves preparing and cooking food using dry heat in an oven.
Phase 3: Basic Skills		"I can follow a recipe and bake some simple treats."	The learner has learned the basic techniques and skills required for baking. A learner can now follow recipes and make simple baked goods like cookies or muffins.

(*continued*)

Table 2-1. (*continued*)

Phases	Progression	Define	Describe
Phase 4: Analyze and Experiment		"I can analyze recipes, adjust ingredients, and experiment with different flavors."	A learner has developed a deeper understanding of baking. A learner can now analyze recipes to understand their components, adjust ingredients to customize their preferences, and experiment with different flavors and techniques.
Phase 5: Mastery		"I have mastered the art of baking."	At this stage, a learner has acquired extensive knowledge, skills, and experience in baking. A learner can confidently create a wide range of complex baked goods, customize recipes to your liking, and resolve any issues.

Cognitive Load of Mind

Cognitive load refers to the mental effort or capacity to process and manipulate information during learning or problem-solving tasks. It is the number of mental resources, such as attention, working memory, and processing power, that an individual needs to dedicate to a particular cognitive task.

Cognitive load can be categorized into Intrinsic Load, Extraneous Load, and Germane Load. To create a fantastic learning experience, instructional designers must aim for an **MOP** approach:

- **Managing** intrinsic cognitive load

- **Optimizing** the extraneous cognitive load

- **Promoting** germane cognitive load

Table 2-2 provides a brief overview of all the types of cognitive loads.

Table 2-2. *Types of Cognitive Loads*

Parameters	Intrinsic Load	Extraneous Load	Germane Load
Definition	The inherent complexity of learning materials or tasks is presented to a learner.	Unnecessary cognitive effort and mental strain are imposed on learners due to irrelevant and distracting elements.	Mental efforts learners actively invest in understanding and constructing their knowledge, contributing to meaningful learning and the development of long-term memory.

(continued)

Table 2-2. (*continued*)

Parameters	Intrinsic Load	Extraneous Load	Germane Load
Reasons that increase the load	• Complexity of content • Lack of prior knowledge of the learner • Lack of meaningful connections to prior knowledge or real-life examples • Abstract and unfamiliar concepts • Insufficient time allocated for processing and understanding the material	• Confusing or cluttered visuals • Irrelevant information and excessive multimedia effects • Inconsistent formatting or presentation styles • Irrelevant and repetitive content that does not contribute to the learning objectives • Excessive use of complex vocabulary or technical terms	• Promote active learning strategies • Provide engaging problem-solving tasks • Provide opportunities for reflection and elaboration • Support meaningful connections • Emphasize critical thinking skills
Required in the learning process?	No	No	Yes

(*continued*)

Table 2-2. (*continued*)

Parameters	Intrinsic Load	Extraneous Load	Germane Load
Ways to balance the load	• Chunking of information • Use visual aids • Scaffold learning • Provide real-world examples	• Design clear and concise materials • Offer worked examples • Remove unnecessary distractions • Provide effective guidance • Organize and chunk information correctly • Optimize multimedia use • Use the familiar format and tools	• Applying knowledge to real-life situations or problem-solving scenarios • Elaborating on concepts by exploring related examples or creating new examples • Participating in discussions or collaborative activities with peers • Analyzing and critically evaluating information or arguments • Asking questions and seeking a more profound understanding

Cognitive Load Theory (CLT)

John Sweller proposes Cognitive Load Theory (CLT) in the late 1980s. According to him, our human brain is divided into sensory memory, working memory, and long-term memory. Let's understand this with the help of an analogy.

Imagine you are in a study room with a bulletin or notice board, a study table, and a bookshelf. In this analogy:

A bulletin or notice board, which represents your sensory memory, quickly captures and displays important messages, timetables, reminders, and visual stimuli that catch your attention. However, just like you remove the items from the bulletin board after some time, sensory memory holds onto information only for a short time before it either disappears or moves into working memory.

A study table, which represents your working memory. It is like the surface of your table where you can place a few items and work with them at a time.

A bookshelf, which represents your long-term memory. It is where you store all your books, knowledge, and information. It has a vast capacity to hold many books, just like your long-term memory can keep a vast amount of data over a lifetime.

So when you start studying, new information enters your sensory memory, like looking at the timetable on your notice board (Figure 2-1). Then, you choose the relevant materials according to the timetable.

Figure 2-1. *Sensory Memory Analogy*

To effectively manage the table's surface (cognitive load), you would place the relevant books and notebooks on the study table, ensuring easy access and immediate availability. However, like your study table has limited space (Figure 2-2), your working memory can only hold a limited number of items simultaneously. So you must be selective and consciously decide which information to focus on.

Figure 2-2. *Working Memory Analogy*

As you continue studying, you may need to refer to additional books. The connection between your study table (working memory) and the bookshelf (long-term memory) becomes crucial here. When you no longer need certain books on the study table, you put them back on the bookshelf for future reference (Figure 2-3). Similarly, when you have processed and understood the information in your working memory, you transfer it to your long-term memory for later recall.

Figure 2-3. *Long-Term Memory Analogy*

Renowned psychologist George Miller once famously characterized the capacity of working memory as 7±2, indicating that individuals typically hold around seven items in their working memory. However, it is prudent to err on the side of caution and assume a more conservative estimate of 3±1 when it comes to learning. This approach ensures that we account for potential worst-case scenarios and do not overwhelm learners with an excessive cognitive load. It is worth noting that working memory capacity can be even lower for younger children, necessitating further consideration and adaptation in instructional design to accommodate their cognitive limitations.

To transfer the information from working to long-term memory, you need to elaborate on the information you have processed. In our analogy, consider a bookshelf in that study room, in which all the books and notebooks are arranged logically.

Elaboration is taking a new book from the study table and storing it on the bookshelf. Just like organizing books on a bookshelf, you find the right spot for the new book, placing it alongside related books or within a specific category. This establishes connections and associations between the new book and books sharing similar themes or concepts.

The more you engage in elaboration, the more connections you create between books on the study table and the bookshelf. Each time you make connections, you strengthen the relationships and increase the likelihood of retrieving that knowledge when needed.

The following conclusions are drawn based on the discussion:

- **Sensory memory** is short-lived, automatic, and unconscious and can hold vast amounts of incoming sensory memory information from different modalities (visual, audio, tactile, etc.).

- **Working memory** is fast and has limited capacity; it can only handle small amounts of information and tends to vanish after a few seconds.

- **Long-term memory** holds a vast amount of information and is logically structured.

Applying CLT to Improve Learning Outcomes

As an instructional designer, you strive to create an unforgettable learning experience for learners. To mark the success of the outcome of the learning journey, the learners must reflect upon the learnings from the journey. To make learners reflect upon the same, you, as an instructional designer, should design the experience while considering the cognitive load. The following principles will suffice to design the same.

Principle 1 – Reducing Problem Space: The problem space refers to the difference between the current situation and the desired outcome of the instruction. When the problem space is extensive, it can overwhelm learners' working memory. This often occurs when learning complex concepts simultaneously requires memorizing numerous details. A more effective approach is to decrease the problem space by breaking down information into smaller parts or providing worked examples with solutions. This strategy enhances learning outcomes by reducing the problem space and relieving learners of excessive cognitive load.

Principle 2 – Reducing Split Attention Effect:
The split attention effect refers to a cognitive phenomenon that occurs when individuals are required to divide their attention between multiple sources of information that are not integrated or presented. It happens when learners have to mentally integrate information from separate and disconnected sources, which can result in cognitive overload and reduced learning outcomes.

To mitigate the split attention effect, instructional designers can use strategies such as integrating text and visuals into a single presentation, aligning information spatially, or providing explicit guidance to help learners connect and integrate the different pieces of information. By reducing the need for mental integration and minimizing split attention, learners can focus their cognitive resources on understanding and learning the content more efficiently.

Principle 3: Using Practical Instructional Approaches Tailored to the Nature of the Content:
By adopting instructional methods like employing metaphors and analogies to relate new information to familiar concepts, utilizing chunking techniques

to break down complex material into manageable units, encouraging rehearsal to reinforce learning, facilitating mental visualization of material or concepts through imagery, and incorporating mnemonic devices to aid memory retention, instructional designers can optimize the learning experience and enhance knowledge acquisition.

Learning Theories

Learning theories serve as a fundamental framework for instructional solutions to effectively attain desired learning outcomes. These theories enable instructional designers to gain insight into the cognitive processes involved in information retention and retrieval and the factors that foster motivation and engagement in learning.

Various scientists and psychologists (Figure 2-4) have made significant contributions to the development of learning theories over time. One of the oldest theories is behaviorism, which employs programmed instruction to reinforce specific learner behaviors. Conversely, cognitivism emphasizes that learning is not solely based on stimulus and response but also involves the learner's cognitive abilities. Both behaviorism and cognitivism prioritize the role of the instructor in guiding the learning process. Another prominent theory is constructivism, which asserts that learning occurs when learners actively build upon their existing knowledge and apply their newfound understanding to real-life situations. Humanism, another noteworthy theory, posits that learners can attain self-actualization, the highest level in the hierarchy of needs, when they derive enjoyment and passion from the learning process, surpassing their physiological and safety requirements. Constructivism and humanism underscore the significance of learner-centered approaches in fostering meaningful and fulfilling learning experiences. Lastly, connectivism asserts that learning is enhanced when learners engage and connect with their peers, allowing for collaborative discussions and the development of new skills. This theory acknowledges technology's growing influence and networked learning's role in the digital age.

Behaviorism

Jhon B watson
1913

Cognitivism

Jean Piaget
1930

Constructivism

Lev Semyonovich
Vygotsky
1968
Jean Piaget
1973

Humanism

Abraham
Maslow
Early 1900s

Connectivism

George
Siemens
2005

Figure 2-4. *Fathers of the Learning Theories*

Let's start understanding the five learning theories – behaviorism, cognitivism, constructivism, humanism, and connectivism – with the help of an analogy that incorporates elements from all the learning theories (Figure 2-5), specifically within the context of the baking industry.

Analogy: Mastering the Art of Baking

Scenario. Imagine you are aspiring to become a master baker and create exquisite pastries. Here's how each learning theory can be incorporated sequentially.

Behaviorism
Following Tried-and-Tested Recipes

To start your baking journey, you follow well-established recipes that provide precise measurements and step-by-step instructions. You observe and replicate the behaviors outlined in the recipe, such as mixing ingredients in a specific order or setting the oven temperature accurately. You learn the foundational techniques and principles of baking through repetition and reinforcement.

Constructivism
Experimenting and Innovating

With knowledge and experience, you begin experimenting with new flavors, textures, and combinations. You create unique recipes, modify existing ones, and adapt techniques to suit your preferences and creative vision. Through trial and error, you construct your understanding of flavor profiles, ingredient combinations, and baking methods, leading to the development of your distinct baking style.

Connectivism
Sharing and Learning from the Baking Community

As you progress on your baking journey, you connect with fellow bakers, pastry chefs, and enthusiasts through workshops, forums, or social media platforms. You share your experiences, seek advice, and learn from the collective wisdom of the baking community. Exchanging knowledge, tips, and techniques broadens your understanding and inspires continuous growth.

Cognitivism
Understanding Ingredients and Techniques

As you gain experience, you dive deeper into the knowledge behind baking. You study the properties of different ingredients, such as flour, sugar, and butter, and their roles in the baking process. You also explore baking techniques, like folding, creaming, or tempering chocolate, understanding how each method impacts the final product. This cognitive understanding enhances your ability to adapt and troubleshoot when faced with challenges.

Humanism
Personalization and Artistic Expression

In your baking endeavors, you express your individuality and personal touch. You infuse your creations with creativity, aesthetic appeal, and unique presentations. You take pleasure in baking, deriving satisfaction and fulfillment from the artistry and the joy it brings to others. Baking becomes a reflection of your self-expression and passion.

Figure 2-5. *Learning Theories Analogy*

Evolution of Learning Theories

The corresponding section combines the collection of theories by pioneering psychologists and scientists throughout history. Each theory is evidence of the enduring pursuit to unravel the mysteries surrounding how knowledge is acquired, processed, and applied.

As we traverse the chronological path, the theories unfold like chapters in a grand saga, with each subsequent theory building upon the foundations laid by its predecessors. This progression reflects the ongoing quest for knowledge and a deeper understanding of how we learn. From the early days of behaviorism to the cognitive revolution, the rise of constructivism, and the emergence of connectivism, each theory represents a step forward in our understanding of the learning process.

The chronological arrangement of these theories helps us trace their historical development and highlights the ongoing conversation and exchange of ideas. As theories are refined and challenged, our understanding of learning deepens. This evolution reflects the continuous pursuit of knowledge and the pivotal moments when new perspectives reshape our understanding of the learning process.

Note   The implementation examples given in this section will vary depending on the specific instructional context and subject matter. These examples illustrate how the theories can be applied but may need to be adapted to suit the particular learning objectives and target audience.

Behaviorism

Behaviorism is a psychological approach that focuses on observable behaviors and the environmental factors shaping them. The following are some of the influential behaviorism theories:

Classical Conditioning (Ivan Pavlov, 1897)

- Initial theory that explores the association between stimuli and responses

- Focuses on involuntary, reflexive behaviors

- Demonstrates how organisms can learn to associate neutral stimuli with meaningful responses through repeated pairings

Stimulus-Response Theory (Edward Thorndike, 1905)

- Expands on classical conditioning by emphasizing the influence of consequences on behavior

- Introduces the concept of satisfying or unsatisfying consequences shaping behavior

- Focuses on the relationship between stimuli and observable responses, often in the context of a specific task or situation

Behavior Modification (John B. Watson, Early 20th Century)

- Focuses on modifying behavior through environmental factors

- Highlights the use of rewards and punishments to encourage desired behaviors and discourage unwanted behaviors

- Emphasizes the importance of consistent cues, feedback, and reinforcement in behavior change

Operant Conditioning (B.F. Skinner, 1938)

- Further develops the role of consequences in shaping behavior

- Highlights the significance of reinforcement and punishment in influencing behavior

- Introduces the concept of shaping behavior through reinforcement schedules, such as continuous or intermittent reinforcement

Radical Behaviorism (B.F. Skinner, 1957)

- Takes a comprehensive view of behavior, emphasizing the influence of external stimuli and consequences

- Advocates for a deterministic understanding of behavior, where environmental factors entirely determine behavior

- Focuses on operant conditioning principles and reinforcement's role in shaping behavior

Social Learning Theory (Albert Bandura, 1977)

- Expands the behaviorist approach by incorporating social factors into learning

- Introduces the concept of learning through observation, modeling, and self-efficacy

- Emphasizes the role of social interactions, role models, and the cognitive processes involved in learning

Table 2-3 summarizes all the theories discussed previously and provides the way to implement them in instructional materials with examples.

Table 2-3. Overview of Behaviorism Theories

Learning Theory	Key Idea	Pros	Cons	Implementation in Instructional Design	Examples of Implementation
Ivan Pavlov (1897) – Classical Conditioning	Learning through the association of stimuli and responses	Simple and straightforward model of learning	Limited to reflexive and involuntary responses	Incorporate repeated associations of stimuli and responses to create associations between new information and prior knowledge.	Introduce a bell sound before presenting new information in a lesson to associate the sound with the upcoming content.
Edward Thorndike (1905) – Stimulus–Response Theory	Behaviors are influenced by satisfying or unsatisfying consequences	Provides a clear understanding of how consequences shape behavior	Limited in explaining complex human behaviors	Design instructional materials that provide clear goals and objectives, break down complex tasks into smaller, achievable steps, and offer rewards or consequences based on learners' performance.	Break down a complex math problem into smaller steps, provide immediate feedback, and offer a reward or praise for correct solutions.

(continued)

Table 2-3. (*continued*)

Learning Theory	Key Idea	Pros	Cons	Implementation in Instructional Design	Examples of Implementation
John B. Watson (Early 20th Century) – Behavior Modification	Changing behavior by modifying environmental factors	Effective in addressing specific behavioral issues	May neglect the importance of internal thoughts and feelings	Design instructional materials that provide clear and consistent cues, feedback, and rewards to reinforce desired behaviors and discourage unwanted behaviors.	Use a token economy system where learners earn tokens or points for completing tasks and can exchange them for rewards or privileges.
B.F. Skinner (1938) – Operant Conditioning	Behavior is shaped by consequences and reinforcement	Provides a systematic framework for behavior modification	May overlook the influence of internal mental processes	Use rewards and punishments to reinforce the desired behaviors and discourage undesirable behaviors in instructional activities.	Provide a digital badge or virtual reward for completing a task correctly, or deduct points for incorrect responses in an online quiz.

Theory	Key Concept	Strengths	Weaknesses	Application	Example
B.F. Skinner (1957) – Radical Behaviorism	Behavior is determined by external stimuli and consequences	Offers a comprehensive and deterministic view of behavior	Ignores cognitive processes and internal experiences	Incorporate clear behavioral objectives and provide immediate feedback and reinforcement based on learners' responses in instructional materials.	Use immediate feedback and praise in digital learning platforms to reinforce correct answers or provide guidance for incorrect responses.
Albert Bandura (1977) – Social Learning Theory	Learning through observation, modeling, and self-efficacy	Emphasizes the role of social factors in learning	Does not fully account for individual differences	Incorporate opportunities for learners to observe and imitate desired behaviors through demonstrations or role models. Encourage collaborative learning and provide opportunities for learners to build self-efficacy.	Assign group projects where learners observe and learn from their peers' presentations or performances. Provide opportunities for learners to showcase their skills and receive positive feedback from their classmates.

Cognitivism

Cognitivism recognizes that learning goes beyond simple stimulus and response associations and acknowledges the significant role of the learner's cognitive abilities in learning. The following theories offer unique insights into how individuals acquire, process, and apply knowledge. These theories have significantly contributed to our understanding of cognitive processes and have informed instructional practices in various educational settings.

Schema Theory (Jean Piaget, 1952)

- Proposes that individuals construct knowledge through the development of mental schemas

- Focuses on assimilation and accommodation as individuals actively interact with their environment

Information Processing Theory (George Miller, 1956)

- Views the mind as an information processing system that encodes, stores, and retrieves information

- Explores how individuals process and manipulate information in a step-by-step manner

Dual Coding Theory (Allan Paivio, 1971)

- Suggests that information is processed and stored in verbal and visual forms

- Highlights the benefits of utilizing multiple modalities for encoding and retrieving information

Social Cognitive Theory (Albert Bandura, 1977)

- Integrates cognitive processes, behavior, and social factors in understanding learning and behavior change

- Emphasizes the role of observational learning, self-efficacy, and reciprocal determinism

Cognitive Load Theory (John Sweller, 1988)

- Explores the limitations of working memory and the impact of cognitive load on learning

- Offers strategies to optimize instructional design by reducing extraneous cognitive load and fostering effective learning

Cognitive Apprenticeship Theory (Collins, Brown, and Newman, 1989)

- Focuses on learning through active participation in authentic tasks and the guidance of an expert

- Emphasizes modeling, coaching, and scaffolding to facilitate the acquisition of complex skills

Table 2-4 summarizes all the theories discussed previously and provides the way to implement them in instructional materials with examples.

Table 2-4. Overview of Cognitivism Theories

Learning Theory	Key Idea	Pros	Cons	Implementation in Instructional Design	Examples of Implementation
Jean Piaget (1952) – Schema Theory	Knowledge organized in mental schemas	Explains how individuals actively construct understanding	May not account for individual differences in schema formation	Design instructional materials that activate and build upon learners' existing schemas, providing opportunities for assimilation and accommodation through meaningful activities.	Present new information within the context of learners' existing knowledge structures.
George Miller (1956) – Information Processing Theory	Mind as an information processing system	Provides a framework for understanding cognitive processes	Simplifies complex cognitive processes	Design instructional materials that align with the cognitive processes of encoding, storing, and retrieving information, and incorporate strategies for improving memory and problem-solving.	Present information in a structured and organized manner, utilizing mnemonic techniques for memorization.

| Allan Paivio (1971) – Dual Coding Theory | Processing and storing information in verbal and visual forms | Enhances learning and memory through multiple modalities | May not apply equally to all types of information | Design instructional materials that present information through both verbal and visual channels, promoting meaningful connections and facilitating better understanding and recall. | Use graphic organizers, diagrams, and multimedia presentations to supplement textual information. |
| Albert Bandura (1977) – Social Cognitive Theory | Interaction between cognition, behavior, and the environment | Emphasizes the role of observational learning and self-efficacy | May overlook the influence of individual internal processes | Design instructional materials that incorporate modeling, provide opportunities for observational learning, and foster a supportive social environment to enhance learning and self-efficacy. | Include videos or demonstrations of desired behaviors and provide opportunities for peer collaboration |

(continued)

Table 2-4. (*continued*)

Learning Theory	Key Idea	Pros	Cons	Implementation in Instructional Design	Examples of Implementation
John Sweller (1988) – Cognitive Load Theory	Optimize instructional design by reducing cognitive load	Enhances understanding of the limitations of working memory	May oversimplify the complexities of cognitive processes	Design instructional materials with appropriate content complexity and support learners in managing cognitive load through clear organization and effective use of multimedia elements.	Divide complex concepts into smaller chunks and provide supporting visuals or examples.
Collins, Brown, and Newman (1989) – Cognitive Apprenticeship Theory	Learning through active participation and guidance	Facilitates skill acquisition and application	Requires skilled instructors for effective implementation	Design instructional materials that provide authentic tasks, modeling, coaching, and scaffolding to support learners' progression from novice to expert in a specific domain.	Pair learners with experts who provide guidance and support in completing real-world tasks.

Constructivism

Constructivism posits that learning occurs through active engagement as learners construct knowledge by building upon their existing understanding and applying it to real-life situations. It emphasizes the learner's active role in learning and the importance of hands-on experiences in facilitating meaningful understanding. The following theories have significantly influenced educational practices and instructional design, promoting learner-centered approaches and meaningful engagement in the learning process.

Cognitive Constructivism (Jean Piaget, 1950)

- Focuses on learners constructing knowledge through cognitive processes such as assimilation and accommodation

- Emphasizes the importance of learners' active engagement and critical thinking in the learning process

Social Constructivism (Lev Vygotsky, 1968)

- Expands on cognitive constructivism by highlighting the social and cultural aspects of learning

- Emphasizes the role of social interactions, collaboration, and language in knowledge construction

- Proposes the concept of the zone of proximal development, where more knowledgeable others support learners

Radical Constructivism (Ernst von Glasersfeld, 1970)

- Builds upon social constructivism by emphasizing the subjective nature of knowledge construction

- Views knowledge as a personal interpretation of experiences and rejects the notion of objective truths

59

- Emphasizes learners' active role in constructing meaning based on their subjective experiences

Constructionism (Seymour Papert, 1980)

- Builds upon constructivist principles by emphasizing the creation of tangible artifacts or projects

- Views learning as enhanced through the active construction of physical objects or digital creations

- Encourages learners to engage in hands-on, active learning and apply knowledge meaningfully

Experiential Learning (David Kolb, 1984)

- Integrates constructivist principles with the idea of learning through concrete experiences

- Emphasizes a cyclical learning process involving substantial experience, reflection, conceptualization, and testing

- Highlights the importance of reflection and active engagement in the learning process

Situated Learning (Jean Lave and Etienne Wenger, 1991)

- Expands on constructivism by highlighting the situated nature of learning in authentic contexts

- Emphasizes the importance of communities of practice and real-world contexts in supporting learning

- Proposes that learning is socially and culturally embedded within specific activities and contexts

Table 2-5 summarizes all the theories discussed previously and provides the way to implement them in instructional materials with examples.

Table 2-5. *Overview of Constructivism Theories*

Learning Theory	Key Idea	Pros	Cons	Implementation in Instructional Design	Examples of Implementation
Jean Piaget (1950) – Cognitive Constructivism	Learners actively construct knowledge through cognitive processes.	Encourages active engagement and critical thinking	May not fully account for social and cultural aspects of learning	Design instructional activities that promote exploration, discovery, and problem-solving to foster learners' cognitive development.	Present open-ended problems or scenarios that require learners to apply their understanding to find solutions.
Lev Vygotsky (1968) – Social Constructivism	Learning is a social process that occurs through collaboration and interaction with others.	Emphasizes the importance of social interactions and cultural context in learning	May require skilled facilitators to support collaborative learning effectively	Design collaborative learning activities that promote peer interaction, discussion, and shared construction of knowledge.	Implement group projects or case studies where learners work together to solve problems and share their understanding.

(continued)

Table 2-5. (*continued*)

Learning Theory	Key Idea	Pros	Cons	Implementation in Instructional Design	Examples of Implementation
Ernst von Glasersfeld (1970) – Radical Constructivism	Knowledge is actively constructed by individuals based on their subjective experiences.	Recognizes learners' active role in constructing their understanding	Critics argue that it can lead to relativism and subjective interpretations of knowledge.	Design learning activities that encourage learners to reflect on their own experiences and make connections to construct meaning.	Assign reflective journals or personal narratives where learners can articulate their understanding and insights.
Seymour Papert (1980) – Constructionism	Learning is enhanced through the active construction of tangible artifacts or projects.	Promotes hands-on, active learning and creativity	Requires access to resources and materials for creating tangible artifacts	Design project-based learning experiences where learners create tangible products or solutions that address real-world problems.	Engage learners in building physical models, creating multimedia presentations, or developing prototypes to demonstrate their understanding.

Theory					
David Kolb (1984) – Experiential Learning	Learning occurs through a concrete experience, reflection, conceptualization, and testing cycle.	Encourages active engagement and reflection on learning experiences	May require additional support for learners who struggle with reflective thinking	Design learning activities that provide opportunities for hands-on experiences, reflection, and application of knowledge in practical contexts.	Incorporate simulations, role plays, or case studies where learners can apply their knowledge and reflect on their experiences.
Jean Lave and Etienne Wenger (1991) – Situated Learning	Learning is situated in authentic contexts and communities of practice.	Promotes real-world application of knowledge and skill development.	May require access to relevant communities of practice for effective implementation	Design authentic learning experiences that connect learners to real-world contexts and provide opportunities for participation in relevant communities of practice.	Incorporate internships, apprenticeships, or fieldwork experiences where learners engage in authentic tasks and interact with professionals in the field.

Humanism

The humanistic learning theories offer unique perspectives on the role of personal agency, intrinsic motivation, individual growth, and the creation of positive learning environments. They prioritize the holistic development of individuals and the promotion of well-being and self-actualization. The following theories have significantly influenced fields such as education, counseling, and positive psychology, highlighting the importance of fostering a learner-centered approach that nurtures learners' intrinsic motivation, autonomy, and personal growth.

Maslow's Hierarchy of Needs (Abraham Maslow, 1943)

- Describes a hierarchy of human needs, from physiological to self-actualization

- Recognizes the importance of addressing basic needs to create an optimal learning environment

Person-Centered Theory (Carl Rogers, 1951)

- Emphasizes the importance of a supportive and empathetic learning environment that promotes personal growth

- Values the learner's subjective experience and emphasizes empathy, unconditional positive regard, and genuineness in the learning process

Experiential Learning (Carl Rogers and David Kolb, 1960)

- Learning occurs through concrete experiences, reflection, conceptualization, and testing.

- Emphasizes the importance of hands-on experiences, reflection, and active experimentation in learning.

Self-determination Theory (Edward Deci and Richard Ryan, 1985)

- Focuses on intrinsic motivation and fulfilling basic psychological needs for optimal learning and well-being

- Emphasizes the importance of autonomy, competence, and relatedness in promoting motivation and engagement in learning

Positive Psychology (Martin Seligman, 1998)

- Focuses on promoting well-being, positive emotions, character strengths, and resilience in learning

- Emphasizes the importance of a positive mindset, character development, and cultivating strengths for optimal learning and growth

Table 2-6 summarizes all the theories discussed previously and provides the way to implement them in instructional materials with examples.

Table 2-6. Overview of Humanism Theories

Learning Theory	Key Idea	Pros	Cons	Implementation in Instructional Design	Examples of Implementation
Abraham Maslow (1943) – Maslow's Hierarchy of Needs	Describes a hierarchy of human needs, from physiological needs to self-actualization	Recognizes the importance of addressing basic needs to create an optimal learning environment	Critics argue that the hierarchy may not apply to all cultures and individuals	Design learning experiences that address learners' basic needs, establish a safe and supportive classroom environment and promote self-actualization	Ensure learners' well-being by providing comfortable learning spaces and foster a sense of belonging through class discussions and team-building activities
Carl Rogers (1951) – Person-Centered Theory	Emphasizes the importance of a supportive and empathetic learning environment that promotes personal growth	Values the learner's subjective experience and emphasizes empathy, unconditional positive regard, and genuineness in the learning process	Critics argue that it may not provide enough guidance or structure for some learners	Create a learner-centered environment that values learners' perspectives, provides constructive feedback, and fosters a positive and trusting relationship	Encourage active listening and empathy in classroom discussions, provide opportunities for self-reflection and self-expression, and tailor instruction to individual needs and interests

Theory					
Carl Rogers and David Kolb (1960) – Experiential Learning	Learning occurs through concrete experiences, reflection, conceptualization, and testing.	Emphasizes the importance of hands-on experiences, reflection, and active experimentation in learning	Critics argue it may not be suitable for all subjects or learning contexts.	Design learning activities that involve real-world experiences, promote reflection and critical thinking, and provide opportunities for application and experimentation.	Engage learners in field trips or simulations, incorporate reflective journals or portfolios, and design problem-solving tasks that require the application of knowledge.
Edward Deci and Richard Ryan (1985) – Self-Determination Theory	Focuses on intrinsic motivation and fulfilling basic psychological needs for optimal learning and well-being	Emphasizes the importance of autonomy, competence, and relatedness in promoting motivation and engagement in learning	Critics argue that it may not fully consider the role of external rewards and punishments in motivation	Design learning activities that provide choices, support learners' autonomy, provide meaningful feedback, and foster collaboration and social connection	Allow learners to choose topics of interest for projects, encourage self-assessment and reflection, and create a supportive and inclusive learning environment
Martin Seligman (1998) – Positive Psychology	Focuses on promoting well-being, positive emotions, character strengths, and resilience in the learning process	Emphasizes the importance of a positive mindset, character development, and cultivating strengths for optimal learning and growth	Critics argue that it may overly focus on positive aspects and neglect the role of negative emotions or challenges	Foster a positive classroom climate, promote character development, and integrate activities that build resilience and a growth mindset	Encourage learners to identify and develop their strengths, incorporate gratitude exercises or mindfulness practices, and promote positivity and support

67

Connectivism

Connectivism learning theories offer unique perspectives on learning in the digital age, emphasizing the interconnectedness of knowledge, networks, and technology. They recognize the importance of digital resources, online communities, and personal networks in facilitating learning and knowledge acquisition. The following theories have significantly influenced the field of online and technology-enhanced learning, highlighting the need for individuals to develop skills in navigating information networks, collaborating in online communities, and leveraging digital tools for learning and knowledge creation.

Personal Learning Networks (PLNs) (Siemens and Downes, 2004)

- Individuals cultivate personalized networks for learning, connecting with resources and communities.

- Learners curate their learning resources and leverage social media to connect with experts and access relevant materials.

Connectivism (Siemens, 2004)

- Learning is creating and maintaining connections with information sources and networks.

- Networks and digital resources play a crucial role in learning.

- Learners navigate and connect with various sources to expand their knowledge.

Networked Learning (Wenger, White, and Smith, 2009)

- Learning occurs through participation in networked communities and collaborative activities.

- Social interaction and community engagement support learning and knowledge exchange.

- Learners participate in online communities, engage in discussions, and share knowledge within a networked structure.

Digital Learning Ecology (Buchem, 2012)

- The interplay between technology, social connections, and physical spaces influences learning.

- The complex relationships between individuals, tools, communities, and environments shape learning experiences.

- Instructional design considers the ecological aspects of digital learning environments, incorporating online and offline interactions.

Table 2-7 summarizes all the theories discussed previously and provides the way to implement them in instructional materials with examples.

Table 2-7. Overview of Connectivism Theories

Learning Theory	Key Idea	Pros	Cons	Implementation in Instructional Design	Examples of Implementation
Siemens and Downes (2004) – Personal Learning Networks (PLNs)	Individuals cultivate personalized networks for learning, connecting with resources and communities	Supports individualized and self-directed learning, enabling learners to curate their learning resources	Requires learners to actively build and maintain their networks, which may require time and effort	Encourage learners to develop and maintain their PLNs, connect with experts via social media and online channels, and access relevant resources	Assign tasks that involve researching and curating resources, participating in online communities, and seeking expertise outside the classroom
George Siemens (2004) – Connectivism	Learning is creating and maintaining connections with information sources and networks	Emphasizes the role of networks and digital resources in learning	Critics argue that it may downplay the importance of individual cognition and internal knowledge construction	Design learning activities that promote online resources, collaborative tools, and networked learning environments	Facilitate online discussions, encourage learners to curate and share relevant resources, and participate in online communities to expand their knowledge network

Wenger, White, and Smith (2009) – Networked Learning	Learning occurs through participation in networked communities and collaborative activities.	Emphasizes the social nature of learning and the importance of community participation	May require careful facilitation to ensure productive engagement within online communities	Design collaborative learning tasks that leverage online platforms, encourage peer interaction, and promote knowledge sharing within a networked structure	Assign group projects that involve online collaboration, use discussion forums for knowledge exchange, and engage learners in joint problem-solving activities
Ilona Buchem (2012) – Digital Learning Ecology	The interplay between technology, social connections, and physical spaces influences learning	Recognizes the complex relationships between individuals, tools, communities, and environments in learning	Challenges include adapting the instructional design to leverage digital environments effectively.	Design learning experiences that leverage technology, incorporate online and offline interactions, and create learning spaces that support collaboration and engagement	Create blended learning environments that combine online activities, face-to-face discussions, and collaborative projects that bridge digital and physical spaces

Evolution of Instructional Design Approaches

As our understanding of learning grew, so did the methods we employed to design effective instruction. The evolution of learning theories played a pivotal role in shaping instructional design approaches, marking significant paradigm shifts along the way. Let us journey through time, observing how these shifts unfolded.

Behaviorism and Programmed Instruction (1920s–1950s)

- Behaviorist theories, particularly those of B.F. Skinner, laid the foundation for instructional design. Behaviorism focuses on the idea that learning results from a stimulus-response relationship.

- Programmed Instruction (PI) emerged during this time, emphasizing self-paced learning through instructional materials that presented content in small, manageable steps. Skinner's teaching machine and the work of Sidney L. Pressey were notable contributions.

Systematic Instructional Design (1950s–1960s)

- In response to the increased demand for effective training during World War II, systematic instructional design models emerged.

- Robert M. Gagne proposed a nine-step instructional design process incorporating elements such as analyzing learner characteristics, specifying objectives, and designing instructional strategies.

Instructional Systems Development (1960s–1970s)

- The advent of computer technology influenced instructional design approaches. The focus shifted from programmed instruction to broader instructional systems.

- The US military played a significant role in developing instructional systems, formulating the Instructional Systems Development (ISD) model. ISD emphasized analyzing training needs, designing instructional materials, developing and implementing instruction, and evaluating effectiveness.

Cognitive Approaches (1970s–1980s)

- Cognitive psychology and theories of learning, such as information processing and constructivism, gained prominence.

- Influential researchers like David Ausubel, Jerome Bruner, and Jean Piaget emphasized the importance of meaningful learning, prior knowledge, and active learner engagement.

- Cognitive instructional design models, such as Merrill's Component Display Theory (CDT) and Reigeluth's Elaboration Theory, focused on organizing and presenting information to enhance learning.

Multimedia and Technology Integration (1980s–1990s)

- The development of multimedia technologies, including computers, graphics, audio, and video, led to multimedia integration into instructional design.

- Richard Mayer's research on multimedia learning principles highlighted the importance of aligning instructional design with cognitive processes.

- Computer-based training (CBT) and multimedia learning environments became popular to enhance learner engagement and interactivity.

Constructivist and Situated Learning (1990s–2000s)

- The constructivist and situated learning perspectives gained prominence, emphasizing the active construction of knowledge and learning in authentic contexts.

- The work of theorists like Jean Lave, Etienne Wenger, and John Dewey influenced instructional design by promoting collaborative learning, problem-solving, and authentic tasks.

- Problem-based learning (PBL) and inquiry-based learning (IBL) approaches became prevalent, focusing on real-world problem-solving and active learner participation.

Online and E-Learning (2000s–Present)

- The widespread availability of the Internet and technological advancements led to the growth of online and eLearning.

- Learning Management Systems (LMS) and rapid eLearning authoring tools make creating and delivering online courses easier.

- Instructional design models like the ADDIE (Analysis, Design, Development, Implementation, Evaluation) and SAM (Successive Approximation Model) frameworks are commonly used in designing online learning experiences.

- It is essential to note that instructional design evolves as new technologies and research emerge. Modern approaches often incorporate learner-centered design, adaptive learning, mobile learning, and data-driven decision-making.

How to Choose a Learning Theory?

After going through all flavors of learning theories and understanding that each learning theory offers a unique taste to balance the journey of the learning experience, it is crucial to decide the learning theories to implement while designing the learning experience. The following factors are to be considered:

1. **The Level of Learners' Knowledge:** This is the foundation upon which the learning experience must be developed. An instructional designer must assess the learners' prior knowledge like an architect does while designing a structure. It functions as a compass, directing us to appropriate theories while ensuring that we build on the pre-existing framework of understanding. By embracing this element, we may avoid overloading learners or leaving them adrift in a sea of complexity, instead creating a growth-friendly environment.

2. **The Learners' Thought Process:** Understanding the learners' mental processes is analogous to solving the complexities of a fine-crafted puzzle. Each mind works in its way, making connections and patterns. Instructional designers can adjust their techniques to the learner's mental patterns by diving into the varied landscapes of cognition. It enables them to deliver information in a way that is consistent with their cognitive framework, increasing comprehension and aiding the internalization of new concepts.

3. **The Desired Outcome (Generation of New Ideas or a Single Answer):** The desired outcome acts as a guiding star for our educational vessel. Do we want to encourage learners' creative wellsprings so that fresh ideas can grow like colorful blooms? Is our goal to find a single solution, like a prospector finding gold in the depths of a mine? By identifying the final aim, instructional designers can modify their techniques to create an environment that fosters the intended outcome. Aligning techniques with the desired outcome improves the success of the educational journey, whether through encouraging brainstorming sessions that stimulate divergent thinking or engaging in focused conversations that lead to convergent solutions.

By embracing these essential considerations, educators can create an environment where knowledge blooms, minds thrive, and the pursuit of learning becomes an artful masterpiece.

Summary

- Human learning involves establishing initial neural connections, refining and interconnecting those connections, acquiring basic skills, analyzing and experimenting with the concept, and reaching a state of mastery.

- Cognitive load refers to the mental effort required for learning or problem-solving tasks, including attention, working memory, and processing power. It can be categorized into intrinsic, extraneous, and germane load.

- John Sweller proposed Cognitive Load Theory (CLT), which identifies the human brain's sensory, working, and long-term memory components.

- Learning theories provide a framework for instructional solutions to achieve desired learning outcomes by understanding cognitive processes, motivation, and engagement.

- Behaviorism focuses on reinforcing specific behaviors, while cognitivism emphasizes cognitive abilities. Constructivism highlights active learning and application in real-life situations, and humanism prioritizes learner enjoyment and self-actualization. Connectivism emphasizes the importance of collaboration and technology in networked learning. These theories guide instructional design for meaningful and fulfilling learning experiences in the digital age.

- Different theories such as behaviorism, systematic instructional design, cognitive approaches, multimedia and technology integration, constructivist and situated learning, and online and eLearning have shaped instructional design over time.

- Factors to consider when choosing a learning theory for instructional design include the level of learners' knowledge, understanding the learners' thought process, and determining the desired outcome.

Let's Brainstorm

These mini-scenarios will make you scratch your head and scribble on your pad.

Mini-scenario 1

Assume you are designing an instructional program for teaching complex mathematical concepts. How would you apply principles from both behaviorism and cognitivism to ensure effective learning outcomes?

Mini-scenario 2

In an instructional design project, you are tasked with reducing the split attention effect for learners. Provide any three working strategies you would employ to integrate information sources and reduce cognitive overload.

Mini-scenario 3

As an instructional designer, you are developing a course on a new technological tool. How would you apply constructivist principles to encourage collaborative learning and problem-solving?

Mini-scenario 4

You are creating an online course for a wide range of learners with varied levels of prior knowledge. How would you assess and utilize their knowledge to create a growth-friendly learning environment?

Mini-scenario 5

When creating an instructional program, you want to promote creativity and generate new ideas among learners. Describe any two instructional techniques you would employ to stimulate divergent thinking and encourage brainstorming sessions.

CHAPTER 3

Learner Attributes and Learning Modalities

The previous chapter unravels various learning theories and the mysteries of the learner's mind and its ability to process new information and retrieve old information. It is time to ascend to the next level of enlightenment to explore the various traits of learners. By analyzing these traits, instructional designers can tailor impactful and compelling learning experiences crafted to the unique characteristics of each learner.

Gone are the days of one-size-fits-all (generic) training because we now know that customized experiences result in complete immersion and engagement. A mosaic of learners awaits us, each with its colors, shapes, and desire for knowledge. It is our responsibility as instructional designers to help people realize their full potential by sparking their interest and taking them on a transformative educational journey.

We must interpret the mysterious codes of their learning methods and preferences to complete this feature. Are they visual connoisseurs looking for colorful images and fascinating graphics? Or are they audial virtuosos drawn to the melody of spoken words? Some people may hunger for hands-on experiences that combine the physical and intellectual.

© Ankita Jiyani Mangtani 2024
A. J. Mangtani, *Instructional Design Unleashed*, Design Thinking,
https://doi.org/10.1007/979-8-8688-0416-8_3

Examine the kaleidoscope of learners' hopes and dreams. A clear understanding of their goals and desires serves as the compass directing our design decisions. We spark the fires of inspiration by connecting the learning experience with their intrinsic motivations and lofty aspirations, generating a symphony of intellectual fulfillment.

Learner Attributes

When conducting a learner analysis in the instructional design process, gathering information about the traits of adult learners is crucial for making informed instructional decisions. Understanding the learners' needs, preferences, and abilities allows instructional designers to tailor their materials and activities to create a relevant and effective learning experience. This learner-centered approach increases the chances of achieving the desired learning outcomes and facilitating effective knowledge transfer.

The following are some key aspects of learner analysis that can assist instructional designers:

Demographic Information: Gathering basic demographic details such as age, gender, educational background, and work experience helps in understanding the diversity within the learner group and identifying any specific requirements or considerations.

Learning Styles: Analyzing how adults prefer to learn, whether through visual, auditory, or kinesthetic means, helps designers create instructional materials that align with the learners' preferred learning styles. Some prefer hands-on activities, while others learn best through reading or listening.

Learning Preferences: Understanding how learners prefer to receive and engage with instructional material can inform decisions on the format and delivery methods. Some learners prefer online modules, while others benefit more from face-to-face instruction or a blended approach.

Prior Knowledge and Skills: Assessing the existing knowledge and skills of the learners allows designers to build upon their previous understanding and avoid unnecessary repetition. This analysis helps identify knowledge gaps and design appropriate content and activities to bridge them.

Learning Objectives and Goals: Understanding the learners' personal and professional goals helps instructional designers align the content and activities with these objectives. It ensures that the instruction is relevant and meaningful to the learners, increasing their motivation and engagement.

Learning Needs and Challenges: Identifying any specific challenges or constraints that learners may face, such as time limitations or technological barriers, helps designers adapt the instructional approach accordingly. This analysis includes supportive resources, clear instructions, and flexible learning options to accommodate different learner needs.

Motivation and Expectations: Assessing the learners' motivation levels and expectations regarding the learning experience can help

> designers incorporate elements that enhance
> engagement and create a positive learning
> environment. Try including interactive activities,
> real-world examples, or opportunities for
> collaboration and feedback.

Questionnaires, surveys, pre-assessment, registration forms, and self-reflection exercises would suffice to gather information about the characteristics of the learners.

Multiple Intelligence Theory

Before exploring various learning modalities, it is crucial to grasp the concept of multiple intelligence theory. This theory, developed by Howard Gardner in 1983, posits that individuals possess distinct types of intelligence, including naturalistic, linguistic, logical-mathematical, spatial, musical, bodily-kinesthetic, interpersonal, and intrapersonal (Figure 3-1). Understanding multiple intelligences is critical to recognizing learners' diverse strengths and preferences, paving the way for tailored instructional approaches that effectively cater to their unique abilities and learning needs.

In addition to these eight intelligences, there has been a suggestion to include a ninth type, existential intelligence. Existential intelligence involves delving deeper into questions about life, meaning, and existence. Individuals with a strong sense of existential intelligence are profoundly aware of the larger picture and contemplate existential matters with depth and introspection.

While this framework has been influential in educational settings, it is crucial to approach it with a balanced perspective.

- **Oversimplification of Learner's Capabilities:**
  Categorizing learners into dominant intelligence types
  may not fully capture the nuanced and dynamic nature
  of human intellect. Intelligence is often an interplay

of multiple domains rather than a single predominant type, and individuals can develop and demonstrate strengths in various areas over time.

- **Cultural Bias:** Different cultures may emphasize and nurture specific types of intelligence based on their societal values and needs. For instance, while Western cultures might highly value logical-mathematical and linguistic intelligence, other cultures might place greater importance on interpersonal or spatial skills. This cultural variability underscores the importance of considering diverse cultural contexts when applying Gardner's theory.

Figure 3-1. *Theory of Multiple Intelligence*

Table 3-1 offers a comprehensive overview of the various types of intelligence learners can possess, along with their corresponding characteristics and potential career choices.

Table 3-1. *Overview of Multiple Intelligence*

Type of Intelligence	Strengths	Characteristics	Possible Career Choices
Visual-Spatial Intelligence	Visual and spatial judgment	Proficient at visualizing things; skilled with directions, maps, charts, videos, and pictures	Architect, Artist, Engineer
Linguistic-Verbal Intelligence	Words, language, and writing	Able to use words effectively in writing and speaking; strong in storytelling, memorization, and reading	Writer/Journalist, Lawyer, Teacher
Logical-Mathematical Intelligence	Analyzing problems and mathematical operations	Skilled at reasoning, recognizing patterns, and logical analysis; conceptual thinking about numbers, relationships, and patterns	Scientist, Mathematician, Computer Programmer, Engineer, Accountant
Bodily-Kinesthetic Intelligence	Physical movement, motor control	Excellent hand-eye coordination and dexterity; skilled at dancing, sports, and creating with hands	Craftsperson, Dancer, Builder, Surgeon, Sculptor, Actor

(continued)

Table 3-1. (*continued*)

Type of Intelligence	Strengths	Characteristics	Possible Career Choices
Musical Intelligence	Rhythm and music	Proficient in thinking in patterns, rhythms, and sounds; a deep appreciation for music	Musician, Composer, Singer, Music Teacher, Conductor
Interpersonal Intelligence	Understanding and relating to other people	Skilled at understanding and interacting with others; perceptive about emotions, motivations, and intentions	Psychologist, Philosopher, Counselor, Salesperson, Politician
Intrapersonal Intelligence	Introspection and self-reflection	High self-awareness; strong in analyzing strengths, weaknesses, and personal motivations	Philosopher, Writer, Theorist, Scientist
Naturalistic Intelligence	Finding patterns and relationships to nature	Attuned to nature; interested in nurturing, exploring the environment, and learning about other species	Biologist, Conservationist, Gardener, Farmer
Existential Intelligence	The ability to see the big picture	Contemplation of existential questions; consideration of the meaning of life and how actions serve larger goals	Philosopher, Theologian, Pastoral Counselor, Pastor

Instructional Strategies

The instructional strategies mentioned in Table 3-2 align with the respective strengths of each intelligence. These materials can effectively engage learners and cater to their specific learning preferences and strengths, ensuring complete absorption of information.

Table 3-2. *Instructional Strategies*

Type of Intelligence	Strengths	Instructional Strategies
Visual-Spatial Intelligence	Visual and spatial judgment	• Infographics and visual aids • Maps, charts, and diagrams • Videos and multimedia presentations • Pictures and illustrations
Linguistic-Verbal Intelligence	Words, language, and writing	• Textbooks and written materials • Reading assignments and articles • Written exercises and essays • Verbal discussions and debates
Logical-Mathematical Intelligence	Analyzing problems and mathematical operations	• Math problems and exercises • Logical reasoning puzzles • Scientific experiments and simulations • Conceptual models and diagrams
Bodily-Kinesthetic Intelligence	Physical movement, motor control	• Hands-on activities and experiments • Role-playing and simulations • Building and crafting projects • Physical exercises and movement-based learning

(continued)

Table 3-2. (*continued*)

Type of Intelligence	Strengths	Instructional Strategies
Musical Intelligence	Rhythm and music	• Music composition and performance • Listening to and analyzing musical pieces • Singing exercises and vocal training • Incorporating music into lessons and mnemonic devices
Interpersonal Intelligence	Understanding and relating to other people	• Group discussions and collaborative projects • Role-playing and simulations of real-life interactions • Peer feedback and cooperative learning activities
Intrapersonal Intelligence	Introspection and self-reflection	• Self-reflection prompts and journaling • Personal goal-setting exercises and self-assessments • Individual research projects and independent study
Naturalistic Intelligence	Finding patterns and relationships to nature	• Field trips and nature exploration • Environmental observation and data collection • Nature documentaries and documentaries on ecological systems

(*continued*)

Table 3-2. (*continued*)

Type of Intelligence	Strengths	Instructional Strategies
Existential Intelligence	The ability to see the big picture	• Philosophical discussions and debates • Reflective writing assignments and thought-provoking prompts • Studying literature and art that explores existential questions

Learning Modalities: The Concept

The learning modalities play a pivotal role in shaping the design and delivery of effective learning experiences. These modalities encompass learners' strengths and preferences in response to stimuli and information processing. As instructional designers, we must recognize and understand these individual differences, as they serve as a foundation for various models and families of learning styles.

Learning modalities are behavioral patterns encompassing cognitive, affective, and physiological aspects. They indicate how learners approach and engage in the learning process. While different psychologists may define the term uniquely, learning modalities generally revolve around specific characteristics.

These characteristics encompass how learners think, feel, and respond to educational stimuli. They are developed and solidified over time, representing persistent qualities in behavior. Various factors, including past experiences, cognitive processes, environmental demands, and the nature of the current learning task, influence a learner's learning style.

Understanding and acknowledging learning modalities provides valuable insights into the underlying causes of learners' behaviors, shedding light on how their minds relate to and interact with the world. By recognizing and addressing these preferences, instructional designers can create optimal educational conditions that align with learners' styles, enhancing their overall learning experience.

Moreover, learning modalities serve as a gateway to various models and families within the field. These models offer frameworks and categorizations that help classify and understand learners' diverse learning styles. By studying and applying these models, instructional designers gain deeper insights into the complexities of learners' preferences, enabling them to design more targeted and effective instructional strategies.

These modalities offer valuable insights for instructor-led training, and a single modality is not enough to create instructional materials or solutions that effectively cater to a diverse group of learners. It is crucial to integrate multiple modalities to address the varied needs of learners comprehensively. While the modalities discussed in this chapter provide a useful framework, they also have limitations that must be considered when assessing learner styles. Equally important is the emphasis on encouraging learners to develop a broad spectrum of skills and strategies beyond their preferred styles. This multifaceted approach not only fosters adaptability and resilience but also equips individuals to navigate diverse challenges and contexts effectively.

Learning Modalities: The Analogy

Compare learning modalities to a Swiss Army knife – a multi-tool encompassing various functions. Just as a Swiss Army knife offers different tools for different needs, learners have distinct preferences and requirements for absorbing information.

A Swiss Army knife includes various tools such as blades, screwdrivers, and bottle openers. Similarly, learners may prefer interactive experiences, detailed explanations, practical applications, or text-based materials.

Just as a Swiss Army knife user selects the tool that suits the task, learners should be able to access instructional materials that align with their preferred learning styles. This diversity in approach ensures learners effectively engage with the content, leading to a deeper understanding and better knowledge retention.

Education accommodating various learning styles is like a well-equipped Swiss Army knife – versatile, adaptable, and tailored to individual needs, helping learners succeed in their intellectual pursuits.

Learning Modalities: Embracing Diversity in Learning

Let us understand the concept of learning modalities through the eyes of different learners from diverse backgrounds, preferences, and knowledge. In a bustling metropolis like New York, where cultures collide and ideas intertwine, we encounter four K-12 learners who recently enrolled for a four-year undergraduate program in the university, each bringing their unique experiences, values, and family support.

Meet the "Learners Squad," a group of extraordinary learners, each with their dreams, aspirations, and ways of learning. As we progress through the chapter, we will witness their learning journey unfold, discovering how their preferred learning styles evolve. Along the way, we will understand the significance of designing instructions using blended approaches, catering to these learners' individual needs and strengths, which will not only help them perform better in the university exams but also allow them to walk onto their desired career paths.

Aisha: The Multiculturalist

Aisha is a bright-eyed learner hailing from a culturally rich neighborhood in Queens. Her parents are immigrants from different countries and instilled a deep appreciation for diversity in her. She learns about various traditions, languages, and customs at home, forming a solid sense of empathy and open-mindedness. Aisha's family support encourages her to embrace new perspectives, making her a natural bridge between different cultures at school. As she starts her educational journey in this new environment, Aisha's curiosity and enthusiasm for learning will inspire her peers.

Ethan: The Traditionalist

From the close-knit community in the Bronx, we meet Ethan, who holds his family's traditional values close to his heart. His parents emphasize the importance of discipline, hard work, and adherence to established norms. For Ethan, routine and structure provide stability and security, allowing him to thrive academically. However, he might face challenges in this new school setting when presented with unconventional ideas or unfamiliar concepts. Nonetheless, with his family's unwavering support and determination, Ethan is well prepared to tackle any obstacle that comes his way.

Sofia: The Environmentalist

Sofia resides in the green outskirts of Brooklyn, where her family passionately advocates for environmental sustainability and conservation. Growing up with a deep connection to nature, Sofia has developed a strong sense of responsibility toward the planet. Her family's eco-conscious beliefs inspire her to seek knowledge and understand environmental issues. As she joins the new school, Sofia's passion for the environment will likely lead her toward science subjects and projects that explore sustainable practices, making her an eco-warrior.

Amir: The Activist

Amir's family lives in the heart of Manhattan, where activism and social justice are integral parts of their lives. Discussions around inequality and human rights have profoundly impacted Amir's worldview, instilling a drive to bring about meaningful change. At school, Amir's strong sense of justice will make him actively participate in discussions and activities addressing societal issues. His family's unwavering support and encouragement emboldened him to voice his beliefs and stand up for what he believes is right, making him a potent force for social change in the school community.

These four learners from diverse backgrounds, preferences, and knowledge are joining the new college, and their journeys of exploration and growth are about to commence. Each learner's family support and ideology will play a pivotal role in shaping their learning behavior and style, propelling them forward as they navigate the complexities of education and the world beyond. Their unique experiences and perspectives will enrich the college community, fostering an environment that celebrates the beauty of diversity and the power of individual beliefs.

VARK Model

The VARK learning style model, developed by Neil Fleming, incorporates a questionnaire to identify an individual's sensory modality preference in learning. This model categorizes learners into four distinct learning modes, represented by the acronyms VARK. Table 3-3 summarizes the instructional strategies for each learning modality of the VARK model.

Table 3-3. *VARK Model*

Learning Preference	Description	Instructional Strategies
Visual (V)	Relies on visual cues and images	Charts, diagrams, graphs, videos
		Visual representations
Auditory (A)	Learns best through listening and speaking	Verbal explanations, lectures, discussions
		Audiobooks, group discussions
Reading/Writing (R)	Prefers written materials	Reading books, articles, and written instructions
		Note-taking, making lists, organizing information
Kinesthetic (K)	Thrives on hands-on experiences and movement	Experiments, role-playing, interactive demonstrations
		Physical engagement, practical application

Best-Suited Approach for the Learners Squad

At the start of their learning journey in a new college, the administration administered a VARK test to the Learners Squad to better understand their learning preferences. The learning squad members responded to a series of questions (see Appendix A).

Table 3-4 shows the assessment scores from the VARK test; the administration classified the learners into various learner categories. The college formulated a list of subjects they might excel in using these categories and considering their interests.

Table 3-4. *VARK Analysis of the Learners Squad*

Learner Name	Questionnaire Scores	Modality	Subjects They Might Excel In
Aisha	V-9, A-8, R-10, K-9	Multimodal (VARK)	• Intercultural Communication • Language Learning • Social Sciences
Ethan	V-10, A-3, R-4, K-9	Visual and Kinesthetic	• Science • Mathematics
Sofia	V-9, A-4, R-8, K-3	Visual and Read/Write	• Biology • Ecology • Environmental Science
Amir	V-4, A-9, R-10, K-8	Auditory, Read/Write, and Kinesthetic	• Sociology • Political Science • Human Rights • Law

Dunn and Dunn Model

The Dunn and Dunn Learning Style Model, developed by Rita and Kenneth Dunn, is an educational theory that suggests individuals have different preferred ways of learning and processing information. The idea behind the model is that learners learn best when instruction aligns with their unique learning style preferences. The Dunns identified five main categories of learning styles, referred to as "modalities," and also proposed various environmental factors that can influence learning (Figure 3-2). The five modalities are as follows:

Environmental: This refers to the learning environment individuals find most conducive to learning. Some prefer a quiet setting, while others thrive in a more interactive or collaborative environment.

Emotional: This modality focuses on the emotional or affective aspects of learning. Some learners may be more receptive to learning when motivated, while others may prefer a calm and relaxed emotional state.

Sociological: This modality relates to learning preferences in group dynamics. Some learners may prefer to work alone, while others thrive in group settings.

Physical: This refers to using one's physical senses during learning. Some learners may prefer hands-on experiences, while others may be more auditory or visual learners.

Psychological: This modality relates to cognitive preferences. Some learners may benefit from structured and organized learning materials, while others may prefer more exploratory and open-ended approaches.

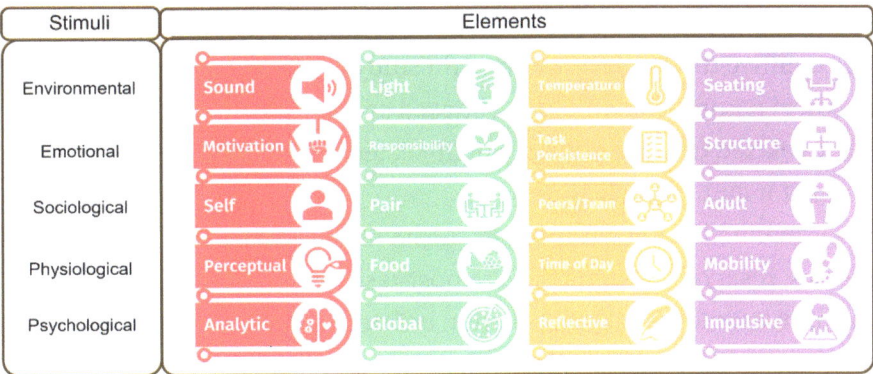

Figure 3-2. *Stimuli and Elements of the Dunn and Dunn Model*

Best-Suited Approach for the Learners Squad

Following four months of VARK test, the learning squad underwent assessment with the different stimuli encompassing all the learning modalities presented by the Dunn and Dunn model. This assessment aimed to offer learners blended learning approaches for excelling in specific fields. The learners responded to a series of questions (refer to Appendix A).

The assessment results led to the identification of distinct stimuli tailored to the learners. Afterward, the administration compiled a list of subjects they might excel in based on these categories and the learners' interests.

Aisha: The Multiculturalist

Aisha's optimal learning environment encompasses serene acoustics, well-lit surroundings, and a comfortably warm setting. Her adaptability extends to seating choices, allowing her to tailor her study space. Motivated by both academic success and a fervent curiosity about diverse cultures, Aisha thrives in collaborative settings, embracing open-mindedness and empathy while working with peers from varied backgrounds. She values guidance from knowledgeable adults, finding structured oversight to be motivating.

Aisha's multifaceted learning approach involves auditory, visual, and textual methods, with narratives and discussions enhancing her retention. Her cognitive rhythm favors late-night study sessions, aligning with her analytical nature. Transitioning between spaces, she effortlessly switches from a relaxed reading chair to a high-seated writing station or a supine posture for auditory learning.

Disciplined in her study habits, Aisha refrains from eating during sessions, reserving sustenance as a post-task reward. Her methodical approach extends to task completion before embarking on new challenges, a practice enhancing her efficiency.

With a penchant for analytical exploration and meticulous note-taking, Aisha excels in fields demanding critical thinking, comprehensive analysis, and a nuanced perspective. This aligns her strongly with areas such as cross-cultural studies or cultural anthropology. Aisha's unique blend of adaptability, intrinsic motivation, analytical acumen, empathy, and disciplined work ethic sets the stage for a promising academic journey and impactful contributions to her chosen field.

Ethan: The Traditionalist

Ethan's optimal learning environment thrives in tranquil surroundings devoid of auditory distractions, complemented by subdued lighting and a comfortable, moderate temperature. Seated at a high table, his preference for structure is evident, fostering an environment of focus and discipline. Ethan draws motivation from academic achievements and his passion-driven pursuits, promoting a balanced drive for excellence.

Disciplined and responsible, Ethan's work ethic is characterized by a diligent approach to task completion before advancing to the next endeavor. He finds solace and effectiveness in solitary learning, eschewing group dynamics that may disrupt his established rhythm and commitment. Guided reminders about his progress serve as a motivational beacon, with teacher encouragement fueling his journey toward milestones.

Ethan's information retention is optimized through visual and auditory channels, mainly video content and storytelling. He finds the practical application of concepts intriguing, often translating theoretical knowledge into real-life scenarios and interactive games. Averse to eating during study sessions, Ethan reserves consumption for post-task accomplishment, reinforcing his disciplined approach.

A morning person, Ethan capitalizes on his early hours for effective learning and his heightened morning memory recall. His preference for a stationary study approach aligns with his analytical inclination, accentuated by his penchant for detailed note-taking and data analysis.

Ethan's blend of discipline, responsibility, and penchant for in-depth analysis positions him well for subjects demanding methodical exploration, structured understanding, and detailed investigation. Fields such as mathematics, physics, or research-based disciplines may align favorably with his strengths, offering a promising academic trajectory and potential for impactful contributions.

Sofia: The Environmentalist

Sofia's optimal learning environment is intertwined with the soothing symphony of nature's whispers and unfiltered human voices. Illuminated by natural light and cooled by a refreshing temperature, Sofia's unconventional choice of floor seating attests to her distinct learning approach. She draws motivation from academic accomplishments and her dedication to environmental stewardship, forming a harmonious balance of aspirations.

Guided by discipline and a profound sense of responsibility for nature, Sofia seeks consistent reminders to propel her task completion. Collaborative learning with like-minded peers, who share her reverence for the environment, resonates deeply with her. An independent spirit, Sofia relishes uncovering solutions on her terms, embodying a progressive learning ethos.

Sofia's appetite for knowledge finds expression through thought-provoking documentaries on environmental shifts. Her learning retention flourishes through lively peer discussions, complemented by movie viewing and article perusal. Sipping a beverage enhances her studying experience, aiding her concentration.

An early riser, Sofia harnesses the early morning tranquility to internalize knowledge. Her penchant for mobility – walking while listening and adopting a seated stance while reading and writing – accentuates her versatility.

Sofia's reflective nature aligns with her propensity for dialogue, relishing idea exchange and input before shaping her viewpoints. This trait suggests her potential to excel in subjects requiring comprehensive analysis and thoughtful syntheses, such as environmental studies, ecology, or sustainable development. Her interactive approach and eco-conscious spirit could pave the way for impactful contributions in these domains.

Amir: The Activist

Amir thrives in an environment that blends the gentle hum of low to moderate sound with dim, focused lighting and a comfortably cool temperature. Nestled in a cozy chair, his ideal study setting echoes his pragmatic and insightful approach to learning. His motivation stems from academic achievements and a profound passion for assisting others, particularly in problem-solving capacities.

An embodiment of discipline and a fervent advocate for justice, Amir's sense of responsibility extends to individuals and society. Frequent reminders, often initiated by a guiding adult, serve as his compass for task completion. Collaborative learning with peers under the watchful eye of a mentor resonates with his dedication to progress, while learning from seasoned experts in the realm of justice and law fuels his intellectual curiosity.

Amir's learning journey is characterized by a thirst for direction and a hunger for discovery. He thrives when navigating legal intricacies, guided by documentaries, films, and case studies. His method of retention flourishes in dialogue with peers, reflecting his inclination to engage, dissect, and deliberate.

Empowered by beverage sips and gum-chewing, Amir's study sessions mirror his dynamic approach. He finds solace in the serene ambiance of late-night reading, contrasting his morning disposition. Stationary during the study, he channels his impulsive energy into multitasking, a trait that sets him on various paths of exploration.

Amir's learning style harmonizes with both impulsive and reflective tendencies. His penchant for idea exchange and input-seeking underscores his enjoyment of collaborative brainstorming, mirroring his adaptable approach. This blend of traits suggests his potential to excel in law, social justice, political science, or even conflict resolution, where he can channel his passion for helping others and his flair for navigating intricate matters into tangible contributions.

Gregorc Learning Style Delineator

The Gregorc Learning Model, originating in 1969 as the energic model and evolving into the mind style model in 1984, was formulated by Anthony F. Gregorc, a distinguished international lecturer, researcher, author, and consultant. This model introduces a systematic learning approach, categorizing individuals into four cognitive styles that reflect their information processing, perception, and organizational abilities (Figure 3-3). Understanding these styles allows individuals to identify strengths and weaknesses, facilitating strategic career advancement.

The Gregorc Learning Model centers on two essential qualities: perceptual quality and ordering ability. Perceptual qualities, categorized as concrete or abstract, influence how individuals engage with sensory information and intuitive understanding. Ordering abilities, sequential or random, affect how individuals organize thoughts and ideas, impacting their approach to tasks and problem-solving.

The learning styles within the model include the following:

> **Abstract Random (AR):** Learning through visualization and imagination, preferring flexible guidelines and forming strong relationships

Abstract Sequential (AS): Using planned visualization and logical analysis, making decisions thoughtfully, and communicating diplomatically

Concrete Sequential (CS): Thriving with detailed, step-by-step instructions, learning through sensory experiences and hands-on learning

Concrete Random (CR): Learning through senses and trial and error, excelling in competitive environments, risk-taking, and independent problem-solving

By discerning their learning style within the Gregorc framework, individuals can optimize their educational strategies and chart an informed career trajectory. Table 3-5 summarizes different parameters for each learning modality of the Gregorc Learning Style Delineator.

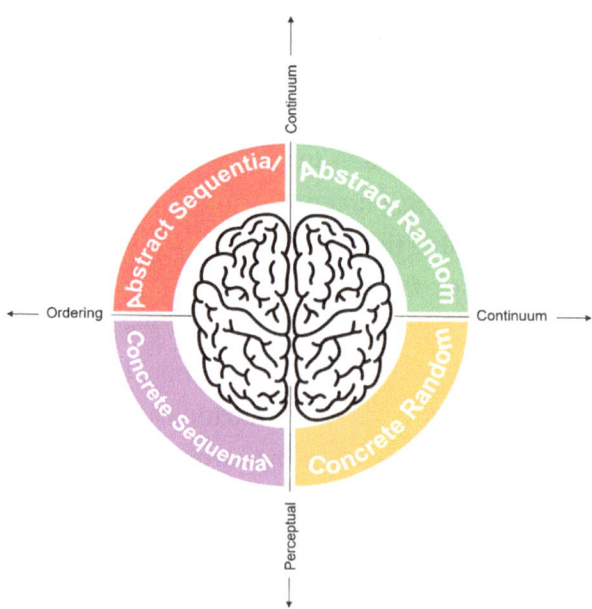

Figure 3-3. *Styles of Gregorc's Delineator*

Table 3-5. *Overview of Gregorc's Learning Style Delineator*

Types of Learning	Key Characteristics	Weakness	Instructional Strategies
Abstract Random	• Visual and imaginative learning • Thrives in unstructured settings • Values broad guidelines	• Explain emotions • Competitive context • Authoritarian interactions • Restrictive setting • Unfriendly peers • Focused attention • Detailed information • Embracing critique	• Provide creative and open-ended assignments • Encourage exploration and varied perspectives
Abstract Sequential	• Planned visualization and logical analysis • Thoughtful decision-making • Diplomatic communication	• Differing views collaboration • Limited time for depth • Repetitive tasks • Abundant rules • Emotional thinking • Expressing feelings • Diplomatic persuasion • Shared conversation	• Offer structured, step-by-step tasks • Encourage logical planning and organization

(*continued*)

Table 3-5. (*continued*)

Types of Learning	Key Characteristics	Weakness	Instructional Strategies
Concrete Sequential	• Detailed, step-by-step learning • Hands-on and sensory experiences • Practical and direct approach	• Group work • Aimless discussions • Unorganized setting • Unclear Instructions • Unpredictable colleagues • Abstract concepts • Imagination required • Ambiguous questions	• Provide clear, detailed instructions • Incorporate hands-on activities and real-world applications
Concrete Random	• Sensory-based and trial-and-error learning • Thrives in competitive and independent situations • Risk-taking and experimentation	• Limits and rules • Official reports • Fixed routines • Repetition • Detailed records • Displaying process • Single choice • No alternatives	• Introduce challenges and problem-solving activities • Encourage independent projects and creative exploration

Best-Suited Approach for the Learners Squad

After four months of the Dunn and Dunn test, the administration assessed the learning squad to examine their cognitive style for information processing. This test aimed to comprehend their cognitive structure and offer the most suitable instructional strategies to enhance their learning experience. The learners responded to a series of questions (see Appendix A). Table 3-6 summarizes the preferred modality of the squad.

Table 3-6. *Gregorc's Analysis for the Learners Squad*

Learner	Modality	Subject Best Learned
Aisha – The Multiculturalist	Abstract Random	Creative Arts, Sociology
Ethan – The Traditionalist	Concrete Sequential	Mathematics, Engineering
Sofia – The Environmentalist	Concrete Random	Environmental Science, Biology
Amir – The Activist	Abstract Sequential	Law, Political Science

Myers-Briggs Type Indicator

As far as we understand, recognizing the diverse ways individuals learn is crucial for achieving optimal learning outcomes. Each learner possesses unique strengths, preferences, and weaknesses that influence their learning experience. This understanding has led to the development of tools like the Myers-Briggs Type Indicator (MBTI), a personality evaluation tool that can significantly enhance the effectiveness of training programs.

The Role of the Myers-Briggs Type Indicator

The Myers-Briggs Type Indicator (MBTI) offers valuable insights into individuals' personality traits and learning preferences. Before designing a training course, instructional designers can leverage this tool to better understand learners' styles and tastes. Instructional designers can maximize engagement, participation, and knowledge retention by tailoring instructional methods to align with learners' learning styles.

Aligning Instructional Design with Learner Preferences

To illustrate the importance of aligning instructional design with individual learning styles, let's consider a training program focused on enhancing presentation skills. The course utilizes interactive activities, practice sessions, and feedback to develop participants' public speaking abilities.

In this scenario, learners who thrive in dynamic and hands-on environments, enjoy collaborating with peers, and readily embrace opportunities to practice and receive feedback are likely to excel in the course. Their confidence grows as they actively engage in group discussions, refine their delivery skills, and adapt to audience feedback.

On the other hand, individuals who prefer individual reflection and a more structured approach to learning may face challenges in the course. They may need help with the fast-paced nature of group activities or feel overwhelmed by the need for spontaneous responses. These learners might benefit from additional support, such as dedicated reflection time or structured presentation frameworks.

The Myers-Briggs Type Indicator (MBTI) deconstructs learner personalities into 16 distinct types (Figure 3-4), each characterized by a combination of four dichotomies.

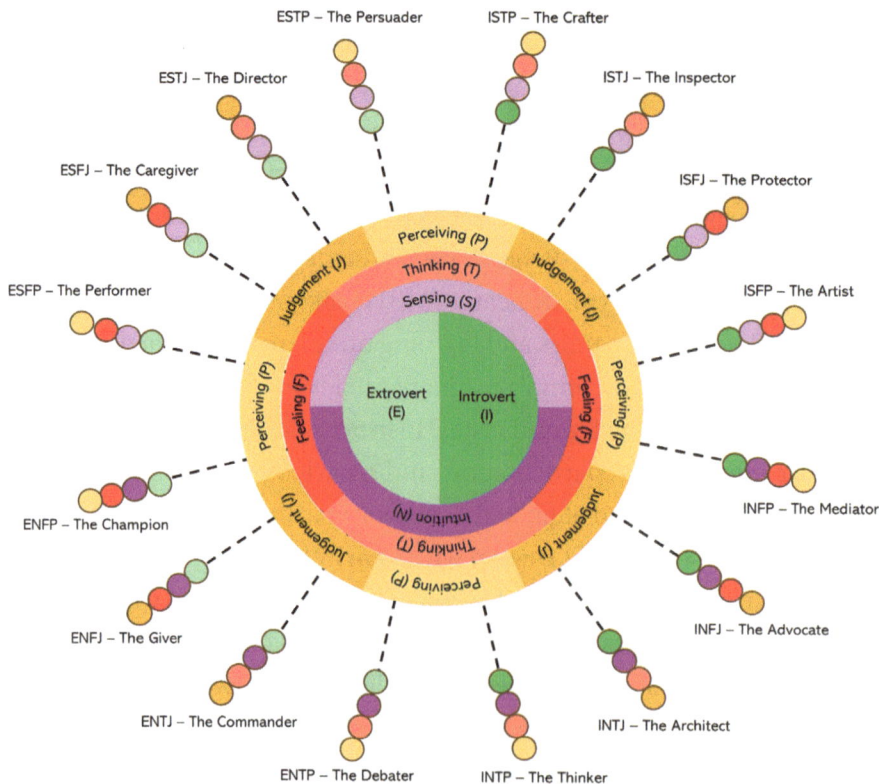

Figure 3-4. *Myers-Briggs Personality Types*

These dichotomies, including Extroverted (E) vs. Introverted (I), Sensing (S) vs. Intuition (N), Feeling (F) vs. Thinking (T), and Judging (J) vs. Perceiving (P), provide valuable insights into learners' preferences, motivations, and learning styles. Table 3-7 summarizes the different parameters for all pairs of MBTI dichotomies.

Table 3-7. *Dichotomies of MBTI*

Dichotomy	Description	Representation	Instructional Strategies
E-I dichotomy	It captures learners' tendencies toward internal or external influences	Extroverted learners draw inspiration from external sources	• Design highly interactive content • Include hands-on activities and exercises • Change topics frequently
		Introverted learners are motivated by internal factors	• Provide individual work opportunities • Include optional supplementary materials
S-N dichotomy	It highlights how learners perceive and grasp new information	Sensing learners prefer concrete and direct learning experiences	• Focus on memorizing key facts • Use visual elements such as diagrams and videos

(*continued*)

Table 3-7. (*continued*)

Dichotomy	Description	Representation	Instructional Strategies
		Intuitive learners excel at understanding concepts in broader and abstract categories	• Illustrate theoretical concepts • Utilize metaphors and symbols • Encourage pattern recognition and associations
F-T dichotomy	It sheds light on how learners relate to and engage with learning content	Feeling learners connect deeply with subjects that resonate with their emotions	• Design collaborative and interactive courses • Highlight the societal impact of concepts
		Thinking learners approach learning with a logical and analytical mindset	• Present concepts logically • Foster critical questioning and debating • Provide clear performance criteria

(*continued*)

Table 3-7. (*continued*)

Dichotomy	Description	Representation	Instructional Strategies
J-P dichotomy	It reflects learners' preferences for structured or flexible learning environments	Judging learners thrive in well-organized and systematic settings	• Create well-structured course schedules • Communicate expectations and outcomes
		Perceiving learners excel in more spontaneous and adaptable learning contexts	• Incorporate open-ended statements and questions • Include diverse sources of information • Implement deadline-driven deliverables

Table 3-8 summarizes all sixteen personality types of MBTI.

Table 3-8. Overview of MBTI Personalities

Four-Letter Code	Full Form	Key Characteristics	Strengths	Weaknesses	Career Paths
ISTJ – The Inspector	Introverted Sensing Thinking Judging	• Reserved • Practical • Loyal • Orderly • Traditional	• Dependable • Organized • Responsible • Detail oriented	• Inflexible • Overly cautious • Struggle with change or new ideas	• Accountant • Engineer • Lawyer • Administrator • Project Manager
ISTP – The Crafter	Introverted Sensing Thinking Perceiving	• Independent • Adaptable • Practical skills	• Resourceful • Logical • Adaptable • Problem-solving	• Dislikes routine • Impulsive • Not good at expressing emotions	• Engineer • Technician • Detective • Pilot • Athlete
ISFJ – The Protector	Introverted Sensing Feeling Judging	• Warm hearted • Dedicated • Supportive • Protective	• Empathetic • Dependable • Responsible • Good at anticipating the needs of others	• Reluctant to change • Overly self-sacrificing • Avoid conflict	• Nurse • Counselor • Teacher • Social Worker • Librarian

Type				
ISFP – The Artist	Introverted Sensing Feeling Perceiving	• Easygoing • Artistic • Expressive • Value personal freedom	• Creative • Sensitive • Compassionate • Tune with their emotions	• Struggle with decision-making • Overly private • Avoid conflict	• Musician • Writer • Artist • Counselor • Veterinarian
INFJ – The Advocate	Introverted Intuitive Feeling Judging	• Idealistic • Insightful • Creative • Analytical • Driven to make a positive impact	• Insightful • Empathetic • Passionate • Committed to personal growth	• Overly sensitive • Perfectionistic • Struggle with setting boundaries	• Counselor • Writer • Psychologist • Nonprofit Work • Social Justice Advocate
INFP – The Mediator	Introverted Intuitive Feeling Perceiving	• Idealistic • Values driven • Empathetic • Desire to make a positive difference	• Compassionate • Creative • Open-minded • Dedicated to personal growth	• Overly sensitive • Prone to self-doubt • Struggle with practical tasks	• Writer • Counselor • Social Worker • Artist • Human rights advocate

(continued)

Table 3-8. (*continued*)

Four-Letter Code	Full Form	Key Characteristics	Strengths	Weaknesses	Career Paths
INTJ – The Architect	Introverted Intuitive Thinking Judging	• Visionary • Strategic • Logical • Analytical • Goal oriented	• Independent • Innovative • Organized • Efficient problem solvers	• Arrogant or cold • Overly critical • Struggle with expressing emotions	• Scientist • Engineer • Entrepreneur • Strategist • Academic
INTP – The Thinker	Introverted Intuitive Thinking Perceiving	• Analytical • Introspective • Value intellect • Logical reasoning	• Rational • Objective • Open-minded • Adept at complex problem-solving	• Detached • Unemotional • Overthinking • Struggle with decision-making	• Scientist • Researcher • Philosopher • Computer Programmer • Analyst
ESTP – The Persuader	Extraverted Sensing Thinking Perceiving	• Outgoing • Action oriented • Center of attention • Thrive in the present moment	• Energetic • Adaptable • Confident • Skilled at engaging with others	• Impulsive • Risk-taking • Maybe insensitive • Struggle with long-term planning	• Salesperson • Entrepreneur • Athlete • Detective • Paramedic

Type	Personality	Traits		Challenges	Careers
ESTJ – The Director	Extraverted Sensing Thinking Judging	• Assertive • Organized • Responsible • Strong sense of responsibility	• Efficient • Dependable • Natural leaders • Excel in structured environments	• Overly controlling • Inflexible • Struggle with expressing emotions	• Manager • Administrator • Military Officer • Project Manager • Police Officer
ESFP – The Performer	Extraverted Sensing Feeling Perceiving	• Outgoing • Spontaneous • Enjoy the spotlight • Thrive in social situations	• Fun-loving • Adaptable • Energetic • Skillful at engaging and entertaining others	• Seek immediate gratification • Have difficulty with long-term planning • Sensitive to criticism	• Actor • Entertainer • Sales Representative • Event Planner • Coach
ESFJ – The Caregiver	Extraverted Sensing Feeling Judging	• Warm-hearted • Outgoing • Caring • Generous • Supportive	• Compassionate • Dedicated • Cooperative • Attentive to the needs of others	• Overly self-sacrificing • Seek approval from others • Avoid conflict	• Teacher • Counselor • Nurse • Social Worker • Event Coordinator

(continued)

113

Table 3-8. (*continued*)

Four-Letter Code	Full Form	Key Characteristics	Strengths	Weaknesses	Career Paths
ENFP – The Champion	Extraverted Intuitive Feeling Perceiving	• Charismatic • Energetic • Enthusiastic • Creative • Driven by ideals	• Optimistic • Imaginative • Passionate • Skilled at inspiring and motivating others	• Struggle with practical details • Prone to being over-idealistic • Have difficulty with routine tasks	• Teacher • Counselor • Writer • Journalist • Social Worker
ENFJ – The Giver	Extraverted Intuitive Feeling Judging	• Loyal • Sensitive • Understanding • Generous • Supportive	• Empathetic • Charismatic • Supportive • Skilled at interpersonal relationships	• Overly idealistic • Prone to burnout • Struggle with setting boundaries	• Teacher • Counselor • Social Worker • Human Resources Manager • Event Planner

Type	Traits				Careers
ENTP – The Debater	Extraverted Intuitive Thinking Perceiving	• Innovative • Inquisitive • Enjoy exploring ideas and possibilities	• Quick-witted • Adaptable • Excellent problem solvers • Thrive in challenging environments	• Have difficulty with practical tasks • Easily bored • May appear argumentative	• Scientist • Engineer • Entrepreneur • Strategist • Lawyer
ENTJ – The Commander	Extraverted Intuitive Thinking Judging	• Assertive • Confident • Strong leadership skills • Excel at planning and organizing projects	• Strategic thinkers • Efficient • Logical • Skilled at making tough decisions	• Controlling • Impatient • Struggle with emotions	• CEO • Manager • Entrepreneur • Lawyer • Executive • Consultant

115

Best-Suited Personality for the Learners Squad

After four months of the Gregorc test, the administration assessed the learning squad to examine their personality styles. This test aimed to identify the personality types and provide suitable approaches to help the learners excel in specific fields. The learners responded to a series of questions (refer to Appendix A).

Aisha

Aisha's preference for a quiet environment, bright light, and flexibility in seating options indicates an "I" (Introverted) nature. Her motivation driven by performing well and her interest in exploring diverse traditions align with an "N" (Intuitive) and "F" (Feeling) personality. Aisha's inclination to engage with groups of peers from various backgrounds could suggest a preference for "E" (Extroverted) interactions. Therefore, Aisha might fit into the ENFP (Extroverted, Intuitive, Feeling, Perceiving) or INFP (Introverted, Intuitive, Feeling, Perceiving) personality type.

Ethan

Ethan's disciplined approach to completing tasks before moving on indicates a strong "J" (Judging) trait. His preference for learning alone and his methodical study habits align with an "I" (Introverted) nature. Ethan's enjoyment of watching videos and discussing concepts with others suggests he could be more aligned with "N" (Intuitive) and "F" (Feeling) traits. Therefore, Ethan might fit into the INFJ (Introverted, Intuitive, Feeling, Judging) or ISFJ (Introverted, Sensing, Feeling, Judging) personality type.

Sofia

Sofia's preference for a serene environment with natural sounds and her enjoyment of unconventional seating options align with the "I" (Introverted) trait. Her affinity for discussing ideas with others and engaging in real-life applications suggests an "N" (Intuitive) and "F" (Feeling) nature. Sofia's strong sense of responsibility toward nature and the environment resonates with the "J" (Judging) trait. Therefore, Sofia could be categorized as an INFJ (Introverted, Intuitive, Feeling, Judging) or an ISFJ (Introverted, Sensing, Feeling, Judging).

Amir

Amir's disciplined and responsible demeanor, robust sense of justice, and advocacy for assisting others harmonize effectively with the "J" (Judging) trait. His preference for collaborative learning and seeking guidance from knowledgeable adults suggests a "P" (Perceiving) trait. Amir's enjoyment of discussing ideas with others and his adaptability in brainstorming reflect traits associated with the "E" (Extroverted) nature. Therefore, Amir could be categorized as an ESTJ (Extroverted, Sensing, Thinking, Judging) or an ESFJ (Extroverted, Sensing, Feeling, Judging).

Jackson's Learning Styles Profiler

Learning Styles Profiler (LSP) is a hybrid model of learning developed by Charles J. Jackson and Lawty-Jones in 1996. The LSP incorporates personality, self-efficacy, goal orientation, emotional intelligence, and conscientiousness to measure functional and dysfunctional learning associated with high performance.

The LSP identifies five different learning styles, each reflecting various aspects of the learning process. Table 3-9 summarizes the different parameters for the five learning styles of Jackson's Learning Style Profiler.

Table 3-9. Overview of Jackson's Learning Styles

Learning Style	Description	Implication	Instructional Strategies
High Sensation Seeker	Prefers new and challenging activities and consciously works toward achieving challenging goals	Flourishes in settings characterized by novelty and excitement More open to taking risks and embracing new learning opportunities	Provide hands-on and experiential learning opportunities Encourage exploration and experimentation in the learning process
Goal-Oriented Achiever	Sets specific and challenging learning goals, has the self-confidence to achieve them, and has control over goals and direction	Driven and motivated to achieve learning objectives Likely to excel in structured and goal-oriented learning environments	Provide clear and achievable learning goals Offer structured learning materials and resources
Emotionally Intelligent Achiever	Learns from mistakes and approaches learning with scientific, detached, and autonomous reasoning	Adaptable and responsive to feedback, enhancing emotional intelligence Effective at learning from experiences and improving problem-solving skills	Provide opportunities for self-reflection and self-assessment Encourage collaboration and group discussions to learn from peers

Conscientious Achiever	Responsible and rational learners use complex higher-level learning strategies to comprehend difficult information	Diligent and focused on achieving positive learning outcomes Likely to excel in academically challenging environments	Provide challenging and intellectually stimulating learning tasks Foster a supportive learning environment to reduce anxiety and stress
Deep Learning Achiever	Positive inclination toward learning, open to new ideas, seeks to understand knowledge	Motivated by intellectual curiosity Likely to excel in academic settings that encourage critical thinking and exploration	Encourage self-directed learning and exploration Use real-world examples and applications to make learning relevant

Best-Suited Personality for the Learners Squad

After four months of the MBTI test, the administration assessed the learning squad's personality using Jackson's Learning Style Profiler. This test aimed to identify a new personality dimension that complements the personality traits identified through MBTI. Identifying these interconnected dimensions seeks to determine the most effective instructional strategy for enhancing their learning efficiency. The learners responded to a series of questions (refer to Appendix A). Table 3-10 summarizes the modality and subject for the squad.

Table 3-10. *Jackson's Learning Style Analysis of the Learners Squad*

Learner	Modality	Subjects They Might Excel In
Aisha	Conscientious Achiever, Emotionally Intelligent Achiever, and High Sensation Seeker	Literature, Psychology, Sociology, or Art, where she can engage in deep discussions, analyze emotions, and explore different perspectives
Ethan	Goal Oriented Achiever, Conscientious Achiever, and Emotionally Intelligent Achiever	He could excel in Mathematics, Engineering, Computer Science, or Economics, leveraging his methodical study habits and analytical skills as valuable assets
Sofia	Conscientious Achiever and Emotionally Intelligent Achiever	She could contribute effectively in fields such as Environmental Science, Psychology, Anthropology, or Sustainable Development, leveraging her dedication to the environment and her skill in empathizing with people's emotions, which are valuable assets.
Amir	Goal Oriented Achiever, Conscientious Achiever, and Emotionally Intelligent Achiever	His strong sense of justice, responsibility, and collaborative ability would be assets in law, Political Science, Social Work, and Business Management

Apter's Motivational Style Profile

Apter's (2001) reversal theory revolves around the study of individual motivational styles. The Apter Motivational Style Profile (AMSP) is a comprehensive tool used for personality profiling. Its primary focus lies in understanding the impact of an individual's motivational factors on their life, personal, or organizational.

The theory proposes that people experience eight core motivational states, organized into four pairs of opposing states. Within these pairs, individuals constantly undergo dynamic shifts or reversals. Psychological well-being and effectiveness are optimized when individuals can experience all eight states and adapt their state to meet the demands of various situations.

Central to the AMSP is the concept that our experiences are shaped by different ways of perceiving the world, each grounded in a fundamental value or motive. The profile identifies four pairs of opposite motivational states through which individuals frequently switch or "reverse" in their everyday lives and circumstances. Table 3-11 summarizes the instructional strategies for all pairs of AMSP.

Table 3-11. Overview of Apter's Motivational Style Profile

Domain	State	Description	Instructional Strategies
Means-Ends	Telic	Motivation in this state comes from goal achievement and accomplishment, with value placed on an activity's future consequence or benefit (the ends).	Encourage setting clear and achievable goals. Emphasize the long-term benefits of their efforts and how they align with their values and aspirations.
	Paratelic	Motivation in this state comes from enjoying an activity's process	Promote activities that they find intrinsically enjoyable and fulfilling. Encourage them to explore different hobbies and engage in creative pursuits.
Rules	Conforming	Motivation comes from belonging and doing what is expected, with rules and norms experienced as supportive.	Emphasize the importance of societal norms and rules. Provide a supportive and inclusive environment that values compliance and fosters a sense of belonging.
	Negativistic	Motivation comes from breaking out of the rules and expectations. In this state, the motive is to act outside the status quo.	Encourage critical thinking and questioning of the status quo. Offer opportunities for creativity and innovation while maintaining a respectful approach.

Transactions	Mastery	Motivation comes from power, control, and strength – for oneself or another.	Provide challenging tasks that require problem-solving and leadership skills. Provide personal growth and skill development opportunities.
	Sympathy	Motivation comes from providing or receiving personal care and emotional support for oneself or another.	Promote empathy and emotional intelligence. Encourage acts of kindness and opportunities to support and care for others.
Relationships	Autic (self)	Motivation comes from filling one's needs and focusing on personal success or nurturing.	Encourage self-care and personal development. Provide resources for setting and achieving personal goals and pursuing individual interests.
	Alloic (other)	Motivation comes from identifying with others and filling their needs.	Foster a sense of community and teamwork. Encourage collaboration and understanding of the needs of others.

(continued)

Table 3-11. (*continued*)

Domain	State	Description	Instructional Strategies
Transaction Pairs	Autic Mastery	Motivation comes from having power, skills, or strength and feeling capable or strong.	Provide opportunities for skill-building and personal empowerment. Encourage self-confidence and assertiveness in pursuing their goals.
	Autic Sympathy	Motivation comes from caring for oneself and wanting support and caring from others.	Provide opportunities for them to reflect on their progress and celebrate their accomplishments, fostering a sense of self-appreciation.
	Alloic Mastery	Motivation comes from wanting to give power and knowledge to others, to give others ability or skill.	Encourage mentorship and teaching others. Provide platforms to share knowledge and empower others to develop their abilities.
	Alloic Sympathy	Motivation comes from wanting to care for others and to give them love and nurturing.	Cultivate an environment of care and support. Encourage acts of compassion and provide love and nurturing to others.

Best-Suited Personality for the Learners Squad

After four months of Jackson's profiler test, the administration assessed the learning squad to examine their motivational sources. This test aimed to identify the learners' motivations for academic performance and learning. The learners responded to a series of questions (see Appendix A). Table 3-12 summarizes the preferred AMSP pairs of the squad.

Table 3-12. *Analysis of AMSP of the Learners Squad*

Learners	Means-Ends	Rules	Transactions	Relationships	Transaction Pairs
Aisha	Telic (Aisha's motivation is driven by achieving goals and anticipating future benefits.)	Conforming (Aisha's interest in exploring diverse traditions aligns with conforming to established norms.)	Sympathy (Aisha's inclination to engage with groups of peers and focus on personal success aligns with sympathy.)	Autic (self) (Aisha's preference for engaging with various backgrounds suggests an autic orientation.)	Autic Sympathy (Aisha's interest in exploring diverse traditions and engaging with others aligns with autic sympathy.)
Ethan	Telic (Ethan's disciplined approach and methodical study habits align with goal achievement.)	Conforming (Ethan's preference for learning alone and adhering to established guidelines.)	Mastery (Ethan's disciplined approach and methodical study habits suggest a mastery orientation.)	Autic (self) (Ethan's preference for learning alone and focusing on personal success aligns with autic orientation.)	Autic Mastery (Ethan's methodical study habits and desire for personal success align with autic mastery.)

	Telic	Conforming	Sympathy	Autic (self)	Autic Sympathy
Sofia	Telic (Sofia's strong sense of responsibility toward nature aligns with achieving future benefits.)	Conforming (Sofia's enjoyment of unconventional seating and responsibility toward nature.)	Sympathy (Sofia's affinity for discussing ideas and engagement in real-life applications aligns with sympathy.)	Autic (self) (Sofia's preference for a serene environment and affinity for discussing ideas align with autic orientation.)	Autic Sympathy (Sofia's engagement in real-life applications and desire for support align with autic sympathy.)

	Telic	Conforming	Mastery	Alloic (other)	Alloic Mastery
Amir	Telic (Amir's disciplined and responsible nature aligns with goal achievement.)	Conforming (Amir's strong sense of justice and advocacy align with conforming to rules.)	Mastery (Amir's disciplined and responsible nature suggests a mastery orientation.)	Alloic (other) (Amir's preference for collaborative learning and seeking guidance aligns with alloic orientation.)	Alloic Mastery (Amir's advocacy for helping others and enjoying discussions aligns with alloic mastery.)

Allinson and Hayes Cognitive Style Index

The Allinson and Hayes Cognitive Style Index (CSI) is a self-report psychometric measure of cognitive style that assesses preference-related differences in information processing according to intuition and analysis. Graham Allinson and John Hayes developed it in 1996. The CSI comprises 38 items, each rated using a 3-point scale (true, uncertain, false). The items are designed to assess a person's information processing preferences linked with the two cognitive styles of intuition and analysis.

- Intuitive style emphasizes feelings, open-mindedness, and a global perspective. People with an intuitive style tend to be quick to see patterns and connections and are often good at coming up with creative solutions to problems.

- Analytical style emphasizes reasoning, detail, and structure. People with an analytical style tend to be more cautious and deliberate in their decision-making, and they are often good at solving problems systematically and logically.

Table 3-13 summarizes the instructional strategies for learning styles of CSI.

Table 3-13. *Overview of CSI Learning Styles*

Types of Learner	Characteristics	Instructional Strategies
High Intuitive (0–28)	Makes decisions primarily by relying on gut instincts and intuition.	Encourage activities that allow for creative and imaginative thinking.
	Prefers to focus on the bigger picture and may overlook finer details.	Provide opportunities for brainstorming and ideation sessions.
	Often makes quick decisions without extensive analysis.	Foster an environment where risk-taking and experimentation are encouraged.
	Comfortable with ambiguity and uncertainty.	Incorporate real-life case studies and scenarios to enhance their problem-solving skills.
Quasi Intuitive (29–38)	The individual shows a mixture of intuition and analytical thinking.	Provide a mix of open-ended tasks and data-driven projects to appeal to their blended approach.
	Achieves a balance between relying on instinctual feelings and engaging in a certain degree of analysis.	Encourage self-assessment and reflection on their decision-making processes.
	Adapts decision-making approach based on the complexity of the situation.	Promote a supportive learning environment that values both intuitive and evidence-based thinking.
	Open to exploring possibilities before making a final decision.	Incorporate activities that encourage them to explore multiple solutions before settling on one.

(*continued*)

Table 3-13. (*continued*)

Types of Learner	Characteristics	Instructional Strategies
Adaptive (39–45)	Adapts cognitive style based on the demands of the situation.	Provide diverse learning experiences to accommodate their flexible approach.
	Able to switch between intuitive and analytical thinking as needed.	Offer self-paced and collaborative learning opportunities to cater to their adaptive nature.
	Comfortable with both ambiguity and structured situations.	Encourage them to develop metacognitive skills to identify each context's most appropriate thinking style.
Quasi Analytical (46–52)	Shows a mixture of analytical thinking and some level of intuition.	Provide activities that encourage data analysis while allowing for intuitive insights.
	Values evidence and data-driven decision-making.	Propose guidance on how to strike a balance between logic and intuition in various situations.
	May incorporate some intuitive insights into analytical processes.	Encourage self-assessment to identify their strengths in both analytical and intuitive domains.
	Tends to consider various options before concluding.	Provide opportunities for group problem-solving tasks to foster collaboration and exchange of ideas.

(continued)

Table 3-13. (*continued*)

Types of Learner	Characteristics	Instructional Strategies
Highly Analytical (53–76)	Primarily relies on logical and analytical thinking.	Provide structured learning materials and well-organized content to cater to their preference for logic.
	Thoroughly examines data and evidence before making decisions.	Encourage critical thinking and problem-solving through data analysis and evidence-based reasoning.
	Prefers a systematic and methodical approach to problem-solving.	Offer opportunities for in-depth research projects and organized procedures for their learning tasks.
	May be cautious in taking risk and avoids impulsive decisions.	Provide a supportive learning environment that values careful analysis and encourages them to gradually step out of their comfort zone.

Best-Suited Personality for the Learners Squad

After four months of the Apter's motivation test, the administration assessed the learning squad to examine their cognitive styles and information processing capabilities in alignment with intuition and analysis. The learners responded to a series of questions (see Appendix A). Table 3-14 summarizes the CSI scores of the squad.

Table 3-14. *CSI Analysis of the Learners Squad*

Learner	Scores	Modality
Aisha	18	High Intuitive
Ethan	31	Highly Analytical
Sofia	37	Adaptive with Quasi Intuitive tendencies
Amir	62	Adaptive with Quasi Intuitive tendencies

Herrmann Brain Dominance Instrument

The Herrmann Brain Dominance Instrument (HBDI) is a model developed by Ned Herrmann that categorizes cognitive preferences into four quadrants based on the idea of one brain part being dominant (Figure 3-5). It divides the brain into four sections, each representing distinct thinking and learning styles. The HBDI illustrates that engaging different quadrants leads to varied learning and thinking processes.

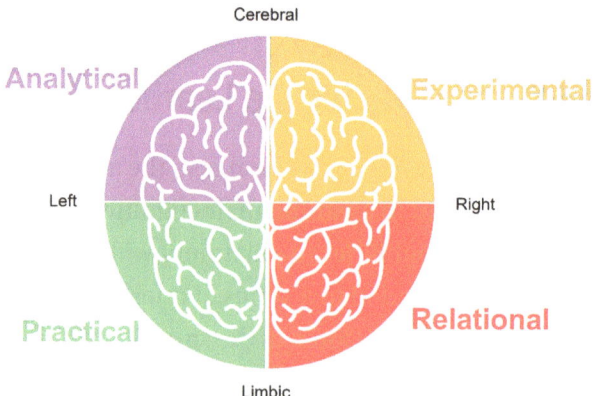

Figure 3-5. *Quadrant of HBDI*

HBDI Preferences

Brain dominance shapes thinking style preferences, influencing attention and optimal learning approaches. Each quadrant corresponds to preferred learning/thinking activities:

> **A Quadrant (Analytical):** Quantifying, analyzing, theorizing, logical processing
>
> **B Quadrant (Practical):** Organizing, sequencing, evaluating, practicing
>
> **C Quadrant (Relational):** Sharing, internalizing, moving, involvement
>
> **D Quadrant (Experimental):** Exploring, discovering, conceptualizing, synthesizing

Brain Dominance

Individuals exhibit varying levels of dominance within these quadrants. Typically, one of these four areas holds greater prominence, often aligning with the individual's personality traits. Nevertheless, research indicates that individuals frequently employ multiple styles or quadrants. Primary and secondary quadrant preferences are shared among most people. A primary preference pertains to a brain area readily and comfortably utilized. Concurrently, secondary preferences involve brain regions accessible as needed. Furthermore, certain preferences may prove challenging to access or deliberately avoided. Table 3-15 summarizes the different parameters for each quadrant of HBDI.

Table 3-15. *Overview of HBDI Quadrants*

Quadrants	Key Characteristics	Weakness	Instructional Strategies	Personality Type
A Quadrant	Analytical thinking, logical, factual, rational, critical, technical, mathematical, quantitative	May overanalyze, lack creativity	Lecture, critical thinking, logical reasoning, factual-oriented learning	Theorists (analytical)
B Quadrant	Sequential thinking, conservative, controlled, structured, organized, detailed, planned	May resist change, overlook broader context	Outlining, checklists, structured exercises, problem solving with steps	Organizers (sequential)
C Quadrant	Interpersonal thinking, kinesthetic, emotional, spiritual, musical, sensory, feeling	May rely on feelings, lack objectivity	Brainstorming, metaphors, illustrations, mind mapping, holistic approaches	Humanitarians (interpersonal)
D Quadrant	Imaginative thinking, visual, holistic, intuitive, innovative, artistic, spatial, conceptual	May struggle with practicality, lack focus	Cooperative learning, group discussion, role-playing, visual and holistic approaches	Innovators (imaginative)

Best-Suited Personality for the Learners Squad

After four months of the CSI test, the administration assessed the learning squad to determine their dominance in using their brain to store and process information. The learners responded to a series of questions (see Appendix A). Table 3-16 indicates squad preference for each quadrant.

Table 3-16. *HBDI Analysis of the Learners Squad*

Learners	Category
Aisha	Quadrant C
Ethan	Quadrant A
Sofia	Quadrants C and D
Amir	Quadrants A and C

Felder-Silverman Learning Style Model

The model consists of four dimensions, each delineating a spectrum of preferences (Figure 3-6). Individuals have the capacity to locate themselves along each continuum, potentially embodying a blend of preferences. Table 3-17 summarizes instructional strategies for each dimension of the Felder-Silverman Learning Style Model.

Table 3-17. *Overview of the Felder-Silverman Learning Style Model*

Dimension	Description	Characteristics	Instructional Strategies
Sensing vs. Intuitive Learning	It determines how the learner prefers to perceive or take in information	Sensing learners: concrete, practical, direct observation, concerned with facts and procedure	Provide real-world examples and hands-on applications
		Intuitive learners: innovative, abstract, patterns/ connections, concerned with theories and meanings	Encourage discussions to explore underlying concepts
Visual vs. Verbal Learning	It describes how learners prefer information to be presented	Visual learners: images, spatial relationships, flowcharts, diagrams	Use visual aids, diagrams, and videos
		Verbal learners: words, language	Provide written and spoken explanations
Active vs. Reflective Learning	It determines how the learner prefers to process the information	Active learners: hands-on, experimentation, group work	Engage in interactive activities and group discussions
		Reflective learners: internal processing, thinking, introspection	Encourage personal reflection and independent study
Sequential vs. Global Learning	It determines how learners prefer to organize and progress toward understanding information	Sequential learners: linear, step-by-step learning, learns in small incremental steps	Present information in a logical order in step-by-step format
		Global learners: holistic, big-picture understanding, learn in large leaps	Provide overarching concepts before details; use mind maps

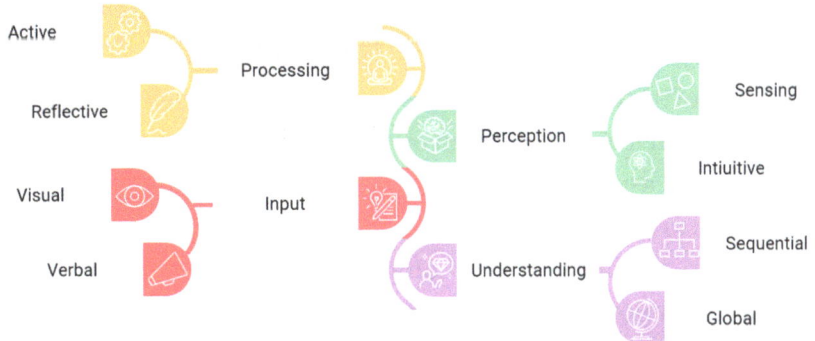

Figure 3-6. *Learning Styles of the Felder-Silverman Model*

Best-Suited Personality for the Learners Squad

After four months of the HBDI test, the administration assessed the learning squad to determine their preferred learning styles. The learners responded to a series of questions (refer to Appendix A). Table 3-18 summarizes the preferred dimension of the squad.

Table 3-18. *Felder-Silverman Analysis of the Learners Squad*

Dimensions	Aisha	Ethan	Sofia	Amir
Sensing vs. Intuitive Learning	Intuitive	Sensing	Intuitive	Intuitive
Visual vs. Verbal Learning	Verbal	Verbal	Both	Verbal
Active vs. Reflective Learning	Active	Reflective	Active	Active
Sequential vs. Global Learning	Global	Sequential	Global	Global

Kolb's Learning Styles

Kolb's Experimental Learning Theory

Kolb proposed that acquiring abstract concepts is essential for learning, enabling their flexible application across diverse scenarios. As a result, knowledge emerges from the transformative nature of experience.

At the heart of Kolb's Experiential Learning Theory lies a two-fold framework. Firstly, it outlines a four-stage cycle that governs the progression of the learning process. By navigating through these stages, Kolb asserts that learners can convert their experiences into knowledge.

The second aspect focuses on learning styles and learners' cognitive mechanisms to attain knowledge. The theory emphasizes the ability of individuals to demonstrate their comprehension and learning when they successfully employ abstract concepts in novel situations.

Four-Stage Kolb's Learning Cycle

Let us understand the four-stage cycle with the help of an analogy of baking chocolate chip cookies to analyze the learning process.

Imagine you are an enthusiastic baker who decides to bake a batch of flavorful chocolate chip cookies for the first time. You diligently follow the recipe, measuring the ingredients precisely and preheating the oven to the recommended temperature **(concrete experience)**. However, as you eagerly retrieve the cookies from the oven, you notice that they are burnt on the edges but still raw in the center. Oh no!

Curious about the outcome, you take a moment to reflect on your baking process. You review the steps you took during baking, carefully comparing them to the recipe to ensure you followed the instructions accurately **(reflective observation)**.

Upon reflection, you realize the oven temperature was too high, causing the cookies to burn on the edges. Additionally, you understand that the baking time might have been insufficient, resulting in the cookies being undercooked **(abstract conceptualization)**. Understanding the importance of learning from mistakes, you decide to discard the first batch of cookies and embark on a second attempt.

This time, you actively experiment and make adjustments. You lower the oven temperature and extend the baking time based on your newfound knowledge **(active experimentation)**. As you proceed with the second attempt, you engage in a new concrete experience, putting your adjusted methods into practice.

In line with Kolb's Experiential Learning Theory, the learning process has two key goals. Firstly, you aim to acquire specific knowledge about baking chocolate chip cookies. Secondly, you learn about your own learning process through trial and error, experimenting with different factors and reflecting on the results to achieve your desired outcome.

By embracing the iterative learning cycle through concrete or practical experience, reflection, conceptualization, and active experimentation, you gain insights into baking specifics and develop a deeper understanding of how you acquire and expand your baking knowledge.

Kolb's Learning Styles

In 1984, Kolb expanded upon his learning cycle and introduced a model that delved into various learning styles. By combining these models, Kolb formulated the Experiential Learning Theory, which aims to delve into the inner cognitive processes of learners.

While Kolb's four stages of learning interact harmoniously to form a comprehensive learning process, individuals may exhibit preferences for specific components. Some individuals may gravitate toward concrete and reflective experiences while dedicating less time to the abstract and active stages. This recognition led Kolb to identify four distinct learning styles corresponding to the earlier four-stage learning cycle.

Kolb posited that our learning style preference results from two variables representing two distinct "choices" we make. To represent each stage of the learning cycle, Kolb depicts these stages along the intersecting axes of these two variables.

Kolb introduces two fundamental axes in his framework: the Processing Continuum, represented horizontally, and the Perception Continuum, represented vertically (Figure 3-7). According to Kolb's perspective, learners cannot simultaneously engage in both variables on a single axis (e.g., thinking and feeling), as this can create an internal conflict. Consequently, learners unconsciously make a choice that determines their preferred learning style.

The Processing Continuum relates to our emotional response and how we attribute meaning to information. Learners opt for a specific approach to transform and process their experiences. This can involve active experimentation through hands-on involvement or reflective observation through careful observation and analysis.

On the other hand, the Perception Continuum focuses on our approach to tasks. It represents how learners approach and grasp information. This can involve an emphasis on concrete experience through a sensory and immersive approach or abstract conceptualization through analytical thinking and interpretation.

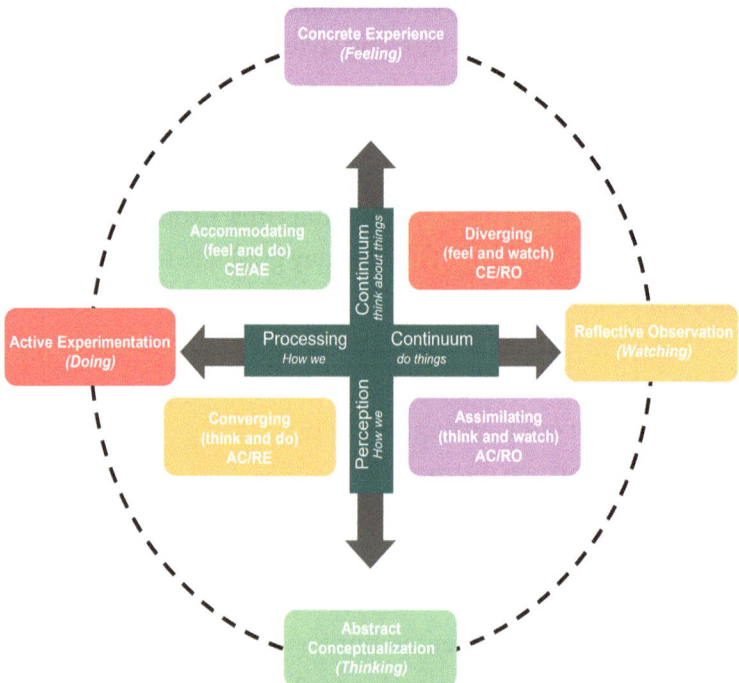

Figure 3-7. *Kolb's Learning Styles*

These distinct learning styles highlight a preference for specific learning cycle phases. Table 3-19 summarizes the different parameters for each learning style of Kolb's Learning Cycle.

Table 3-19. Overview of Kolb's Learning Styles

Learning Styles	Key Characteristics	Strengths	Weaknesses	Instructional Strategies
Diverging (CE/RO)	Idea generation, multiple perspectives	Creative thinking, gathering information, broad cultural interests	Difficulty making decisions, may struggle with practical application	Group work, open-ended discussions, visual aids, brainstorming
Assimilating (AC/RO)	Logical approach, emphasis on ideas and concepts	Strong analytical skills, understanding complex information, organizing data	Less emphasis on interpersonal interactions, limited practical application	Clear explanations, demonstrations, lectures, theoretical frameworks, logical models
Converging (AC/AE)	Practical problem-solving, applying ideas to new situations	Strong problem-solving skills, technical competence, application of theories	Limited focus on interpersonal activities, less comfortable with ambiguity	Hands-on activities, case studies, simulations, real-world applications
Accommodating (CE/AE)	Hands-on experiences, intuitive decision-making	Active experimentation, adaptability, goal driven	Less emphasis on analytical thinking, reliance on trial and error	Practical exercises, experiential learning, field trips, role-playing

Note   The table provides a concise overview of each learning style. It is important to remember that individuals may combine these styles to varying degrees. Instructional strategies can be tailored to cater to the diverse learning preferences of group or individual learners.

Best-Suited Personality for the Learners Squad

After four months of the Felder-Silverman test, the administration assessed the learning squad to examine their approaches to learning new information and experimenting with their learning styles. The learners responded to a series of questions (see Appendix A). Table 3-20 summarizes the score of the squad obtained.

Table 3-20. *Kolb's Analysis of the Learners Squad*

Learners	Scores	Modality
Aisha	40	Diverging Style
Ethan	45	Assimilating Style
Sofia	43	Diverging Style
Amir	46	Converging Style

4MAT Learning Styles Model

The 4MAT Learning Styles Model offers a unified framework to comprehend how individuals and groups engage with knowledge, progressing through interpretation, assimilation, action, and integration stages. It is based on David Kolb's four learning styles. This model elucidates learning as a process of observing and processing information.

Observation involves translating personal experiences into conceptual forms, bridging emotions and ideas, and crucially connecting learners' values with experts'. Processing entails reflecting on and structuring new knowledge and applying and testing ideas in real-world contexts. This interplay between reflection and action enhances learning, promoting adaptation across diverse situations.

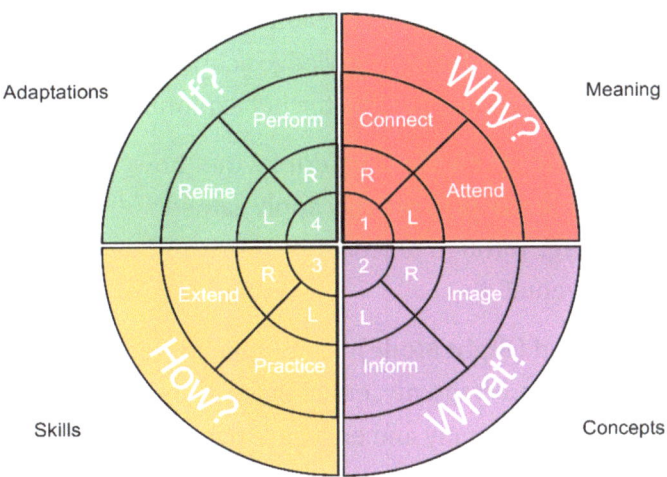

Figure 3-8. *An Eight-Step Learning Process of 4MAT*

The 4MAT Learning Styles Model (Figure 3-8) outlines an eight-step learning process aligned with distinct brain functions. Each quadrant is a learning style influenced by left and right brain modes. A balanced interplay between these modes enhances creative thinking and problem-solving abilities. The learning cycle begins as follows:

1.1 Connect (Right Mode): Establish the "why" of learning, engaging learners through personal experiences to create meaning and motivation.

1.2 Attend (Left Mode): Shift to objective discussion, encouraging reflection and dialogue to integrate the experience and understand its significance.

2.1 Image (Right Mode): Address the "what" by delving deeper, transitioning from concrete to abstract, and forming a mental image to reinforce focus.

2.2 Inform (Left Mode): Present core conceptual information, fostering knowledge acquisition and conceptualization, guiding further exploration.

3.1 Practice (Left Mode): Address the "how," allowing learners to actively apply gained knowledge through practical exercises and experimentation.

3.2 Extend (Right Mode): Encourage personal synthesis and creativity, enabling learners to contribute uniquely and engage actively with the content.

4.1 Refine (Left Mode): Analyze and improve the applied learning, refining concepts and strategies used in exercises.

4.2 Perform (Right Mode): Conclude by having learners explain their learning and achievements, reflecting on limitations, summarizing, and evaluating.

The model's structure integrates diverse learning styles and brain functions to optimize learning, fostering comprehensive understanding and practical application.

The four categories of learners have different learning techniques, with each asking a different inquiry regarding the information they learn. Table 3-21 summarizes the different parameters for each learning type of the 4MAT Learning Styles Model.

Table 3-21. *Overview of 4MAT Learning Styles*

Learner Type	Key Characteristics	Weaknesses	Instructional Strategies
Imaginative Learner – Why should I learn this?	• Engages in feelings, reflection • Desires personal meaning, involvement	May prioritize personal meaning over depth of factual knowledge.	Encourage personal connections, reflection, and involvement.
Analytical Learner – What should I learn?	• Thinks critically, values facts • Enjoys independent research	May overanalyze and lack practical application of knowledge.	Provide factual information and encourage critical thinking.
Common Sense Learner – How should I learn?	• Practical, enjoys experiments • Likes building, designing	May struggle with abstract concepts and theoretical understanding.	Include hands-on activities, experiments, and real-world applications.
Dynamic Learner – What if I learn this?	• Experimental, seeks possibilities • Values self-discovery, creativity	May need more structured learning and consistent focus.	Encourage trial and error, exploration, and original adjustments.

Best-Suited Personality for the Learners Squad

After four months of the Kolb test, the administration assessed the learning squad to examine how they interact with knowledge. The learners responded to a series of questions (refer to Appendix A). Table 3-22 summarizes the score of the squad obtained.

Table 3-22. *4MAT Analysis of the Learners Squad*

Learner	Score	Modality
Aisha	7	Imaginative Learner
Ethan	9	Analytical Learner
Sofia	7	Common Sense Learner
Amir	8	Dynamic Learner

Honey and Mumford Learning Styles

Honey and Mumford propose four distinct learning styles based on Kolb's experimental learning theory: Activist, Theorist, Pragmatist, and Reflector. Each style is a meaningful indicator of individual preferences and strengths within the learning process. They encompass unique combinations of task approaches and emotional responses to learning.

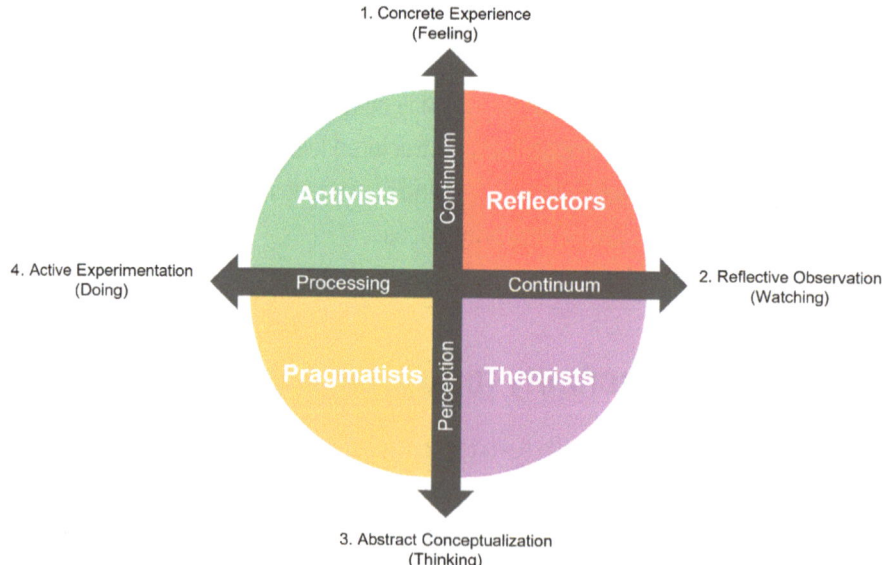

Figure 3-9. *Honey and Mumford Learning Styles*

The model (Figure 3-9) also proposes that most individuals prefer one or two of these distinct learning styles. To assist in identifying one's preferred learning style, they have developed a questionnaire. Table 3-23 summarizes the different parameters for each Honey and Mumford Model learning style.

Table 3-23. *Honey and Mumford Learning Styles Overview*

Learner Style	Key Characteristics	Strengths	Weaknesses	Instructional Strategies
Activist	Learn best through doing and hands-on experiences	Enthusiastic, adaptable, and willing to take on challenges	Learn least through reading/ thinking alone or listening to lectures	Brainstorming, group discussions, role play, puzzles, hands-on problem-solving
Theorist	Learn best by understanding underlying theories	Analytical, logical, value facts, and rational thinking	Learn least when conclusions are ambiguous, or emotions involved	Models, facts and figures, applying theories, storytelling
Pragmatist	Learn best when they see practical applications	Action-oriented, practical, seek to apply learning immediately	Learn least when they can't see practical application	Case studies, problem-solving, discussions on theory-to-practice, practical examples
Reflector	Learn best by observing and carefully thinking	Thoughtful, thorough, consider multiple perspectives	Learn least when forced into rushed or unprepared situations	Observing others, paired discussions, receiving feedback from others

In reality, individuals may exhibit traits from multiple styles or have different preferences depending on the learning context.

Best-Suited Personality for the Learners Squad

After four months of the 4MAT test, the administration assessed the learning squad to determine their preferences and strengths within the learning process. The learners responded to a series of questions (see Appendix A). Table 3-24 summarizes the score of the squad obtained.

Table 3-24. *Honey and Mumford Analysis of the Learners Squad*

Learner	Score	Modality
Aisha	48	Activist
Ethan	43	Theorist
Sofia	46	Pragmatist
Amir	47	Reflector

Reliability and Validity of Learning Styles

Reliability and validity are essential concepts in research and measurement. Reliability pertains to the consistency of measurements over time, while validity concerns the accuracy of measurements in capturing what they intend to measure. Internal consistency reliability assesses the consistency of items within a measurement tool, often evaluated using Cronbach's alpha. Construct validity examines how well measurements align with theoretical constructs.

Internal Consistency Reliability

Internal consistency reliability gauges the consistency among items within a measurement tool. It ensures that these items collectively measure the same underlying concept. Cronbach's alpha, a statistic ranging from 0 to 1, is commonly employed to measure this reliability. For instance, if a math test contains questions on various mathematical operations, like addition, subtraction, and multiplication, they should all relate to mathematics and not diverge into unrelated topics.

Kaiser-Meyer-Olkin (KMO) Measure

The KMO measure evaluates the strength of partial correlations among items of the questionnaire, indicating their suitability. Values closer to 1.0 are ideal, while those below 0.5 are unacceptable.

Bartlett's Test of Sphericity

It determines whether the correlation matrix is an identity matrix; a statistically significant result (typically below 0.05) rejects the null hypothesis of unrelated variables.

Cronbach's Alpha

Cronbach's alpha assesses the reliability of a score when multiple items measure the same construct. Ranging from 0 to 1, a value over 0.7 is generally acceptable, indicating strong internal consistency. A higher alpha implies a strong correlation among test items.

Construct Validity – Convergent and Divergent

Construct validity ensures that measurements accurately represent intended theoretical constructs. Convergent validity establishes a positive correlation between scores of different measures evaluating the same construct. Divergent validity, also known as discriminant validity, demonstrates the lack of correlation between scores from theoretically unrelated measures, ensuring distinctiveness.

Convergent Validity and Discriminant Validity

Convergent validity is confirmed when the Average Variance Extracted (AVE) and Composite Reliability (CR) are both above 0.5 and 0.7, respectively. This affirms convergent validity.

Application and Testing

The reliability and construct validity of various learning style models were evaluated using IBM's SPSS software. The resulting values are presented in Table 3-25, providing insights into the consistency and accuracy of the measurement tools applied to learners in the study.

Table 3-25. *Analysis of Learning Style Models*

Learning Style Models	Reliability					Construct Validity	
	KMO	Bartlett's Sphericity Test			Cronbach's Alpha	Convergent	
		Approx. Chi-Square	df.	Sig.		CR	AVE
VARK	0.51	7123.234	153	.00	0.82	0.72	0.51
Dunn and Dunn	0.46	6534.235	125	.04	0.65	0.68	0.50
Gregorc	0.62	4567.254	111	.02	0.77	0.69	0.48
MBTI	0.48	5412.253	143	.03	0.65	0.59	0.41
Jackson	0.69	6543.243	165	.00	0.78	0.78	0.56
Apter	0.72	7542.542	154	.00	0.83	0.73	0.59
CSI	0.82	6891.321	152	.00	0.92	0.71	0.51
HBDI	0.48	4564.345	123	.03	0.64	0.67	0.49
Felder-Silverman	0.59	3452.343	143	.03	0.57	0.62	0.51
Kolb	0.55	2345.165	132	.04	0.55	0.66	0.48
4MAT	0.68	5432.231	121	.02	0.68	0.70	0.52
Honey and Mumford	0.72	6342.145	156	.00	0.91	0.79	0.61

Evaluating the Performance

The assessment of Learners Squad performance at the culmination of each academic year is an integral component of the educational process, wherein university exams serve as a benchmark to gauge academic progress (Figure 3-10). In parallel, meticulously evaluating instructional material alignment underscores the commitment to delivering effective and tailored educational experiences (Figure 3-11).

The evaluation of instructional material alignment involves a comprehensive assessment encompassing a multitude of factors, such as identifying learners' needs, setting clear objectives, analyzing content, accommodating learning styles, conducting formative assessments, seeking feedback, connecting materials to real-world contexts, employing differentiation and blended learning, and continuously refining the approach.

By understanding individual learner characteristics, instructional designers can tailor instructional materials to suit diverse needs. Clear and specific learning objectives guide the creation of materials, while content analysis ensures alignment with objectives. Adapting to different learning styles and preferences enhances engagement, and ongoing assessment informs instructional adjustments. Learner feedback informs continuous improvement, while contextualizing materials in real-world scenarios enhances practical understanding. Differentiation addresses diverse learner abilities; blended learning combines traditional and technological approaches. Encouraging learners to explore and expand their learning approaches can lead to more comprehensive and adaptable learning experiences.

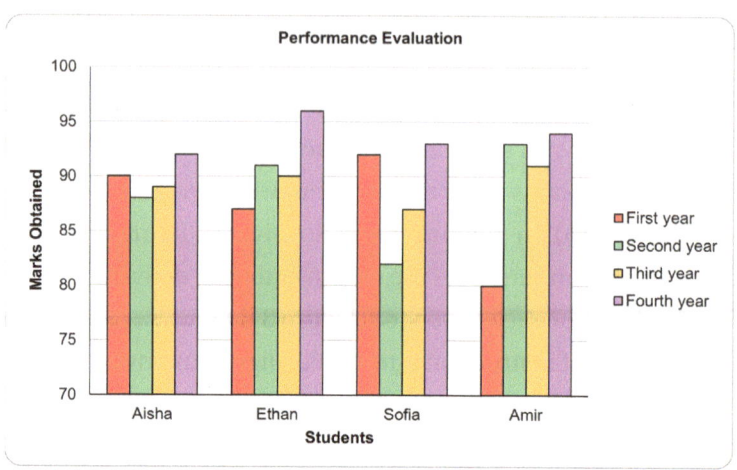

Figure 3-10. *Performance Evaluation of the Learners Squad*

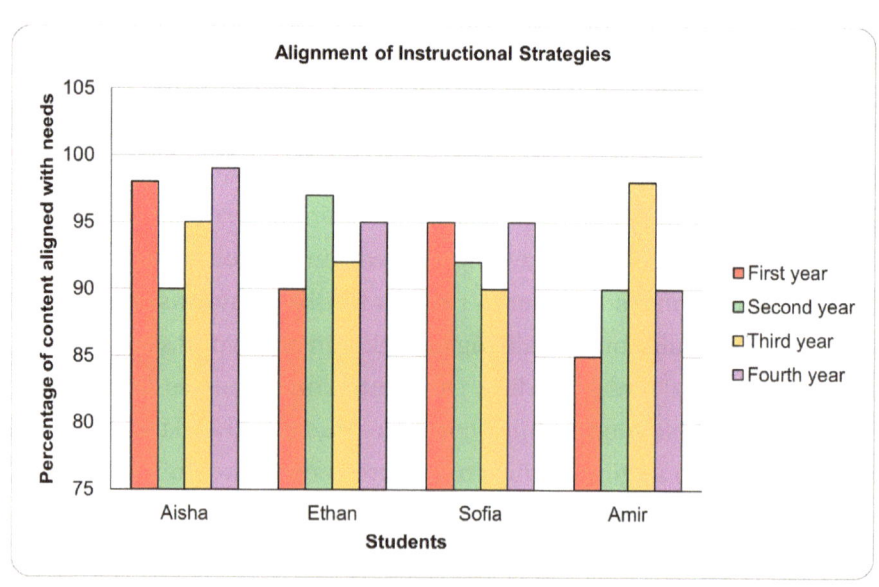

Figure 3-11. *Analysis of Instructional Material Alignment*

Summary

- Key aspects of learner analysis:

    - **Demographic Information:** Age, gender, education, work experience.

    - **Learning Styles:** Visual, auditory, kinesthetic preferences.

    - **Learning Preferences:** Format, delivery methods (online, in-person, blended).

    - **Prior Knowledge and Skills:** Build on existing knowledge and bridge gaps.

    - **Learning Objectives and Goals:** Align content with personal/professional goals.

    - **Learning Needs and Challenges:** Adapt based on constraints and provide support.

    - **Motivation and Expectations:** Enhance engagement and a positive environment.

- According to the theory of multiple intelligence, individuals have distinct types of intelligence: naturalistic, linguistic, logical-mathematical, spatial, musical, bodily-kinesthetic, interpersonal, and intrapersonal.

- Learning modalities shape the design and delivery of effective learning experiences.

- Various learning modalities can be used to analyze the learners' needs and provide them with suitable instructional strategies.

- The learning modalities and styles should be reliable and valid to measure the given hypothesis.

Let's Brainstorm

These mini-scenarios will make you scratch your head and scribble on your pad.

Mini-Scenario 1:

Imagine you are an instructional designer working on an online course. How would you gather information about learners' demographics, preferences, and prior knowledge? How would this information impact your design decisions?

Mini-Scenario 2:

You are tasked with designing a training program for employees with diverse learning styles. How would you leverage the concept of learning modalities to create a training program that accommodates different ways of thinking, feeling, and responding to educational stimuli?

Mini-Scenario 3:

Consider a scenario where a learner has a strong bodily-kinesthetic intelligence. How might you design a learning experience that caters to this learner's preference for hands-on activities? How could this approach enhance their engagement and understanding of the content?

Mini-Scenario 4:

You are creating a blended learning course involving online modules and face-to-face sessions. How would you use your understanding of learning preferences to decide which content should be delivered in which format? How might this contribute to a more effective learning experience?

Mini-Scenario 5:

Reflect on learning modalities as a foundation for various learning style models. Choose one of these models and explain how it can be applied to design instructional strategies that match learners' cognitive, affective, and physiological preferences. Provide an example scenario to illustrate your point.

CHAPTER 4

Pedagogical, Andragogical, and Heutagogical Approaches

In the dynamic landscape of education, one undeniable truth remains constant: learners come in diverse forms, each with unique needs, preferences, and ways of acquiring knowledge. Recognizing this diversity, the field of instructional design has evolved to accommodate various learning styles and approaches by implementing diverse instructional design strategies to increase engagement. In our previous chapter, we explored the intricacies of these learning styles and their corresponding implementation strategies. In this chapter, we embark on a new journey that delves into pedagogical approaches and their pivotal role in shaping the educational experience.

As we dive deeper into pedagogy, we shall explore this foundational concept and navigate the wider landscape of andragogy and other instructional methodologies. We aim to equip you with a comprehensive understanding of these approaches and guide you in selecting the most effective methods for online and offline learning environments.

© Ankita Jiyani Mangtani 2024
A. J. Mangtani, *Instructional Design Unleashed*, Design Thinking,
https://doi.org/10.1007/979-8-8688-0416-8_4

The heart of any successful learning experience lies in the harmonious interplay between two essential elements: the instructional process and the instructional material. To achieve this harmony, it is essential to appreciate the significance of both content and pedagogy.

Content experts excel in their knowledge of the subject matter, which is the "what" of learning. Nevertheless, their expertise in the "how" of learning – the methodology – may not be as extensive, often resulting in content-centric instructional materials. On the other hand, the instructional design process commences with a keen understanding of the learner, placing learner-centricity at its core.

Throughout this chapter, we will explore the very essence of pedagogy – an academic discipline dedicated to the study of teaching methods, the goals of education, and how these goals are realized. This exploration will be grounded in insights derived from educational psychology and enriched by scientific learning theories.

History of Pedagogy

Pedagogy, the art and science of teaching, has undergone a remarkable evolution over time, profoundly influencing the field of instructional design. Understanding this historical context is vital for instructional designers as they craft materials and courses that cater to diverse learners in the modern era.

Throughout history, pedagogical evolution has been driven by societal, technological, and philosophical shifts. Ancient civilizations focused on elite education, emphasizing subjects like mathematics and philosophy. The Middle Ages brought religious-centered learning, while the Renaissance revived classical education. These historical shifts highlighted the importance of adapting pedagogical methods to meet the needs and values of their times.

The Enlightenment era introduced a focus on reason and individual rights, challenging traditional educational norms. The Industrial Revolution necessitated mass education to support industrialization. Progressive educators like John Dewey promoted experiential learning, leading to a shift toward learner-centered approaches.

Behaviorism, constructivism, and the advent of educational technology in the 20th century further transformed pedagogy. Behaviorism emphasizes observable behaviors, while constructivism highlights learners' active role in knowledge construction. The digital age introduced new tools and mediums for teaching and learning, facilitating personalized and distance education.

Today, instructional designers draw from this rich history of pedagogical evolution. They blend traditional and modern approaches, incorporating technology, learner-centered principles, and inclusive practices. Instructional materials and courses are designed with a deep understanding of how pedagogy has evolved, aiming to engage and empower learners in a rapidly changing world.

In this dynamic landscape, instructional designers stand as architects of learning experiences, utilizing the wisdom of pedagogical history to create innovative, effective, and inclusive educational materials and courses that prepare learners for the challenges and opportunities of the 21st century.

Types of Pedagogy

Pedagogy encompasses diverse teaching approaches. Social pedagogy integrates social skills development, preparing learners for collaborative environments. Critical pedagogy promotes questioning and critical analysis, empowering learners to challenge dominant beliefs. Culturally responsive pedagogy tailors education to diverse backgrounds, fostering inclusivity. Socratic pedagogy encourages critical thinking and dialogue. These various pedagogical methods cater to different learning needs and help instructional designers create engaging and effective learning experiences.

Social Pedagogy

As instructional designers, we recognize education's pivotal role in imparting knowledge and shaping individuals into socially adept and well-rounded members of society. Social pedagogy places a strong emphasis on the idea that education is not just about academic learning; it's also critical for the social development of learners. Table 4-1 illustrates an overview of social pedagogy.

Table 4-1. Overview of Social Pedagogy

Parameters	Examples	Challenges	Benefits	Instructional Strategies
Emphasis on Social Development	• Integrating social skills development into education. • Fostering socialization among learners. • Recognizing the significance of soft skills alongside academic knowledge.	• Balancing social development with academic goals. • Measuring and assessing social growth. • Adapting to diverse learner needs and backgrounds.	• Creates socially adept individuals. • Enhances interpersonal skills. • Prepares learners for real-world social interactions.	• Include activities that encourage socialization. • Promote group discussions and collaborative projects. • Offer mentorship and guidance on social skills development.
Integration of Social Skills	• Teaching respectful communication and active listening. • Encouraging teamwork and cooperation. • Fostering empathy and kindness among learners.	• Addressing resistance to change or social skill development. • Ensuring all learners are included and feel valued. • Balancing academic content with social skill instruction.	• Creates a supportive and inclusive learning environment. • Enhances learners' ability to work effectively in teams. • Develop empathy and emotional intelligence.	• Incorporate respectful communication exercises into lessons. • Assign group projects to encourage teamwork. • Promote discussions on kindness, empathy, and cultural awareness.

(continued)

159

Table 4-1. (*continued*)

Parameters	Examples	Challenges	Benefits	Instructional Strategies
Collaboration and Peer Support	• Encouraging learners to seek assistance and collaborate with peers. • Creating a culture of support and cooperation in the classroom. • Recognizing the value of peer learning and mentoring.	• Addressing potential resistance to seeking help or collaborating. • Ensuring that peer interactions remain constructive and respectful. • Balancing individual and collaborative learning needs.	• Facilitates a network of mutual support among learners. • Fosters a sense of belonging and community. • Enhances learning through peer perspectives and experiences.	• Assign collaborative projects and peer-review activities. • Create opportunities for learners to share their experiences and insights. • Offer mentorship programs and peer-led workshops.

Practical Application in Group Projects	• Utilizing group projects to reinforce social skills. • Emphasizing teamwork, communication, and conflict resolution in group work. • Encouraging learners to apply social skills in a real-world context.	• Managing group dynamics and conflicts effectively. • Ensuring that all learners participate and contribute to group projects. • Assessing individual contributions to group work.	• Prepares learners for future collaborative work environments. • Enhances problem-solving and negotiation skills. • Encourages self-reflection and growth through group interactions.	• Assign group projects with clear guidelines on roles and responsibilities. • Facilitate regular group discussions on teamwork and communication. • Offer guidance on conflict resolution and peer evaluation.

161

Critical Pedagogy

As instructional designers, we are committed to creating educational experiences beyond mere transfer of information. Critical pedagogy, inspired by critical theories and radical philosophies, offers a powerful educational approach. It strongly emphasizes empowering learners to question, challenge domination, and critically evaluate beliefs and practices that may be perceived as dominant within society. Table 4-2 illustrates an overview of critical pedagogy.

Table 4-2. Overview of Critical Pedagogy

Parameters	Examples	Challenges	Benefits	Instructional Strategies
Emphasis on Critical Inquiry	• Encouraging learners to question assumptions. • Promoting critical analysis of ideas, beliefs, and practices. • Fostering an environment of intellectual curiosity.	• Balancing critical inquiry with an established curriculum. • Managing resistance to questioning deeply held beliefs. • Ensuring a respectful and open classroom culture for debates.	• Develops critical thinking skills. • Encourages independent and analytical thought. • Prepares learners for informed decision-making.	• Pose open-ended questions that provoke critical thinking. • Facilitate Socratic dialogues to challenge perspectives. • Encourage research projects that require critical analysis.

(continued)

163

Table 4-2. (*continued*)

Parameters	Examples	Challenges	Benefits	Instructional Strategies
Deconstruction of Dominant Narratives	• Analyzing historical events from multiple perspectives. • Examining societal norms and their impact on individuals. • Critiquing media representations and biases.	• Navigating sensitive topics and potential conflicts. • Balancing the deconstruction of dominant narratives with providing alternative viewpoints. • Managing discussions on potentially divisive issues.	• Fosters empathy and cultural sensitivity. • Encourages a deeper understanding of complex issues. • Promotes tolerance and open-mindedness.	• Incorporate diverse perspectives and materials into the curriculum. • Encourage debates on controversial topics with respectful guidelines. • Use multimedia resources to showcase different viewpoints.

164

| Integration of Real-World Applications | • Applying theoretical knowledge to practical scenarios.
• Engaging in projects that address real-world issues.
• Connecting classroom learning to societal challenges. | • Finding opportunities for practical application within the curriculum.
• Ensuring that real-world projects align with learning objectives.
• Managing the logistics of community engagement or fieldwork. | • Enhances problem-solving and critical thinking skills.
• Prepares learners for active citizenship and civic engagement.
• Fosters a sense of social responsibility. | • Design projects that require critical analysis and problem-solving.
• Foster partnerships with community organizations for real-world experiences.
• Encourage learners to reflect on the societal impact of their work. |

(continued)

165

Table 4-2. (*continued*)

Parameters	Examples	Challenges	Benefits	Instructional Strategies
Learner-Centered and Collaborative Learning	• Empowering learners to take ownership of their learning. • Encouraging peer-led discussions and debates. • Promoting collaborative projects and group work.	• Balancing learner autonomy with curriculum requirements. • Ensuring active participation of all learners in collaborative activities. • Managing potential conflicts in group work.	• Fosters self-motivation and independence. • Enhances communication and teamwork skills. • Encourages diverse perspectives and collective problem-solving.	• Allow learners to choose research topics aligned with their interests. • Encourage peer-led discussions and debates. • Provide guidelines for effective teamwork and conflict resolution.

Culturally Responsive Pedagogy

As instructional designers, our commitment to effective teaching goes hand in hand with recognizing the rich tapestry of cultures that our learners bring to educational institutions. The educational landscape has witnessed the emergence of culturally responsive pedagogy in response to this diversity. This approach acknowledges that in our culturally diverse society, there are three vital dimensions – institutional, personal, and instructional – each working in tandem to address and embrace the cultural differences among learners. Table 4-3 illustrates an overview of culturally responsive pedagogy.

Table 4-3. *Overview of Culturally Responsive Pedagogy*

Parameters	Examples	Challenges	Benefits	Instructional Strategies
Acknowledgment of Cultural Diversity	• Inclusion of diverse cultural perspectives in the curriculum. • Recognition of learners' cultural backgrounds and identities. • Celebration of cultural heritage through classroom activities.	• Balancing diverse cultural needs and perspectives. • Avoiding stereotypes or biases in teaching materials. • Finding resources representing various cultures.	• Fosters a sense of belonging among learners. • Enhances cross-cultural understanding. • Encourages appreciation of diversity and inclusion.	• Infuse diverse cultural content into the curriculum. • Encourage learners to share their cultural experiences. • Organize cultural heritage events and guest speakers.
Flexibility in Instructional Methods	• Adapting teaching strategies to accommodate different learning styles and cultural preferences. • Incorporating experiential and interactive learning approaches. • Encouraging learner-led discussions and collaborative projects.	• Requires additional time and effort in lesson planning. • Balancing cultural adaptations with curriculum requirements. • Ensuring all learners feel comfortable participating.	• Enhances learner engagement and motivation. • Increases learner retention and success rates. • Fosters a dynamic and inclusive classroom environment.	• Offer various instructional approaches to cater to diverse learning styles. • Encourage peer learning and group projects. • Provide opportunities for learners to share cultural insights.

Inclusive Learning Materials	• Use of textbooks and resources that reflect the diversity of the learner body. • Incorporation of multimedia materials that include diverse voices and perspectives. • Avoiding stereotypes and biases in educational materials.	• Identifying and sourcing culturally inclusive materials. • Ensuring materials align with educational standards. • Evaluating and updating materials regularly.	• Increases learner engagement and interest in the content. • Reduces feelings of exclusion and marginalization. • Enhances critical thinking by exposing learners to diverse viewpoints.	• Regularly review and update educational materials for cultural sensitivity. • Encourage learners to explore and share their own cultural resources. • Promote critical evaluation of learning materials.
Cultural Competency Development	• Encouraging self-reflection and awareness of cultural biases. • Facilitating discussions on cultural awareness and sensitivity. • Providing opportunities for learners to learn from one another's cultural experiences.	• Creating a safe space for open dialogue about cultural differences. • Handling potentially sensitive topics with care. • Assessing and measuring cultural competency effectively.	• Prepares learners for global citizenship and the workforce. • Enhances interpersonal skills and empathy. • Promotes self-awareness and lifelong learning.	• Incorporate cultural competency discussions into the curriculum. • Encourage peer-led workshops or presentations on cultural topics. • Foster a respectful and open classroom culture for discussing cultural differences.

Socratic Pedagogy

As instructional designers, we aim to create engaging and effective learning experiences that empower learners to develop academic knowledge and essential life skills. One pedagogical approach that aligns perfectly with this objective is Socratic pedagogy. Rooted in philosophy, Socratic pedagogy offers a unique and transformative way of education that emphasizes intellectual and social growth. Table 4-4 illustrates an overview of Socratic pedagogy.

Table 4-4. Overview of Socratic Pedagogy

Parameters	Examples	Challenges	Benefits	Instructional Strategies
Emphasis on Dialogue	• Open-ended questions promote discussion. • Learners explore ideas collaboratively. • Debate and analysis of multiple perspectives.	• Managing diverse viewpoints. • Ensuring respectful dialogue. • Encouraging active participation.	• Deepens understanding. • Encourages critical thinking. • Enhances communication skills.	• Pose thought-provoking questions. • Encourage peer-led discussions. • Provide guidelines for respectful debate.
Learner-Centered Learning	• Learners take ownership of their learning. • Instructors act as facilitators, not just lecturers. • Learners explore topics of personal interest.	• Requires more active engagement from learners. • Can be challenging for instructors to let go of control. • May not suit all learning objectives.	• Fosters self-motivation and independence. • Personalizes learning experiences. • Encourages curiosity and exploration.	• Allow learners to choose topics for discussion. • Encourage self-directed research. • Provide guidance rather than answers.

(continued)

171

Table 4-4. (*continued*)

Parameters	Examples	Challenges	Benefits	Instructional Strategies
Critical Thinking Development	• Analysis of evidence and logical reasoning. • Learners question assumptions and biases. • Evaluate the credibility of sources.	• Requires time and practice to develop. • Encouraging learners to question deeply held beliefs. • Assessment challenges beyond rote memorization.	• Enhances problem-solving skills. • Prepares learners for real-world decision-making. • Fosters intellectual curiosity.	• Pose questions that challenge assumptions. • Encourage debate on complex issues. • Provide resources for information evaluation.
Ethical and Civic Growth	• Engages learners in discussions about moral dilemmas. • Encourages empathy and understanding of diverse perspectives. • Prepares learners for responsible citizenship.	• Balancing diverse viewpoints without bias. • Handling sensitive topics in a respectful manner. • Measuring ethical growth.	• Create a safe space for discussing ethics. • Encourage perspective-taking exercises. • Facilitate discussions on contemporary issues.	• Facilitate discussions on contemporary societal issues. • Use case studies and scenarios to explore ethical dilemmas. • Implement peer-led ethical debates with guidelines for respectful discourse

Assessment Beyond Traditional Testing	• Assessments focus on critical thinking, communication, and problem-solving. • Peer evaluations and self-assessments are incorporated. • Learning is measured through dialogue and contributions to discussions.	• Developing rubrics for subjective assessment. • Ensuring fairness and consistency in grading. • Sourcing peer evaluations effectively.	• Measures higher-order thinking skills. • Reflects real-world skills. • Provides immediate feedback.	• Develop clear rubrics for discussion participation. • Include self-reflection in assessments. • Encourage peer evaluations with clear guidelines.

Domains of Pedagogy

As instructional designers, we aim to create engaging and effective learning experiences that empower learners to develop academic knowledge. We recognize the importance of considering learners' social development and critical thinking while embracing cultural differences that promote intellectual and social growth. In today's educational landscape, this task becomes even more challenging as we cater to learners from various generations, each with its unique learning preferences and values.

Generational differences significantly affect how individuals approach education and what they value in life. To address these distinctions effectively, instructional designers must look at the four learner-centric domains of pedagogy (Figure 4-1) out of seven through the lens of instructional designers crafting learning experiences for different generations of learners, including Generation X (born in 1965–1980), Generation Y (born in 1981–1996), Generation Z (born in 1997–2010), and Generation Alpha (born after 2010).

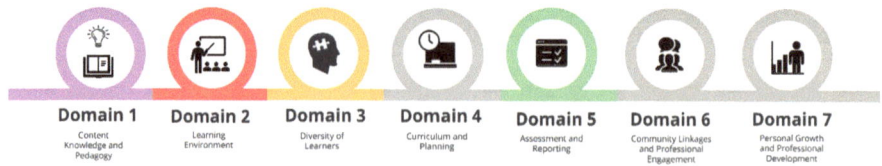

Figure 4-1. *Seven Domains of Pedagogy*

Domain 1: Content Knowledge and Pedagogy

Table 4-5 describes all the strands of domain 1 of pedagogy.

Table 4-5.   Strands of Domain 1

Strand	Generation X	Generation Y	Generation Z	Generation Alpha
Content Knowledge and Application	• Provide comprehensive content knowledge • Emphasize practical applications	• Deliver content through multimedia and technology • Highlight real-world applications	• Offer concise, visually engaging content • Incorporate technology-driven learning	• Introduce interactive and visually stimulating content • Emphasize practical application from a young age
Research-Based Teaching and Learning	• Use evidence-based teaching methods. • Value traditional teaching approaches	• Incorporate research-supported strategies • Blend traditional and modern methods	• Stay updated with emerging educational research • Utilize technology for research-based learning	• Foster inquiry-based learning • Blend technology and hands-on experiences
Positive Use of ICT (Information and Communication Technology)	• Integrate technology cautiously • Focus on practical applications of technology	• Embrace technology for learning and communication • Encourage responsible digital citizenship	• Fully integrate technology for interactive learning • Teach responsible and ethical technology use	• Immerse learners in technology-rich environments • Teach digital literacy from a young age

(*continued*)

175

Table 4-5. (*continued*)

Strand	Generation X	Generation Y	Generation Z	Generation Alpha
Strategies for Promoting Literacy and Numeracy	• Emphasize traditional literacy and numeracy skills • Provide practical applications	• Integrate digital literacy into literacy and numeracy instruction • Utilize technology for practice	• Blend traditional literacy and numeracy with digital skills • Use technology for interactive practice	• Focus on digital literacy alongside traditional literacy and numeracy • Utilize technology for early literacy and numeracy development
Strategies for Developing Higher-Order Thinking Skills	• Encourage critical thinking through traditional methods. Emphasize problem-solving	• Foster critical thinking through interactive discussions • Promote creativity and problem-solving	• Promote critical thinking through digital simulations and problem-based learning • Encourage creativity and innovation	• Emphasize critical thinking and creativity through immersive digital experiences • Provide early exposure to problem-solving activities

Language Use (Mother Tongue, Filipino, English)	• Emphasize proficiency in mother tongue and English • Promote language diversity	• Support proficiency in mother tongue, Filipino, and English • Value multilingualism	• Encourage multilingual proficiency • Leverage technology for language learning	• Foster multilingualism anc digital literacy • Introduce language skills through technology
Classroom Communication Strategies	• Value face-to-face communication in the classroom. • Promote active classroom discussions	• Blend face-to-face communication with digital tools • Encourage online and offline discussions	• Utilize technology for interactive classroom communication • Encourage online collaboration	• Embrace technology for communication from an early age • Promote online collaboration and communication skills

177

Domain 2: Learning Environment

Table 4-6 describes all the strands of domain 2 of pedagogy.

Table 4-6. Strands of Domain 2

Strand	Generation X	Generation Y	Generation Z	Generation Alpha
Learner Safety and Security	• Create a physically safe classroom environment. • Ensure data privacy in digital learning tools.	• Use secure and private digital platforms. • Address online safety and privacy concerns.	• Prioritize online safety in digital learning. • Teach responsible digital citizenship.	• Establish secure online spaces early. • Reinforce physical safety practices.
Fair Learning Environment	• Promote inclusivity and diversity in discussions. • Respect diverse perspectives and backgrounds.	• Encourage open communication and respect for all. • Highlight diverse viewpoints.	• Foster discussions on social issues and fairness. • Include diverse perspectives.	• Highlight global awareness and diversity. • Encourage empathy and respect for others.
Management of Classroom Structure and Activities	• Combine traditional and modern teaching methods. • Maintain a balanced classroom setting.	• Provide structured but flexible learning environments. • Blend technology with in-person interactions.	• Create dynamic, tech-rich learning spaces. • Incorporate VR, AR, and interactive tools.	• Embrace dynamic, tech-rich learning. • Blend online and offline activities.

(continued)

Table 4-6. (*continued*)

Strand	Generation X	Generation Y	Generation Z	Generation Alpha
Support for Learner Participation	• Encourage active participation and discussions. • Facilitate collaborative group projects.	• Foster collaborative learning and peer interactions. • Provide opportunities for group work.	• Use dynamic and interactive learning experiences. • Implement gamification elements.	• Leverage interactive, gamified learning. • Incorporate interactive digital tools.
Promotion of Purposive Learning	• Relate learning to practical applications and careers. • Emphasize relevance to personal goals.	• Connect learning to real-world applications and career benefits. • Highlight practical relevance.	• Emphasize purpose-driven learning. • Show the real-world impact of knowledge.	• Promote purpose-driven learning. • Highlight the global and societal impact of knowledge.
Management of Learner Behavior	• Establish clear behavior expectations. • Encourage respectful behavior in both physical and digital spaces.	• Set clear guidelines for respectful behavior. • Encourage respectful online interactions.	• Promote self-regulation and digital responsibility. • Teach online safety and etiquette.	• Foster self-regulation and digital responsibility. • Begin early education on digital citizenship.

Domain 3: Diversity of Learners

Table 4-7 describes all the strands of domain 3 of pedagogy.

Table 4-7. Strands of Domain 3

Strand	Generation X	Generation Y	Generation Z	Generation Alpha
Learners' Gender, Needs, Strengths, Interests, and Experiences	• Recognize diverse experiences and backgrounds. • Adapt content to cater to diverse interests and needs.	• Acknowledge and respect diverse experiences and identities. • Provide content that relates to their varied interests and experiences.	• Embrace gender and cultural diversity. • Curate content that speaks to their varied interests and experiences.	• Celebrate and accommodate diverse backgrounds, experiences, and identities. • Ensure content resonates with their unique interests and experiences.
Learners' Linguistic, Cultural, Socioeconomic, and Religious Backgrounds	• Promote cross-cultural understanding. • Incorporate diverse perspectives. • Be mindful of socioeconomic differences.	• Embrace multiculturalism and multilingualism. • Integrate diverse cultural perspectives into content. • Address socioeconomic disparities sensitively.	• Celebrate cultural diversity and multilingualism. • Integrate global perspectives into content. • Consider socioeconomic diversity in content creation.	• Encourage global awareness and multilingualism. • Incorporate diverse cultural perspectives and global issues into content. • Be sensitive to socioeconomic disparities and provide equitable opportunities.

Learners with Disabilities, Giftedness, and Talents	• Recognize diverse learning needs. • Provide accommodations for learners with disabilities. • Offer enrichment opportunities for gifted and talented learners.	• Create an inclusive learning environment. • Offer tailored support for learners with disabilities. • Challenge and engage gifted and talented learners.	• Foster inclusivity and accessibility. • Utilize technology for inclusive learning. • Differentiate instruction for diverse needs.	• Prioritize inclusive education. • Incorporate Low floor, High ceiling, and Wide walls for content. • Leverage technology for accessibility and personalization. • Provide enrichment opportunities for talents and gifts.
Learners in Difficult Circumstances	• Be empathetic to learners facing challenges. • Offer support and flexibility for those in difficult circumstances.	• Recognize and address learners' challenges. • Provide resources and emotional support when needed.	• Acknowledge the unique challenges faced by learners. • Provide resources and support to help them overcome difficulties.	• Support learners through challenging circumstances. • Use technology to offer flexible learning options and resources.

183

Domain 5: Assessment and Reporting

Table 4-8 describes all the strands of domain 5 of pedagogy.

Table 4-8. Strands of Domain 5

Strand	Generation X	Generation Y	Generation Z	Generation Alpha
Design, Selection, Organization, and Utilization of Assessment Strategies	• Use a mix of traditional assessment methods like written exams and presentations. • Provide clear grading criteria and rubrics. • Incorporate a balanced approach to assessments. • Allow flexibility in assessment.	• Integrate technology for a blend of traditional and digital assessments. • Offer various assessment options, including online quizzes and collaborative projects. • Allow flexibility in assessment formats to accommodate preferences.	• Prioritize digital assessments, including online quizzes, simulations, and multimedia projects. • Embrace formative assessments and real-time feedback through technology. • Offer a wide range of assessment formats that align with digital natives' skills.	• Exclusively use digital assessments and interactive simulations. • Leverage advanced technology for real-time, adaptive assessments. • Offer innovative assessment formats that align with emerging technologies.

(continued)

Table 4-8. (*continued*)

Strand	Generation X	Generation Y	Generation Z	Generation Alpha
Monitoring and Evaluation of Learner Progress and Achievement	• Use traditional gradebooks and progress tracking systems. • Schedule face-to-face progress discussions when needed. • Maintain in-person communication channels for progress updates.	• Implement digital tools for real-time progress tracking and grade access. • Host virtual progress discussions for convenience. • Prioritize digital communication channels for progress updates.	• Utilize advanced digital platforms for real-time progress tracking and communication. • Foster a culture of self-monitoring with technology. • Promote virtual progress discussions and collaborative goal setting.	• Create a seamless virtual progress tracking system with advanced analytics. • Leverage AI for predictive learning analytics. Promote self-directed, real-time progress tracking and goal setting.
Feedback to Improve Learning	• Deliver feedback through written comments and face-to-face discussions. • Encourage peer review and self-reflection. • Use technology for efficient feedback delivery when necessary.	• Utilize digital platforms for efficient feedback delivery, enabling quick access and response. • Promote peer feedback and self-reflection in digital spaces. - Embrace technology for efficient feedback processes.	• Prioritize digital feedback delivery, including audio and video feedback. • Incorporate real-time feedback in digital spaces. • Leverage technology for automated, personalized feedback.	• Exclusively offer dynamic, multimedia feedback through advanced technology. • Use AI for personalized, real-time feedback. • Foster self-regulated learning with advanced feedback systems.

Communication of Learner Needs, Progress, and Achievement to Key Stakeholders	• Share progress reports with learners and guardians through traditional means like printed reports. • Organize in-person parent-teacher meetings for detailed discussions. • Maintain secure channels for data privacy.	• Utilize digital platforms for real-time progress updates and communication with learners and parents. • Offer virtual parent-teacher conferences for flexibility. • Prioritize data security and privacy in digital communication.	• Foster a strong online school community with real-time progress updates and virtual events. • Leverage social media for parent and community engagement. • Organize virtual school events and webinars.	• Create a seamless virtual school-community ecosystem with AI-driven communication. • Utilize advanced technology for secure, real-time communication with stakeholders. • Promote digital citizenship and data privacy within the virtual community.
Use of Assessment Data to Enhance Teaching and Learning Practices and Programs	• Analyze assessment data for curriculum improvements while maintaining established teaching methods. • Focus on maintaining established teaching methods. Encourage educators to reflect on assessment results.	• Leverage data analytics to identify learning trends and adapt curricula accordingly. • Promote innovative teaching methods based on data insights. • Encourage educators to collaborate and share data-driven best practices.	• Foster a culture of constant reflection and adaptation based on data. • Provide data-driven professional development opportunities. • Encourage educators to learn from real-time data and feedback.	• Instill a commitment to lifelong learning and real-time reflection based on data. • Offer immersive data-driven training in cutting-edge technologies and innovative pedagogy. • Leverage advanced technology for personalized, data-driven professional development.

Pedagogical Approaches

In education, a critical realization is that there's no one-size-fits-all teaching method. Individuals learn differently due to cognitive variances, prior knowledge, cultural backgrounds, and personal experiences. Therefore, instructional designers must employ diverse pedagogical approaches to meet learners' unique needs. This adaptability ensures that every learner can access effective education, fostering inclusivity and personalized growth.

How Is Pedagogy Changing?

It's great to see how pedagogy has been adapting to better suit the needs of 21st-century learners. Traditional classroom lectures have their place but are no longer the sole or most effective means of education. The evolution of pedagogy is driven by several key factors:

Technology Integration: The digital age has brought with it a host of new tools and resources that can be used in education. Online and remote curricula and interactive learning platforms have become integral in facilitating learning both inside and outside the classroom. Technology enables personalized learning experiences and access to a wealth of information.

Active Learning: Pedagogy now strongly emphasizes active and collaborative learning. Instead of passive listening and rote memorization, learners are encouraged to engage with the material through discussions, group projects, hands-on activities, and problem-solving exercises. This fosters a deeper understanding of concepts and better retention of information.

Real-World Relevance: Educators are increasingly incorporating real-world scenarios into their teaching. This approach makes learning more practical and relevant, helping learners connect their learning to their everyday lives and future careers. It also encourages critical thinking by challenging learners to apply knowledge in authentic situations.

Cultural Sensitivity: Acknowledging cultural differences and diversity is essential in modern pedagogy. Recognizing and respecting cultural nuances in the learning process allows learners to feel more included and valued, promoting a more inclusive and equitable educational experience.

Higher-Order Thinking Skills: The focus of education has shifted from mere memorization and regurgitation of facts to the development of higher-order thinking skills. This includes critical thinking, problem-solving, creativity, and effective communication. These skills are considered important for success in the complex and rapidly changing 21st-century world.

Learner Autonomy: Modern pedagogy encourages greater learner autonomy and self-directed learning. Learners are encouraged to take ownership of their education, set goals, and choose how and what they learn. This approach fosters independence and self-motivation.

Incorporating these elements into education fosters a more dynamic and effective learning environment, preparing learners for the challenges and opportunities of the 21st century. This evolution of pedagogy ensures that education aligns with the demands of our rapidly changing world and equips learners with the skills they need to thrive in it.

Inquiry-Based Learning

Inquiry-based learning is like a learning adventure where learners become fearless explorers of knowledge. Instead of simply being handed answers, they're encouraged to be curious detectives, asking questions and delving deep into real-world puzzles. It's a dynamic classroom approach that empowers learners to actively participate in their learning journey, bridging the gap between what they learn in the classroom and the captivating mysteries of the real world. This method not only hones their critical thinking and problem-solving skills but also sparks the flames of creativity that illuminate their path to understanding.

Benefits of Inquiry-Based Learning

- Promotes critical thinking
- Enhances problem-solving skills
- Fosters curiosity and motivation
- Encourages active learning
- Connects learning to the real world
- Supports collaboration
- Develops information literacy
- Boosts creativity
- Promotes lifelong learning
- Prepare for future challenges

Inquiry-Based Learning Examples

- **Case Studies:** Create real-world case studies or scenarios that learners can investigate. Case studies present complex problems or situations that require analysis and critical thinking.

- **Simulations:** Design interactive simulations or virtual experiments that allow learners to explore scientific or practical concepts in a controlled, digital environment.

- **Field Trips:** If applicable, design guides for field trips or site visits, outlining objectives, activities, and questions to consider during the excursion.

- **Presentation-Based Learning:** Offer templates for creating inquiry-based projects, reports, presentations, or multimedia artifacts. These templates ensure learners include key components in their final work.

- **Scaffolding Materials:** Develop scaffolded instructional materials that gradually increase in complexity, providing support as learners progress through the inquiry process.

- **Graphic Organizers:** Create graphic organizers such as concept maps, mind maps, and flowcharts that help learners visually organize their thoughts, questions, and findings.

- **Discussion Prompts:** Offer discussion prompts or debate topics related to the inquiry. These prompts encourage critical thinking and stimulate meaningful class discussions.

Constructivist Learning

The constructivist approach to learning is like a journey where learners are the architects of their own understanding. It's rooted in the belief that we construct our knowledge by actively engaging with the world, drawing upon our unique life experiences to build a mental framework of understanding.

Imagine it as a personal puzzle-solving adventure. Each piece of information we encounter is like a puzzle piece. We don't just passively collect these pieces; we actively arrange them using the pictures we've already created from our prior experiences. In this way, we create a richer, more intricate picture of the world around us.

This approach empowers learners to be the masters of their own learning, encouraging them to use what they already know as a solid foundation for grasping new and complex ideas. It's an approach that celebrates active participation over passive reception, fostering a deeper and more meaningful connection to the learning process.

Benefits of Constructivist Learning

- Fosters active engagement
- Enhances critical thinking
- Improves problem-solving skills
- Understand autonomy and ownership
- Individualized learning
- Increases collaboration
- Fosters lifelong learning
- Fosters creativity and innovation
- Boosts long-term retention
- Enhances intrinsic motivation
- Appreciate diverse perspectives

Constructivist Learning Examples

- **Socratic Questioning:** Pose open-ended questions that promote critical thinking and reflection, encouraging learners to explore and refine their ideas.

- **Storytelling and Narratives:** Use storytelling techniques to present information in a narrative format, making it more engaging and relatable.

- **Collaborative Projects:** Assign group projects, discussions, and collaborative tasks that require learners to work together, share ideas, and construct knowledge collectively.

- **Real-World Scenarios:** Present learners with authentic, real-world situations or case studies that require them to apply their knowledge and problem-solving skills.

- **Problem-Based Learning (PBL) Scenarios:** Develop real-world problems or scenarios that challenge learners to research, analyze, and solve issues by integrating their prior knowledge.

- **Concept Maps and Mind Maps:** Encourage learners to create visual representations of their understanding, showing the relationships between concepts and how they connect.

- **Open-Ended Assessments:** Instead of traditional tests, use open-ended assessments such as essays, projects, and presentations that allow learners to demonstrate their understanding in diverse ways.

Collaborative Learning

Collaborative learning is an educational strategy that prioritizes learners working together to deepen their understanding of a subject. It encourages active participation, teamwork, and exchanging ideas among learners to achieve common learning goals.

Example: Imagine in a science class studying ecosystems. In a traditional approach, learners might read a textbook, listen to the teacher, and take notes individually. However, with the collaborative approach, the class forms small groups, each responsible for researching and presenting a different aspect of the ecosystem.

One group focuses on the rainforest, another on the desert, and so on. You collaborate, share your findings, and discuss the similarities and differences in these ecosystems. This teamwork gives learners a broader perspective and improves their communication and critical thinking skills. It's like assembling a puzzle together, where everyone's piece contributes to the bigger picture of understanding ecosystems more comprehensively.

Benefits of Collaborative Learning

- Enhances understanding

- Understand diverse perspectives

- Improves communication skills

- Develops critical thinking

- Enhances social skills

- Fosters peer support

- Increases efficiency

- Boosts motivation

- Learn to distribute workload

- Fosters creativity

- Promotes cultural awareness

Collaborative Learning Examples

- **Group Assignments and Projects:** Assignments that require learners to work together on a project, presentation, or research paper. These tasks promote teamwork and the sharing of knowledge.

- **Discussion Guides:** Provide structured discussion questions or prompts that guide group discussions. These can be used in class or as part of online forums to facilitate meaningful conversations.

- **Collaborative Online Platforms:** Utilize online platforms, such as discussion boards, forums, or collaboration tools like Google Workspace or Microsoft Teams, to facilitate group communication and document sharing.

- **Peer Assessment Rubrics:** Provide clear criteria and rubrics for learners to evaluate each other's contributions and group work fairly.

- **Synchronous and Asynchronous Activities:** Design both in-class and out-of-class activities to accommodate different schedules and preferences. This includes live discussions, collaborative projects, and asynchronous tasks like forum discussions or group annotations of readings.

- **Role Assignments:** Specify roles within groups, such as a leader, timekeeper, researcher, or presenter, to ensure everyone contributes effectively.

Reflective Learning

The reflective approach to learning is like giving learners the keys to their own educational growth. It's about helping them become the captains of their learning ship. Imagine each learning experience as an adventure; reflection is the compass they use to navigate. They get to assess what's working, adjust their course, and document their discoveries in journals, just like explorers mapping out uncharted territories. This approach transforms learning into an exciting journey of self-discovery and improvement.

Benefits of Reflective Learning

- Improves self-awareness
- Enhances critical thinking
- Ownership of learning
- Continuous improvement
- Increases motivation
- Enhances communication skills
- Better problem-solving
- Fosters emotional intelligence
- Fosters lifelong learning

Reflective Learning Examples

- **Journals and Diaries:** Provide learners with journals or digital diaries where they can regularly record their thoughts, insights, and reflections on their learning experiences.

- **Learning Logs:** Similar to journals, learning logs are structured records where learners can document what they've learned, what they found challenging, and how they plan to improve.

- **Self-assessment Tools:** Create quizzes, self-assessment questionnaires, or rubrics that allow learners to evaluate their understanding and performance. Provide feedback and guidance based on their self-assessment.

- **Reflective Essays or Papers:** Assign reflective writing tasks where learners analyze their progress, discuss their strengths and weaknesses, and propose strategies for improvement.

- **Discussion Forums and Blogs:** Use online platforms for discussions and blogging where learners can share their thoughts, engage in conversations, and reflect on course content.

- **Socratic Seminars:** Organize Socratic seminars or group discussions where learners can engage in deep conversations, challenging each other's ideas and fostering critical thinking.

Integrative Learning

The integrative approach to learning is like connecting the dots in education. Instead of just learning isolated facts, it's about helping learners see how different pieces of information fit together. This approach creates a more comprehensive understanding of the world by showing how topics relate to each other. It keeps learners engaged and curious because they discover the bigger picture, not just memorizing facts.

Benefits of Integrative Learning

- Enhances understanding

- Improves retention

- Fosters holistic learning

- Increases engagement

- Learn with real-world application

- Improves problem-solving abilities

- Fosters interdisciplinary understanding

- Fosters long-term learning

- Preparation for future learning

Integrative Learning Examples

- **Role-Playing and Simulations:** Activities that immerse learners in scenarios where they must apply knowledge from various fields to make decisions.

- **Cross-Disciplinary Projects:** Assignments or projects involving multiple subjects or disciplines encourage learners to integrate knowledge from various sources.

- **Integrated Assessments:** Evaluation methods assess learners' ability to connect and apply knowledge across different topics.

- **Analogies and Metaphors:** Comparisons that draw parallels between different concepts, making abstract ideas more relatable.

- **Concept Maps:** Graphic representations visually display the relationships between concepts and ideas, helping learners see the connections.

- **Interdisciplinary Texts:** Reading materials combining content from multiple subjects encourages learners to extract relevant information from diverse sources.

- **Conceptual Frameworks:** Visual or written frameworks illustrating how different concepts fit together in a larger context.

Implementation of Pedagogical Approaches

Successful implementation of pedagogical approaches requires careful planning, alignment with learning objectives, clear communication, and ongoing support for learners and instructors in both online and offline settings. Additionally, the choice between online and offline approaches should consider the specific goals of the educational program and the resources available. Table 4-9 lists the challenges faced in implementing inquiry-based learning in online and offline settings.

Table 4-9. *Implementation of Pedagogical Approaches in Online and Offline Settings*

Online Setting	Offline Setting
Digital Literacy: Learners and instructors must proficiently use online tools and resources effectively, which may require training.	**Resource Limitations:** Offline inquiry may require physical resources, which can be limited or costly. Access to laboratory equipment, materials, or field sites may be challenging.
Technology Issue: Technical challenges, such as Internet connectivity issues or software compatibility problems, can disrupt the learning process.	**Logistical Constraints:** Scheduling and coordinating activities can be complex, especially involving field trips, experiments, or guest speakers.
Isolation: The absence of face-to-face interactions can lead to a sense of isolation among learners, impacting engagement and motivation.	**Individualization:** It may be challenging to cater to each learner's diverse needs and learning styles in a traditional classroom setting.
Assessment Authenticity: Ensuring the authenticity of online assessments can be challenging, as it may be difficult to verify that learners are completing activities independently.	**Assessment:** Assessing learner progress and understanding in real time can be more challenging, as it often relies on observations and periodic assessments.
Lack of Hands-On Experience: Some inquiry-based activities that require physical manipulation or experimentation may be less feasible online.	**Space Constraints:** Classroom space limitations may restrict the types of hands-on activities that can be conducted.
Privacy and Security: Protecting learner data and privacy in online environments is critical.	**Class Size:** Large class sizes can make it challenging for instructors to provide individualized attention and support to each learner.

Modern Pedagogical Approaches
Project-Based Learning

Project-based learning (PrBL) is a learner-centered teaching approach that encourages learning through real-world, curriculum-related questions or challenges. It allows learners to apply the knowledge and skills they've acquired in their classes and empowers them to develop creative solutions.

The key characteristics of the PrBL model are as follows:

- It centers around significant, open-ended questions, challenges, or issues that prompt learners to investigate, respond to, or resolve.

- It integrates the academic knowledge, understanding, and abilities learners should possess.

- It is inquiry based, fostering intrinsic curiosity and generating questions to guide learners in seeking answers.

- It incorporates 21st-century skills, such as critical thinking, communication, collaboration, and creativity.

- It allows for learner choice in the learning process.

- It offers opportunities for feedback and project revision, mirroring real-world scenarios.

- It mandates learners to present their problems, research methods, and results, similar to scientific research or real-world projects facing peer review and constructive criticism.

Challenges of Implementing PrBL in a Learning Environment

- **Assessment Subjectivity:** One of the prominent challenges in PrBL is subjectivity in assessment. Unlike standardized tests with clear-cut answers, PrBL often requires the subjective evaluation of learner projects. Assessors must consider various factors, including creativity, problem-solving, collaboration, and presentation skills. This can be time-consuming and can raise concerns about grading consistency.

- **Balancing Objectivity:** PrBL values learner projects' open-ended and creative nature. However, finding the right balance between subjectivity and objectivity in assessments can be challenging. Educators must ensure that the assessment criteria are clear and that learners are held to consistent standards while allowing individuality and creativity.

- **Resource Intensity:** Successful PrBL projects often require additional resources in terms of time and materials. This can be a challenge for educators who must carefully plan and allocate resources to support meaningful projects. Not all schools may have the necessary materials and technology to facilitate certain PrBL initiatives.

- **Hyper-focus on Products:** Another challenge is the risk of learners becoming overly fixated on creating a polished product, potentially at the expense of in-depth learning and skill development. Educators must guide learners to balance the process of learning and skill development with the end result.

- **Curriculum Integration:** Integrating PrBL into existing curricula can be challenging, requiring alignment with standards and learning objectives. Teachers must carefully plan how PrBL projects fit into the curriculum to ensure they don't become isolated, stand-alone activities but contribute to comprehensive learning.

- **Teacher Training and Comfort:** Implementing PrBL effectively requires specialized teacher training. They need to be comfortable shifting from traditional teaching methods to a more facilitative role, guiding learners through the learning process. Professional development opportunities and support are crucial in addressing this challenge.

- **Question of Applicability:** In certain subjects, particularly at elementary levels and for skill-based topics like mathematics, the applicability of PrBL may be questioned. There may be doubts about whether learners benefit more from direct instruction and practice or if PrBL can effectively teach these foundational skills.

- **Time Management:** Effective PrBL requires time for learners to explore complex problems, develop solutions, and engage in self-directed learning. Time management can be challenging, as educators must ensure that PrBL projects align with the available instructional time and curriculum requirements.

- **Equity and Accessibility:** PrBL may be less accessible to learners who lack resources or face barriers, for example, limited access to technology or materials. Ensuring equitable access and support for all learners is an ongoing challenge in PrBL implementation.

- **Assessment of 21st-Century Skills:** While PrBL aims to develop skills like critical thinking, collaboration, and communication, assessing these skills can be challenging. Standardized assessments may not adequately measure these soft skills, making it necessary to develop alternative evaluation methods.

Examples of Implementing PrBL

- **Environmental Conservation Project:** In a biology class, learners can undertake a project focused on environmental conservation. They could investigate the impact of pollution on local ecosystems, propose solutions, and design and implement an environmental cleanup initiative. This project involves real-world challenges and encourages the application of biology concepts to address an authentic problem.

- **History-Based Documentary:** In a history class, learners could work on a project to create a documentary about a historical event or figure. This project involves extensive research, critical thinking, and collaboration as learners interview experts and compile information to present a comprehensive view of the topic. The final product, the documentary, becomes a tangible representation of their learning.

- **Math in the Real World:** In a mathematics class, learners can tackle a project that revolves around real-world problem-solving. For instance, they might design a community garden and calculate the necessary materials, costs, and optimal plant arrangements. This project applies mathematical concepts and encourages learners to think critically about spatial relationships and budgeting.

- **Artistic Expression:** In an art class, learners could work on a project to create public art installations. This project involves artistic creativity and requires learners to manage budgets, plan logistics, and engage with the community to ensure their art fits local needs and aesthetics. It's a great example of project-based learning integrating art with real-world application.

- **Social Entrepreneurship Venture:** In a business studies class, learners could embark on a project-based learning journey by creating their social entrepreneurship venture. They would identify a social problem, develop a business plan to address it, and seek funding or support for their venture. This project encourages learners to apply business principles, think critically, and engage with community issues.

- **STEM-Based Sustainable Technology:** In a science, technology, engineering, and mathematics (STEM) class, learners could focus on creating sustainable technology. For instance, they might design a water purification system powered by solar energy designed for regions with restricted access to clean water. This project involves the application of STEM concepts and emphasizes 21st-century skills like innovation and collaboration.

- **Multilingual Cultural Exchange:** In a language class, learners could participate in a project involving a cultural exchange with learners from another country learning their language. Learners improve their language skills and cultural awareness through regular interactions and collaborative projects. This project exemplifies how PrBL can be adapted for language education while promoting global perspectives.

- **Civic Engagement and Advocacy:** In a civics or social studies class, learners could engage in a project that centers around a relevant social or political issue. For example, they might lead a campaign to raise awareness about environmental conservation and advocate for policy changes. This project connects classroom learning with civic engagement, teaching learners the power of advocacy.

Problem-Based Learning

Problem-based learning (PBL) is an approach that utilizes real-world problems to facilitate learner learning of fundamental concepts and principles rather than relying on direct information delivery. It encourages the integration of knowledge from multiple disciplines, promoting a comprehensive understanding of the subject matter. It imparts subject matter knowledge and nurtures critical thinking, problem-solving, and communication skills. Additionally, PBL promotes collaborative group work, research material evaluation, and a commitment to lifelong learning. It is adaptable to various learning contexts, ranging from full semester use as the primary teaching method to incorporation into lab and design courses and initiating focused discussions. The hallmark of PBL is its consistent use of real-world problems as the core learning stimulus. Table 4-10 lists the differences between project-based learning and problem-based learning.

Table 4-10. *Differences Between Project-Based Learning and Problem-Based Learning*

Parameters	Project-Based Learning (PrBL)	Problem-Based Learning (PBL)
Primary Focus	Learning by completing an extended project or task.	Learning through solving complex, real-world problems.
Nature of Task	Learners undertake a broader project, which may include various tasks and phases.	Learners work on solving a specific problem, often presented as a scenario or case.
Problem Complexity	The project can encompass multiple tasks and aspects, offering more complexity.	Emphasis on addressing a single, complex problem or set of problems.
Group vs. Individual Work	Group work is common, but some projects may involve individual work or smaller teams.	Typically involves small group collaboration where learners share responsibilities.
Problem Identification	Learners may be involved in defining the project's scope and objectives.	Learners may identify the problems themselves or receive them from the instructor.
Assessment Criteria	Assessment considers both the process (e.g., project management) and the final product.	Assessment often focuses on the quality of problem-solving and critical thinking skills.
Duration	PrBL projects tend to be more extended, potentially spanning several weeks or even a semester.	PBL may take place over shorter periods, such as a single class session or a few weeks.

(continued)

Table 4-10. (*continued*)

Parameters	Project-Based Learning (PrBL)	Problem-Based Learning (PBL)
Real-World Connection	Also aims for real-world relevance but allows more flexibility in project choice.	Directly connects to real-world problems, making the subject matter practical and relevant.
Teacher's Role	Functions as a guide, assisting with project planning and resources and providing periodic feedback.	Primarily acts as a facilitator, guiding learners in problem-solving and learning.
Examples	Designing a community garden, creating a business plan, or producing a documentary film.	In a medical education context, diagnosing a patient's illness.

Microlearning

Microlearning is an educational strategy that delivers information in small, focused segments, making it easily digestible. It involves breaking down complex topics or skills into concise modules that can be completed within a short timeframe, typically ranging from 15 to 20 minutes. This approach frequently incorporates multimedia elements like infographics, videos, quizzes, or interactive modules to enhance learner engagement and facilitate learning.

In today's dynamic work environment, characterized by shorter attention spans and limited time for dedicated learning, microlearning is a highly advantageous method. It offers benefits such as time efficiency, flexibility, accessibility, continuous learning, improved retention, and targeted learning. These advantages collectively boost learner engagement, increase knowledge retention, and promote the practical application of acquired skills.

The key characteristics of microlearning are as follows:

- **Assess Suitability:** It is crucial to evaluate whether it aligns with your training needs before implementing microlearning. Microlearning may not be a suitable format for every subject matter. Complex tasks or those requiring hands-on experience might not fit within the confines of short, condensed segments. However, microlearning can still serve as a valuable complementary training approach, offering benefits for targeted learning objectives.

- **Repurpose Existing Content:** Efficient microlearning implementation can be achieved by repurposing existing training materials. You can break down components of your current training courses into microlearning segments as long as they logically stand alone. Alternatively, adapt and modify existing content to fit the microlearning format, reducing the effort to create entirely new materials.

- **Leverage Gamification Elements:** Consider integrating gamification elements into your microlearning courses to enhance engagement and motivation. Gamification can transform the learning experience into a game-like scenario, providing learners with a sense of accomplishment upon achieving specific learning goals. This can involve awarding digital badges, points, or status based on course completion. Tangible incentives such as bonuses, extra time off, or certificates can motivate employees to participate actively in microlearning.

- **Incorporate Short Videos and Interactive Components:** Microlearning is designed to be engaging, which can be amplified by integrating attention-grabbing elements like short videos and interactive quizzes. Short videos are particularly effective because they merge visual and auditory elements, capturing viewers' attention. The brevity of these videos aligns with the microlearning approach, ensuring they are engaging and memorable while still delivering relevant information within a suitable timeframe.

- **Emphasize Recurring Content:** Drawing insights from the Ebbinghaus studies, it's evident that revisiting learned materials over time improves knowledge retention. At the conclusion of the microlearning series or within shorter segments, incorporate tools that enable employees to review and reinforce what they've learned. This may involve quizzes assessing retention or summary slides outlining the key takeaways from the course. Alternatively, referencing prior learning modules in subsequent segments refreshes learners' memories and reinforces their comprehension.

- **Ensure Cross-Device Accessibility:** One of the key advantages of microlearning is its flexibility, enabling learners to access content on their own terms. To fully leverage this benefit, it's imperative to ensure that microlearning materials are accessible across various devices, emphasizing smartphones and tablets. Avoid imposing restrictions that confine access solely to work desktop computers. By offering unrestricted access, you can maximize the benefits of microlearning.

- **Promote Social and Collaborative Learning:** The successful implementation of microlearning in an organization can greatly benefit from social and collaborative learning. Social learning involves observing and emulating positive behaviors, while collaborative learning centers on brainstorming and sharing insights with peers. Allowing discussions and comments on microlearning modules actively engages learners, motivates them, and provides valuable insights into areas where additional knowledge may be required. Social and collaborative learning can be powerful allies in a successful microlearning program.

Challenges of Implementing Microlearning in a Learning Environment

- **Resource-Intensive Maintenance:** One common misconception about microlearning is that it's easy to create and maintain. Microlearning programs require significant time, resources, and effort to remain effective. Converting existing courses into microlearning modules is not always straightforward, requiring meticulous planning and relevance assessment for each module. Ensuring that each microlearning session stands alone and aligns cohesively within a broader sequential learning structure can be complex. Furthermore, frequent updates are necessary to keep content current in a rapidly changing business environment, adding to the maintenance workload.

- **Scaling Personalized Content:** The challenge of personalizing microlearning content to cater to individual needs is considerable. Due to the brevity of microlearning courses, a single traditional course can be split into many modules, potentially resulting in an organizational scale of thousands of modules. Managing, personalizing, and effectively directing this extensive volume of content to the right individuals require intricate strategies and substantial effort from instructional designers.

- **Accessibility Issues:** Ensuring that microlearning content is accessible to individuals with disabilities is not straightforward. It involves specific knowledge and technologies to format and optimize courses according to accessibility standards and best practices.

- **Integration with Existing Training:** Integrating microlearning into existing training programs is challenging. Coordinating microlearning with other training modalities and ensuring alignment with broader training objectives require meticulous planning to ensure a seamless transition.

- **Resistance to Change:** Employees and organizations accustomed to traditional training methods may resist adopting microlearning. Overcoming this resistance and ensuring buy-in from all stakeholders can be a significant hurdle in the adoption process.

- **Content Overload:** While microlearning aims to keep content concise, there is a risk of overwhelming learners with many microlearning modules. Striking the right balance and structuring content effectively are crucial to prevent content overload and maintain learner engagement.

Examples of Implementing Microlearning

- **Microcopy for Learning:** Microcopy involves crafting succinct, context-specific messages and hints with the purpose of facilitating user learning.

  Instances of microcopy include guidance in error messages, clarifications within contact forms, and subtle hints that enhance the user experience in e-commerce settings.

- **Microlearning Through Videos:** Microlearning videos are short, purpose-driven video content tailored to attain specific learning objectives.

  They can function as stand-alone knowledge nuggets that offer precise takeaways or seamlessly fit into a more extensive learning journey.

  Illustrative examples encompass explanatory videos, interactive shorts, micro-lectures, whiteboard animations, and kinetic text-based animations.

- **Mobile Apps for Microlearning:** Mobile apps designed for microlearning deliver bite-sized lessons that can be accessed on the go, enabling users to gain knowledge in easily digestible portions.

  Noteworthy applications include Google and YouTube, which offer informative content, Headspace for meditation guidance, Lasting for relationship advice, and "Word of the Day" apps designed to enhance vocabulary skills.

- **Micro-challenges and Gamified Learning:** Micro-challenges and gamified activities present brief scored learning experiences that may include rewards, benefits, badges, or incentives upon successful completion.

  They actively engage learners through elements such as multiple-choice quizzes, polls, flashcards, question-and-response interactions, realistic simulations, and recorded responses from participants.

- **Visual Information with Infographics:** Infographics are visually captivating representations of data, information, or knowledge that often emphasize critical details and numerical values.

  These graphical representations manifest in diverse forms, including statistical infographics, process flowcharts, geographic data maps, and hierarchical diagrams, simplifying the understanding of complex information.

Nano-learning

Nano-learning is an educational approach that focuses on delivering knowledge in ultra-compact, easily consumable formats, such as short videos or multimedia, which can be grasped within a matter of minutes or even seconds. Nano-learning is seen as a promising future-oriented pedagogical approach that prioritizes accessibility, efficiency, and the need for quick, convenient learning experiences, enabling learners to acquire knowledge without significant time or cognitive commitment.

Nano-learning offers a versatile and learner-centric approach, allowing individuals to conveniently access short, engaging modules online, fitting learning seamlessly into their busy lives. This format encourages personalized learning experiences, empowering learners to choose topics and the pace of their education. With its cost-effectiveness, nano-learning becomes an attractive option for organizations seeking efficient and engaging training solutions, making it an ideal choice for modern learners and budget-conscious institutions. Table 4-11 lists the differences between microlearning and nano-learning.

Examples of Nano-learning

- **Micro Videos:** A three-minute YouTube tutorial on basic car maintenance, teaching viewers how to change their car's oil

- **Infographics:** A visually engaging infographic illustrating the steps to create a personal budget, making financial planning easy to grasp

- **Flashcards:** Digital flashcards for learning new vocabulary in a foreign language, like the Duolingo app's flashcard feature

- **Podcasts:** A 5-minute podcast episode discussing the latest trends in digital marketing, perfect for a quick commute update

- **Interactive Quizzes:** An online quiz assessing nutrition knowledge, providing instant feedback on the quality of one's diet choices

- **One-Minute Reads:** A concise blog post explaining the benefits of mindfulness meditation and how to get started with just one minute of practice

- **Social Media Tutorials:** A series of short Instagram stories guiding users through the steps to create eye-catching graphics for their social media posts

- **Product Demos:** A two-minute video showing how to assemble a piece of IKEA furniture with clear, step-by-step instructions

- **Language Learning Apps:** Language learning apps like Babbel, offering daily exercises that take just a few minutes, teaching users new words and phrases in their chosen language

Table 4-11. *Differences Between Microlearning and Nano-learning*

Parameters	Microlearning	Nano-learning
Definition	Short, focused, and concise learning modules	Extremely brief, often just a few seconds
Duration	Typically ranges from a few minutes to 15–20 minutes	Typically consists of just a few seconds
Learning Outcome	Delivers specific learning objectives or skills	Offers quick, highly specific knowledge
Complexity	Can cover moderately complex topics or tasks	Addresses simple, straightforward topics
Delivery Format	Various formats, including videos, infographics, quizzes	Delivered in the form of extremely short texts, images, or videos
Engagement Level	Engages users for a short yet focused learning experience	Grabs attention with ultra-brief, often anecdotal content
Use Cases	Suited for comprehensive topics or skills requiring a brief approach	Ideal for delivering quick tips, reminders, or small knowledge nuggets
Application Examples	Employee training, language learning, skill development	Daily facts, trivia, quick reminders, momentary insights

Gamification

Gamification is the intentional infusion of game elements and mechanics into nongaming situations. The primary objective is to engage participants actively, motivate them, and help them achieve specific goals. This approach taps into our natural tendencies, such as the desire for competition, a sense of accomplishment, teamwork, and altruism, to boost participation, performance, and overall satisfaction.

Gamification involves various strategies, such as introducing scoring systems, awarding achievement badges, and creating leaderboards. These elements add an interactive and rewarding dimension to activities that are not inherently games, like education, marketing, and workplace tasks.

The key characteristics of gamification are as follows:

- **Incorporation of Gaming Elements:** Gamification involves using various gaming features, such as points, badges, leaderboards, levels, challenges, and rewards. These components instill a sense of accomplishment, competition, and progress, drawing upon intrinsic motivators like autonomy, mastery, and purpose.

- **Alignment with Goals:** The success of gamification depends on aligning game elements and mechanics with the overall goals and objectives of the learning context. By connecting learning-related tasks and objectives with these game elements, learners can clearly see how their efforts contribute to desired outcomes.

- **Boosting Motivation and Engagement:** The incorporation of gamification heightens learner motivation and engagement by rendering the learning experience more enjoyable, interactive, and gratifying. It introduces challenges, a sense of achievement, and friendly competition, which, in turn, enhances learner satisfaction and performance.

- **Skill Advancement:** Gamification is a mechanism for cultivating and enhancing specific skills and knowledge areas. By integrating learning elements into game mechanics, learners can acquire new skills, enhance existing ones, and participate in continuous learning within an enjoyable and engaging context.

- **Encouraging Collaboration and Social Interaction:** Gamification can stimulate collaboration and social interaction among learners. Leaderboards, team challenges, and collaborative quests promote a sense of camaraderie, teamwork, and healthy competition, fostering knowledge sharing and collaboration.

- **Providing Timely Feedback and Recognition:** Gamification offers immediate feedback and recognition for learners' efforts and achievements. Real-time feedback, virtual rewards, and badges reinforce positive behaviors, acknowledge accomplishments, and promote desired learning outcomes.

- **Performance Assessment:** Gamification facilitates the monitoring and measuring of learner performance and progress toward learning goals. By collecting data on game-related activities, educational institutions can gain insights into individual and group performance, pinpoint areas for enhancement, and make data-driven decisions.

- **Personalization and Customization:** Effective gamification permits personalization and customization based on individual learner preferences and needs. Allowing learners to select challenges, set goals, and tailor their learning experience heightens their sense of autonomy, engagement, and ownership.

- **Ethical Considerations:** While gamification is potent, addressing ethical concerns is essential. Game mechanics should not generate undue stress, foster unhealthy competition, or result in learner burnout. Ensuring fairness, transparency, and inclusivity in the design and implementation of gamification initiatives is paramount.

- **Continuous Evaluation and Enhancement:** Gamification strategies in the learning domain should undergo continuous evaluation, refinement, and improvement based on feedback, data analysis, and evolving educational requirements. Regular assessment and iteration help maintain relevance, effectiveness, and sustained engagement over time.

Challenges of Implementing Gamification in a Learning Environment

- **Alignment with Objectives:** Ensuring that the gamification strategy aligns with the organization's overarching objectives or learning environment can be challenging. It's important to design the gamification elements to support the desired outcomes.

- **Design Complexity:** Overly complex game mechanics or unclear rules can confuse and frustrate participants. Finding the right balance between engagement and simplicity is crucial.

- **Lack of Engagement:** While gamification is intended to boost engagement, it can sometimes fail to capture the interest of all participants. This may be due to a poor design or a lack of intrinsic motivation for the activities involved.

- **Over-gamification:** Applying gamification to every aspect of an activity or task can lead to an overreliance on rewards and incentives. This can diminish intrinsic motivation and turn participants into "extrinsically" motivated players.

- **Ethical Concerns:** Gamification should be designed with ethical considerations in mind. It should not encourage unhealthy competition or addiction or compromise individuals' well-being.

- **Sustainability:** Sustaining engagement over time can be a challenge. The novelty of gamification can wear off, and participants may lose interest if the gamified elements become repetitive or no longer feel rewarding.

- **Measurement and Analytics:** Measuring the effectiveness of gamification can be difficult. Collecting and analyzing data to assess the impact on engagement, performance, and outcomes is essential.

- **Customization:** Designing gamification that accommodates individual preferences and needs can be complex. A one-size-fits-all approach may not work well for diverse participants.

- **Costs and Resources:** Developing and maintaining gamification systems may require resources in terms of time, money, and expertise. Organizations need to budget for these costs.

- **Technology Integration:** If gamification relies on technology, integrating it with existing systems and ensuring compatibility can be challenging.

- **Resistance to Change**: Some individuals or groups may resist adopting gamification, perceiving it as a distraction or a gimmick rather than a valuable tool.

- **Content Relevance:** Ensuring that the gamified elements are relevant to the content or activity can be tricky. If the connection is too tenuous, it may not effectively enhance the learning or engagement experience.

Examples of Implementing Gamification

- **Embrace Immersive Environments:** Utilize virtual and simulated experiences to help learners hone their skills and tackle real-world challenges effectively. These interactive simulations equip learners to avoid common pitfalls and develop problem-solving abilities. Animated or gamified training options make it easier for learners to enhance their skills.

  Example: Learners can explore a virtual island to learn survival skills, enhancing their problem-solving and decision-making abilities.

- **Conquer Short Attention Spans:** Leverage game mechanics to break down learning into smaller, easily digestible segments that address the "forgetting curve." Employ gamification to divide training into short, engaging sessions that keep learners actively involved.

  Example: Create short, interactive math quizzes that learners can complete in just a few minutes, maintaining their focus and motivation.

- **Elevate Learning Through Gamification:** Activate learning engagement by incorporating active learning techniques.

  - Understand your learners thoroughly.

  - Address their anxieties and concerns.

  - Empower learners to take control of their learning.

  - Inject elements to ward off monotony.

  - Employ familiar characters, case studies, and narratives.

  Example: In a history class, learners can become historical figures in a gamified scenario, making decisions that shaped the course of history.

- **Balanced Reward Systems:** Rewards play a role in gamification, but they shouldn't overshadow the excitement of the learning journey. Maintain equilibrium by focusing on the thrill of competition along with rewards.

  Example: In a fitness app, users can compete on leaderboards for the most steps taken while earning virtual medals for achieving personal fitness goals.

- **Celebrate Participation:** Recognize that everyone's a winner in gamification. Acknowledge participation as a key metric of success.

  Example: In a hobby-related community, participants receive recognition and virtual "achievement" badges for actively contributing to discussions and sharing their knowledge and experiences.

- **Nurture a Culture of Continuous Learning:**
  Encourage ongoing learning by infusing game elements
  into training activities, motivating learners to pursue
  knowledge willingly.

  Example: In a language learning platform, learners
  earn points, unlock levels, and receive virtual trophies
  for consistently practicing and improving their
  language skills.

- **Power of Storytelling:** Enhance engagement by
  incorporating storytelling elements into the learning
  content.

  Example: In a general knowledge quiz app, learners
  embark on a virtual treasure hunt, solving puzzles and
  riddles based on historical events and myths to uncover
  hidden knowledge.

- **Avatars and Personalization:** Elevate personalization
  by allowing learners to create avatars and tailor their
  learning journeys.

  Example: In an online creativity course, learners can
  design and customize avatars representing their artistic
  preferences and explore a personalized gallery of
  creative challenges and projects.

Game-Based Learning (GBL)

Game-based learning is an instructional approach harnessing the
potential of games to shape and bolster educational outcomes.

In a game-based learning environment, these objectives are realized
through educational games that incorporate key elements like learner
engagement, immediate rewards, and friendly competition, all aiming to
sustain and stimulate learning while learners enjoy themselves.

What makes game-based learning so remarkable is its universality; it benefits learners of all ages, spanning from early childhood education to postsecondary studies and even beyond. The mode and context are also highly flexible, allowing learners to acquire knowledge:

- Through digital games on the Internet

- In a physical, face-to-face setting using tangible objects

- Either independently or in collaboration with peers

Successfully integrating game-based learning (GBL) within educational contexts requires several critical components. First and foremost, it's imperative to establish clear learning objectives to provide direction for GBL's inclusion in the curriculum. Developing engaging games harmonizing with the subject matter and educational goals is fundamental to sustaining learner engagement. Implementing assessment tools and offering timely feedback guarantee that learners are progressing, and adapting the learning experience to individual needs is also essential. Ensuring inclusivity, providing educators with training and support, and promoting collaborative learning are all integral elements of a comprehensive GBL approach. Analyzing learner data and maintaining a balanced approach between screen time and other educational methods further enrich the learning environment. Continuous evaluation and enhancement and providing necessary resources and infrastructure are crucial for effectively implementing GBL. Responsible supervision ensures that GBL is used in a safe and productive manner. Table 4-12 lists the differences between gamification and game-based learning.

Examples of Implementing Game-Based Learning

- **Simulations:** Simulations engage learners actively in their learning by having them assume roles to experience events or processes realistically. They help learners understand and transfer knowledge to new situations.

- **Hybrid Inquiry Projects:** These projects integrate inquiry-based and problem-based learning. Teachers pose a problem to learners and ask them to complete tasks as they work toward a solution. Learners earn points and level up as they progress through tasks, advancing their knowledge and problem-solving skills.

- **Digital Game-Based Learning:** Utilize digital tools like "Minecraft for Education" and "Kahoot!" to facilitate game-based learning. Minecraft offers specific assessments and lessons aligned with educational standards and allows learners to collaborate, solve problems, and gather evidence of their learning. Kahoot turns assessment and review into a game, tracking learners' points as they play against each other.

Table 4-12. *Differences Between Gamification and Game-Based Learning*

Parameter	Gamification	Game-Based Learning
Definition	Using game elements and principles in nongaming contexts enhances engagement and motivation.	The incorporation of educational games as a core component of the learning experience.
Core Element	Game elements (e.g., points, badges, leaderboards) are integrated into the existing curriculum or activities.	Actual games are used to deliver educational content and achieve learning objectives.

(*continued*)

Table 4-12. (*continued*)

Parameter	Gamification	Game-Based Learning
Motivation	Designed to motivate and engage learners through competition, rewards, and recognition.	Inherently motivating as learning occurs through gameplay and interactive experiences.
Interactivity	Encourages learner interaction with nongame content, often through rewards and competition.	Requires active engagement with the game itself, where learners apply knowledge and skills within the game environment.
Integration	Applied as an overlay to an existing curriculum, often as a supplement or enhancement.	Integrated as an essential element of the curriculum, serving as a primary means of content delivery.
Learning Outcomes	Primarily focuses on improving engagement, participation, and motivation. May not directly impact a deep understanding of the subject matter.	Directly aligned with educational goals, with games designed to facilitate knowledge acquisition, skill development, and critical thinking.
Examples	Awarding badges for completing course modules, using leaderboards to track progress, and incorporating quizzes with points.	Using educational games to teach subjects such as math, science, history, and language, where learning occurs through gameplay.
Assessment	Gamification often involves limited learning assessment and may focus more on performance in gamified activities.	Game-based learning typically includes a comprehensive assessment of learners' understanding and application of the subject matter.

(*continued*)

Table 4-12. (*continued*)

Parameter	Gamification	Game-Based Learning
Real-World Application	Widely used in business, marketing, and employee training to enhance engagement and motivation.	Commonly applied in K-12 education, higher education, and corporate training to facilitate active learning and knowledge retention.
Adaptability	Can be integrated relatively quickly and easily within existing educational or business frameworks.	Often requires more planning and resources for developing and integrating educational games.
Use of Game Design	Gamification may not necessarily involve game designers; it often leverages game mechanics and elements.	Game-based learning requires game design expertise to create effective, engaging educational games.
Immersion	Typically less immersive as it uses game elements to supplement traditional learning.	Offers a high level of immersion as it centers on gameplay and interactive experiences.

Scaffolding Learning

Scaffolding learning is an instructional approach utilized by educators to facilitate and guide learners in their acquisition of new knowledge and skills. It is a pedagogical strategy that offers initial, structured assistance to learners, with the intent of progressively diminishing this support as learners gain expertise. Drawing inspiration from the construction industry, where temporary structures aid workers in constructing buildings, educational scaffolding is analogous by furnishing temporary support to learners as they construct their understanding and capabilities.

The key characteristics of scaffolding are as follows:

- **Assessment of Learner Needs:** Instructional designers must understand their learners' current knowledge and abilities. This information helps in designing appropriate scaffolds to match the learners' needs.

- **Zone of Proximal Development (ZPD):** Scaffolding is most effective when it operates within a learner's ZPD. A learner can accomplish this range of tasks with support but cannot do it independently. Instructional designers must identify activities that fall within this zone.

- **Gradual Reduction of Support:** Scaffolding should gradually decrease as learners become more competent. The goal is for learners to take on more responsibility for their learning as they progress.

- **Clear Learning Objectives:** Designers should establish clear and achievable learning objectives and break them down into smaller, manageable steps. These steps can be scaffolded to ensure learners achieve each one.

- **Timely Feedback:** Regular feedback is essential in scaffolding learning. Designers should provide feedback to help learners understand their progress and make necessary adjustments.

- **Use of Tools and Resources:** Instructional designers can introduce various tools, resources, and materials to support learners in learning. These might include educational technology, visual aids, or instructional guides.

- **Promotion of Metacognition:** Encouraging learners to reflect on their learning process and develop metacognitive skills can enhance the effectiveness of scaffolding.

Challenges of Implementing Scaffolding in a Learning Environment

- **Identifying the Zone of Proximal Development (ZPD):** Determining the exact level of support a learner needs in their ZPD can be challenging. It requires a thorough understanding of the learner's prior knowledge and current abilities.

- **Differentiation:** Learners have varying needs, learning styles, and paces. Designing scaffolds that cater to these differences can be complex. It may require creating multiple paths or offering various types of support.

- **Over-scaffolding:** Providing too much support for too long can hinder a learner's independence. Over-scaffolding may lead to learners becoming overly reliant on support and failing to develop essential problem-solving skills.

- **Under-scaffolding:** Conversely, providing insufficient support can leave learners feeling frustrated and overwhelmed. This can lead to disengagement or a sense of failure.

- **Monitoring and Adjustment:** Constantly monitoring learners' progress and adjusting the level of support can be time-consuming and resource-intensive. It requires a keen eye for assessing when learners are ready to take on more responsibility.

- **Scaffolding with Technology:** Incorporating technology as a scaffold can be challenging, as educators and instructional designers must be proficient in various educational tools and platforms. Additionally, technology may not always be accessible to all learners.

Examples of Implementing Scaffolding Learning

- **Mathematics Problem-Solving:** A teacher introduces a complex word problem in a mathematics class. They start by solving part of the problem step by step on the board while explaining their thought process. Then, they ask learners to solve the remaining parts of the problem in pairs. The teacher circulates the room, offering support and guidance as needed. As learners become more confident, they work on similar problems independently.

- **Science Experimentation:** In a science class, learners are conducting an experiment. The teacher provides a detailed procedure to follow, including measurements and data recording. As learners gain experience, they are given a more general procedure and asked to design and carry out experiments independently, using their developed skills.

- **Information Research and Evaluation:** In a research project, learners are given a list of pre-approved sources for their research. The teacher guides them in evaluating the credibility of sources and extracting relevant information. Over time, learners are encouraged to independently identify and evaluate their own sources.

- **Art and Creativity:** The instructor introduces a specific art technique (e.g., shading with charcoal) in an art class. They demonstrate the technique and provide step-by-step guidance to create a particular drawing. As learners become more skilled, they are encouraged to experiment with the technique uniquely.

- **Programming and Coding:** A teacher may provide starter code for a basic program when teaching coding to beginners. Learners can modify and extend this code to accomplish specific tasks. As they become more proficient, they begin writing code from scratch to create their programs.

- **Physical Education and Skill Development:** In a physical education class, learners learn to play a sport like tennis. The instructor initially breaks down the skills into fundamental components and guides learners through each aspect (e.g., grip, stance, swing). As learners progress, the teacher gradually reduces the guidance, allowing them to practice and refine their skills more independently.

Spiral Learning

Spiral learning, also known as spiral curriculum, is an educational method that involves revisiting and expanding upon previously learned concepts, topics, or skills multiple times over an extended period. This approach aims to reinforce and deepen learners' understanding by gradually increasing the complexity and sophistication of the material as they progress through their education. One of the notable proponents of the spiral curriculum is Jerome Bruner, who emphasized the importance of revisiting and deepening learning as a means to facilitate meaningful education. The spiral learning approach is often used in subjects where foundational knowledge is crucial, such as mathematics and science, but can also be applied effectively in other disciplines.

The spiral curriculum method involves revisiting topics at different stages of a learner's education, deepening their understanding through incremental complexity. This approach is cyclical, introducing fundamental concepts and periodically revisiting them to reinforce learning and promote mastery. Educators use scaffolding to provide tailored support, and the approach encourages interdisciplinary connections, offering adaptability to suit various subjects and age levels. Table 4-13 lists the differences between scaffolding and spiral learning.

Examples of Implementing Spiral Learning

- **Geography:** In a geography curriculum, learners could begin with an introduction to world maps and basic geographical features. They revisit these concepts at higher grade levels to explore more advanced topics like regional geography, geopolitics, and cartography.

- **Health Education:** In a health education program, learners might first learn about personal hygiene and basic nutrition. As they advance, they revisit these topics to delve into more complex health issues, such as mental health, substance abuse, and disease prevention.

- **Art History:** In an art history course, learners could start with an overview of major art movements and famous artists. As they progress, they revisit these movements and artists but with a deeper understanding, allowing them to analyze artwork more critically and appreciate the historical context.

- **Environmental Science:** In an environmental science program, learners may initially learn about environmental issues at a local level. Over time, they revisit these topics and explore global environmental challenges, sustainability, and in-depth ecological principles.

- **Literacy Development:** In early literacy, learners begin by learning the alphabet and basic reading skills. As they move up in grade levels, they revisit these skills to explore more advanced reading comprehension, critical analysis of texts, and literary genres.

- **Economics:** In an economics course, learners may begin with basic economic principles like supply and demand. These principles are revisited as they progress, allowing learners to understand macroeconomics, financial markets, and global economic systems.

Table 4-13. *Differences Between Scaffolding Learning and Spiral Learning*

Parameter	Scaffolding Learning Approach	Spiral Curriculum/Learning
Definition	Provides structured support to learners as they acquire new knowledge and skills. The support is gradually reduced as learners become more proficient.	Involves revisiting and deepening previously learned concepts or topics multiple times, with increasing complexity each time.

(continued)

Table 4-13. (*continued*)

Parameter	Scaffolding Learning Approach	Spiral Curriculum/Learning
Timing of Revisiting	Typically focuses on the here-and-now learning, providing immediate support and guidance as needed.	Revisits topics over an extended period, often spanning multiple grade levels or educational stages.
Focus on Individual Learners	Tailors support to individual learners' needs, providing assistance as required for each learner.	Generally applies the same curriculum structure to all learners, revisiting topics for the entire class.
Scope of Curriculum	Primarily concerned with individual learning tasks and objectives, offering targeted support as needed.	Concerned with the overall structure and progression of the curriculum across grade levels or educational stages.
Learning Progression	The focus is on adjusting the level of support based on individual learner progress within a specific learning task.	The focus is on learners' long-term development, revisiting topics to deepen understanding over time.
Goal	To facilitate immediate learning tasks and help learners master specific skills or concepts.	To promote deeper, long-term retention and comprehension of subject matter over the course of an educational journey.
Application	Commonly used in individual lessons, assignments, or learning tasks.	Applied to the overall curriculum structure and educational planning.

(*continued*)

Table 4-13. (*continued*)

Parameter	Scaffolding Learning Approach	Spiral Curriculum/Learning
Adaptability	Flexible and adjustable to individual learners' needs and the specific learning task at hand.	Needs careful planning and designing to ensure that topics are revisited coherently and progressively across grade levels.
Assessment	Often involves ongoing, formative assessment to gauge individual progress and adapt support.	Assessment may involve summative evaluations that measure learners' cumulative understanding of topics revisited over time.
Subject Variety	Can be applied to a broad spectrum of subjects and learning tasks.	Typically applied to subjects where foundational knowledge and comprehension are essential.
Teaching Method	Utilizes a range of teaching methods, including direct instruction, feedback, and guidance as needed.	Emphasizes curriculum design and planning to ensure that topics are revisited and deepened over time.

Andragogy

Andragogy is a theory of adult learning that places a strong emphasis on self-directed, experience-driven, and problem-centric education. It recognizes that adult learners are distinct from children and have unique needs and motivations. More precisely, andragogy is an adult learning theory that prioritizes self-directed and problem-oriented learning. Malcolm Knowles is an American educator who first tried to distinguish between pedagogy and andragogy. These differences laid the foundation of andragogical principles. These differences are the assumptions made by Knowles about adult learning. The key assumptions are as follows:

- **Self-concept:** Adult learners are autonomous and self-directed individuals.

- **Learning from Experience:** Prior experiences serve as a valuable resource for learning.

- **Readiness to Learn:** Adult learners are more inclined to engage when the material is relevant and immediately applicable.

- **Immediate Applications:** The focus is on real-life problem-solving and practicality.

- **Intrinsic Motivation:** Adults are primarily motivated by internal factors and personal growth.

- **Need to Know:** Understanding the relevance and benefits of learning is crucial for adult engagement.

Instructional Strategies to Implement Knowles' Assumptions

- **Facilitate Self-directed Learning**

  - Encourage learners to set their learning goals and objectives.

  - Provide resources and tools for self-assessment and self-paced learning.

  - Offer opportunities for learners to choose topics or projects of personal interest.

Example: In a professional development course, participants set their own learning objectives based on their career goals. They have access to a variety of resources and can choose which topics to explore in depth.

- **Tap into Prior Experience**

  - Start by eliciting the learners' prior experiences related to the topic.

  - Use case studies, problem-solving scenarios, or group discussions to encourage the sharing of experiences.

  - Connect new content to the learners' existing knowledge and real-life situations.

  Example: In a leadership training workshop, participants are asked to share past leadership experiences, both successful and challenging. The facilitator then relates these experiences to the course content.

- **Create Problem-Centered Learning**

  - Design learning activities that revolve around real-world problems or challenges.

  - Encourage learners to apply their new knowledge and skills to solve practical issues.

  - Provide opportunities for critical thinking and problem-solving.

Example: In a project management course, participants work in small groups to solve a real-world project management challenge presented by a local company. They apply course concepts to develop a solution.

- **Offer Immediate Relevance**

  - Clearly articulate the practical applications and benefits of what is being taught.

  - Ensure that the content is applicable to the learners' current needs and situations.

  - Use case studies and real-life instances to demonstrate the immediate relevance of the material.

  Example: In a financial literacy class, the instructor explains how the principles of budgeting can immediately help participants manage their personal finances, demonstrating the direct benefits of the learning.

- **Promote Intrinsic Motivation**

  - Cultivate positive and supportive learning environment that encourages self-motivation.

  - Allow learners to choose their learning paths or projects.

  - Recognize and reward achievements to boost learners' self-esteem.

  Example: In a language learning program, learners choose a personal goal, such as traveling or connecting with family members in another country. Their progress and achievements are celebrated, boosting their motivation.

- **Provide Constructive Feedback**

  - Give feedback that is specific, timely, and constructive to help learners improve.

  - Create opportunities for self-assessment and reflection.

  - Encourage peer assessment and self-evaluation.

  Example: In a writing workshop, participants share their work with peers and receive feedback on specific aspects like clarity, structure, and style. The feedback helps them refine their writing skills.

- **Encourage Collaborative Learning**

  - Facilitate group discussions, collaborative projects, and peer teaching.

  - Learning from others' experiences and insights can enhance the learning process.

  - Collaboration fosters a sense of community among adult learners.

  Example: In a sales training course, participants engage in role-playing exercises where they take on different roles, such as the salesperson and the customer, to practice negotiation and communication skills.

- **Utilize Technology and Online Learning**

  - Offer eLearning options that allow for self-directed study.

  - Use online forums, discussion boards, and digital resources to support asynchronous learning.

- Provide access to up-to-date and unbiased online materials.

Example: In an online coding course, learners have access to a library of coding tutorials, forums for asking questions, and coding challenges that allow them to learn at their own pace and convenience.

- **Flexibility in Learning Environment**

  - Accommodate various learning styles and preferences, allowing for individualized learning paths.

  - Provide a blend of both in-person and online learning options to accommodate busy schedules.

Example: In a healthcare certification program, learners have the option to attend in-person lectures, participate in online webinars, or complete self-study modules to accommodate their busy work schedules.

- **Encourage Reflection and Self-assessment**

  - Offer chances for learners to engage in self-reflection regarding their educational experiences and assess their progress.

  - Encourage the development of metacognitive skills for better self-directed learning.

Example: After a leadership seminar, participants are given a journal to reflect on their leadership experiences and set personal development goals. They periodically review their progress and adjust their goals accordingly.

Heutagogy

Heutagogy is an educational paradigm that underscores a learner-centered approach, granting individuals a central role in directing and managing their own learning experiences. This approach represents a pivotal departure from traditional teaching methodologies, which tend to be teacher-centric, primarily focused on transmitting knowledge to learners.

In essence, heutagogy embodies a dynamic evolution from conventional instructional practices, necessitated by the readily accessible and ever-evolving information landscape of the contemporary world. It is recognized as a response to the insufficiency of discipline-based knowledge in preparing individuals for success within modern communities and professional environments.

Within this redefined context, the concept of heutagogy accentuates the proactive involvement of adult learners and employees in shaping their educational journeys. As we navigate an era characterized by the prominence of distance and eLearning platforms, heutagogy emerges as a natural progression from andragogical approaches, serving as a mechanism to better align education with the needs of the future.

Instructional Strategies to Implement Heutagogy

- **Choice and Autonomy:** Provide options for learners to choose topics, resources, and methods of learning that align with their interests and preferences. This empowers them to make decisions about their education.

  Example: In a history class, learners are given the option to choose a specific historical period or event to research and present, allowing them to explore their interests.

- **Problem-Based Learning:** Use real-world problems and scenarios that require critical thinking and problem-solving. Encourage learners to explore solutions independently or collaboratively.

  Example: In a medical school, learners are presented with a patient case and are tasked with diagnosing and proposing a treatment plan, requiring them to research and apply their knowledge.

- **Project-Based Learning:** Assign open-ended projects that allow learners to research, plan, and execute based on their interests and goals.

  Example: In a business course, learners are assigned the task of creating a business plan for a startup, including market research, financial projections, and marketing strategies.

- **Self-paced Learning:** Offer materials that learners can access at their own pace, allowing them to decide when and how they engage with content.

  Example: Learners in an online coding course can progress through modules at their own speed, with access to all course materials from day one.

- **Peer Learning and Collaboration:** Foster collaboration among learners, allowing them to exchange knowledge, experiences, and insights. This can happen through discussion forums, group projects, or peer review.

Example: In a group project for a science class, learners work together to conduct an experiment and present their findings, with each member contributing based on their strengths and interests.

- **Resource Exploration:** Promote exploration of diverse learning resources, such as books, articles, videos, podcasts, and online courses. Encourage learners to evaluate the credibility and relevance of these resources.

  Example: In a research methods course, learners are encouraged to explore a variety of academic databases and sources to find relevant literature for their research projects.

- **Metacognitive Skill Development:** Teach learners metacognitive skills, such as goal setting, time management, self-assessment, and self-regulation. These skills help learners become more effective self-directed learners.

  Example: A study skills workshop teaches learners techniques for effective time management, note-taking, and self-assessment.

- **Feedback and Assessment:** Emphasize formative feedback and self-assessment. Encourage learners to evaluate their own progress and make adjustments accordingly.

  Example: Learners in an art class regularly assess their own work, comparing it to specific criteria, and then meet with the instructor to discuss progress and areas for improvement.

- **E-Portfolios:** Have learners create digital portfolios to showcase their accomplishments, skills, and reflections throughout their learning journey.

  Example: As part of a graduate program, learners create e-portfolios to showcase their academic achievements, projects, and personal growth throughout their studies.

- **Technology Integration:** Use technology tools and platforms that facilitate self-directed learning, including online libraries, educational apps, and content curation tools.

  Example: Learners use an online learning platform that allows them to access a variety of multimedia resources, such as videos, interactive simulations, and discussion forums.

- **Adaptive Learning Platforms:** Leverage adaptive learning platforms that tailor content to individual learner needs and progress.

  Example: An adaptive language learning app adjusts the difficulty of exercises based on the learner's performance, ensuring a personalized learning experience.

Differences Between Pedagogy, Andragogy, and Heutagogy

Pedagogy, andragogy, and heutagogy are three distinct approaches that shape the landscape of teaching and learning. Now, armed with a thorough understanding of each approach, we can dissect the key differences between them and appreciate how they cater to the evolving needs and contexts of education.

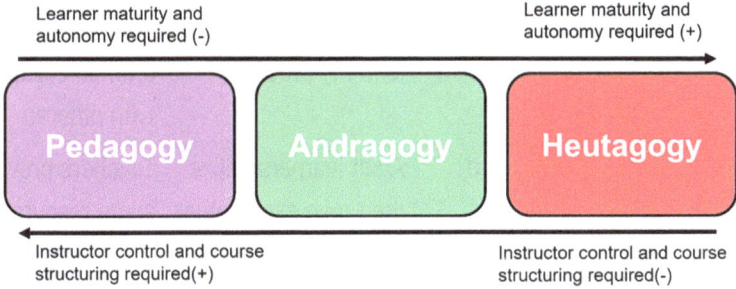

Figure 4-2. *Learner Maturity and Instructor Control in Pedagogy, Andragogy, and Heutagogy*

The fundamental distinction among these three educational paradigms lies in the degree of instructor involvement (Figure 4-2). In pedagogy, the instructor plays a central role in structuring the learning experience and directing content delivery in a systematic manner. On the contrary, in heutagogy, a higher level of learner maturity is a prerequisite. Here, individuals are empowered to autonomously determine their preferred learning styles and adapt their education to align with their specific needs, thereby reducing the necessity for substantial instructor intervention. Table 4-14 lists the differences between all three educational paradigms.

Table 4-14. *Differences Between Pedagogy, Andragogy, and Heutagogy*

Aspect	Pedagogy	Andragogy	Heutagogy
Learner Orientation	Emphasizes the dependent nature of learners. In this approach, teachers dictate what, how, and when learners learn.	Recognizes adult learners as independent and self-directed. They actively seek autonomy in their learning process.	Acknowledges interdependence among learners. They naturally identify and manage their own learning, often in collaboration with others.
Learning Resources	Few resources are given to learners; teachers employ transmission techniques to convey knowledge.	Adult learners utilize their own experiences and the experiences of others as valuable learning resources.	Teachers provide some resources, but learners have autonomy in shaping their learning paths through negotiation.
Motivation for Learning	Learners primarily motivated to progress through predefined curriculum stages. External sources, such as parents, teachers, and competition, drive motivation.	Adult learning is motivated by intrinsic factors like the desire to acquire needed knowledge or enhance performance. Increased self-esteem and self-recognition play a significant role.	Motivation in heutagogy stems from self-efficacy, creativity, and the ability to apply these qualities in various situations, as well as from collaborative learning experiences.

(*continued*)

Table 4-14. (*continued*)

Aspect	Pedagogy	Andragogy	Heutagogy
Purpose of Learning	Learning is directed toward progressing through a fixed subject-centered curriculum.	Adult learning is problem centered, often focused on addressing specific tasks or challenges.	Learning in heutagogy is exploratory and adaptable, driven by the identification of potential learning opportunities in various contexts.
Role of the Teacher	Teachers are central in designing the learning process and determining the content. They are assumed to have superior knowledge.	Teachers act as facilitators or enablers, fostering a collaborative and open learning environment.	The teacher's role is to develop learners' capabilities. Heutagogical educators expect learners to possess the skills to self-direct, be creative, and collaborate effectively.

Summary

- Pedagogy, andragogy, and heutagogy are three distinct approaches that shape the landscape of teaching and learning.

- There are four types of pedagogy: social pedagogy, critical pedagogy, culturally responsive pedagogy, and Socratic pedagogy.

- There are seven domains of pedagogy:
    - Domain 1: Content knowledge and pedagogy
    - Domain 2: Learning environment
    - Domain 3: Diversity of learners
    - Domain 4: Curriculum and planning
    - Domain 5: Assessment and reporting
    - Domain 6: Community linkages and professional engagement
    - Domain 7: Personal growth and professional development
- There are various modern pedagogical approaches such as project-based learning, problem-based learning, microlearning, nano-learning, gamification, game-based learning, scaffolding learning, and spiral learning.
- Andragogy is a self-directed, experience-driven, and problem-centric educational approach.
- Malcolm Knowles gave six key assumptions about adult learning: self-concept, learning from experience, readiness to learn, immediate applications, intrinsic motivation, and need to know.

Let's Brainstorm

These mini-scenarios will make you scratch your head and scribble on your pad.

Mini-scenario 1

You are tasked with developing a comprehensive marketing campaign for a new educational app that aims to cater to learners from various age groups, including Gen X, Gen Y (Millennials), Gen Z, and Gen Alpha. Each generation has distinct preferences and characteristics. How would you design the content and promotional strategies to effectively reach and engage these diverse age groups, considering their unique needs and expectations?

Mini-scenario 2

Imagine you are tasked with designing an educational program for a group of middle-aged professionals who are transitioning into new, unfamiliar careers. These individuals have varying degrees of technological proficiency, and their motivations, prior experiences, and desired career paths differ significantly. How would you apply heutagogical principles in a way that caters to their diverse needs while ensuring successful career transitions?

Mini-scenario 3

Consider you are responsible for developing a continuing education program for a group of mid-career professionals in the healthcare industry. These professionals are seeking to acquire the skills and knowledge necessary to stay in a rapidly evolving field. They have varying schedules, learning preferences, and levels of prior knowledge. How would you design an andragogical learning experience that recognizes and addresses these individual differences, thereby maximizing their engagement and success in the program?

Mini-scenario 4

Imagine you are designing a training program for a group of entry-level employees in a fast-paced retail environment. They need to quickly grasp essential customer service skills and product knowledge. How would you incorporate microlearning techniques to deliver this critical information effectively, considering the fast-paced and dynamic nature of the retail industry?

Mini-scenario 5

Imagine you are an instructional designer for a multinational corporation with a workforce scattered across the globe. You are tasked with providing essential compliance training to employees on a tight schedule. The challenge is that employees have limited time for training due to their busy work schedules. How would you design a nano-learning program that delivers crucial compliance information in short, highly focused modules, ensuring maximum engagement and retention among the employees?

CHAPTER 5

Universal Design of Learning and Development

In the ever-evolving landscape of instructional design, we find ourselves standing on the precipice of a profound understanding of the science that underpins it. We have traversed the terrain of instructional design, delving into the intricacies of creating effective learning experiences. Now, as instructional designers, our gaze shifts toward a crucial aspect that transcends individual differences and embraces inclusivity – the realm of Universal Design for Learning (UDL).

Throughout our journey, we've discovered that learning theories serve as our guiding light, shedding light on the intricacies of acquiring knowledge and forming new connections. As we delved deeper, we explored learning styles and modalities, revealing the various factors influencing the learning process. Motivations, emotions, and necessities intricately shape the learner's experience.

Integrating pedagogical approaches has allowed us to craft learning experiences that seamlessly align with diverse learning styles, making the educational journey enjoyable and engaging. The foundation has been laid, and the stage is set as we stand at the crossroads of pedagogy and inclusivity.

© Ankita Jiyani Mangtani 2024
A. J. Mangtani, *Instructional Design Unleashed*, Design Thinking,
https://doi.org/10.1007/979-8-8688-0416-8_5

As instructional designers, the onus is on us to consider the principles of universal design, ensuring that the learning experience extends its reach to encompass individuals from varied backgrounds, cultures, genders, statuses, ages, linguistic backgrounds, socioeconomic standings, abilities, and, crucially, diverse cognitive and thinking preferences.

In this exploration of Universal Design for Learning, we shall navigate through the dimensions of accessibility and inclusivity, recognizing that our responsibility extends beyond the mere transmission of information. We are tasked with cultivating an environment where the richness of cognitive diversity is not just acknowledged but embraced, making learning an equitable and empowering experience for all.

IDEA Initiative

Inclusion, Diversity, Equity, and Accessibility (IDEA) have emerged as integral to creating inclusive educational environments, building upon the foundation of Diversity and Inclusion (D&I), and emphasizing equitable opportunities for all learners.

Within the educational context, these four components – inclusion, diversity, equity, and accessibility – contribute distinct elements to a comprehensive framework:

> **Inclusion:** Within the educational landscape, inclusion fosters a culture where every learner feels welcome and encouraged to contribute and participate actively. An inclusive educational environment supports, respects, and embraces differences among learners. While diversity ensures the presence of individuals from various backgrounds, inclusion goes a step further by actively promoting their engagement and involvement in the learning community.

Diversity: Diversity in education encompasses the representation of individuals from various backgrounds. This includes different genders, races, ethnicities, abilities and disabilities, religions, cultures, ages, and sexual orientations. Additionally, it extends to people with diverse backgrounds, experiences, skills, and expertise.

Equity: Equity in education underscores the importance of the same treatment, access, opportunity, and advancement for all learners. A commitment to equity involves identifying and dismantling barriers that may impede certain groups and individuals from fully participating in educational opportunities or pursuing specific academic and career paths.

Accessibility: Accessibility in education emphasizes creating an environment that ensures all learners, regardless of their physical or cognitive abilities, have equal access to educational resources, facilities, and opportunities. This involves designing curriculum materials, physical spaces, and technological platforms to accommodate diverse learning needs, promoting an inclusive educational experience for everyone.

In essence, the integration of inclusion, diversity, equity, and accessibility in educational settings is essential for cultivating an environment that not only reflects the richness of human experiences but also ensures accessible and equitable learning opportunities for all learners. The commitment to equity and accessibility in education involves addressing and removing barriers that might hinder the full participation of diverse groups, fostering a truly inclusive, enriching, and accessible educational experience.

Universal Design

As per the Center for Universal Design, Universal Design (UD) is defined as crafting products and environments to be accessible to everyone to the greatest extent possible, eliminating the necessity for adaptation or specialized design. To refine the focus, a modification of this definition is possible. For instance, when applying Universal Design (UD) to learning and development activities, the basic definition can be adapted to state, "the design of learning and development products and environments to be usable and accessible by all learners, to the greatest extent possible, requiring minimal need for adaptation or specialized design."

The essential characteristics of any Universal Design product or environment lie in its accessibility, usability, and inclusivity (Figure 5-1).

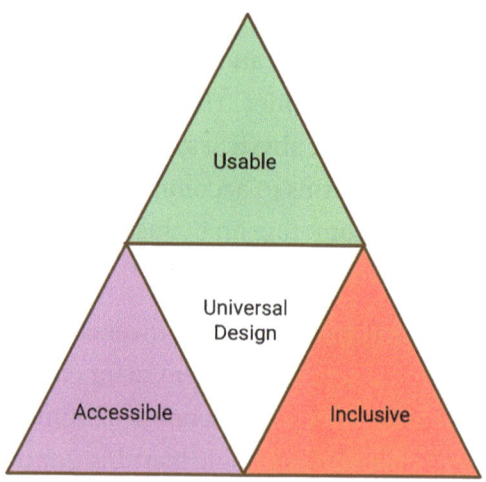

Figure 5-1. *Characteristics of Universal Design*

Seven Principles of Universal Design

The Seven Principles of Universal Design are fundamental guidelines that aim to make environments and products universally accessible. These principles provide a practical framework for creating designs that accommodate various abilities and needs, fostering inclusivity in various aspects of life.

1. **Equitable Use**

    Aim for a design that caters to diverse abilities, ensuring usefulness and marketability. An exemplar is a workplace communication platform that incorporates customizable font sizes, colors, and contrast options to ensure inclusivity for users with diverse visual abilities.

2. **Flexibility in Use**

    Design should be a fit for a wide range of preferences and abilities. A historical monument, for instance, offering visitors the choice to read or listen to display descriptions and history associated with the place, embodies this principle.

3. **Simple and Intuitive**

    Strive for a design that is easy to comprehend, irrespective of the user's background or concentration level. A smartphone with an intuitive user interface that incorporates recognizable symbols and gestures, making it easily navigable for users of all ages and technological backgrounds.

4. **Perceptible Information**

   Ensure the design effectively communicates information, overcoming sensory barriers. Video captioning is a prime example of employing this principle.

5. **Tolerance for Error**

   Minimize risks and adverse consequences of unintended actions. A financial app that incorporates built-in alerts and prompts to prevent users from making potentially costly mistakes in their transactions.

6. **Low Physical Effort**

   Design for efficiency and comfort, reducing physical strain. A smart home system that utilizes voice recognition technology to control lighting, temperature, and security, minimizing the physical effort required for daily tasks.

7. **Size and Space for Approach and Use**

   Provide appropriate size and space for accessibility, accommodating users of varying body sizes, postures, and mobility. A co-working space designed with modular and adjustable workstations accommodates users of diverse body sizes and allows them to personalize their work environment for optimal comfort and productivity.

Universal Design for Learning

Universal Design for Learning (UDL) represents a pedagogical approach to teaching and learning that ensures every student has an equitable opportunity to succeed.

It is important to clarify what UDL isn't in order to grasp its essence. Despite the term "universal," UDL isn't centered on a one-size-fits-all teaching approach. On the contrary, UDL adopts an inclusive strategy.

The overarching objective of UDL is to employ diverse teaching methods to eliminate any barriers to learning. It prioritizes integrating flexibility in instructional design, allowing for personalized adjustments tailored to individual strengths and needs. Consequently, UDL stands out as a framework that universally benefits all learners.

Three Principles of UDL

Universal Design for Learning (UDL) derives its foundation from extensive neuroscience research, pinpointing three fundamental neurological networks pivotal in influencing the learning process. UDL intricately examines human variability by delving into the complexities of the brain. It zeroes in on three specific regions – the affective network, recognition network, and strategic network – that respectively handle the "why," "what," and "how" aspects of learning (Figure 5-2).

UDL Guidelines		
Affective Networks	Recognition Networks	Strategic Networks
The "why" of learning	The "what" of learning	The "how" of learning
Diverse Means of Engagement	Diverse Means of Representation	Diverse Means of Action and Expression
Present alternatives for: • Capture Attention • Foster Long-Term Commitment • Self-control	Present alternatives for: • Perception • Language and Symbols • Comprehension	Present alternatives for: • Physical Action • Expression and Communication • Executive Functions

Figure 5-2. *UDL Guidelines of Three Principles*

Diverse Engagement Approaches: This principle is aligned with the affective network and explores the "why" of learning by connecting course material with students' personal goals. It centers on fostering interest and motivation and sustaining effort for all learners through recruitment, persistence, and self-regulation options.

Aligning with Universal Design for Learning (UDL) principles, the instructional strategy involves accessing ways to capture attention, building engaging options for fostering long-term commitment, and internalizing self-control to ensure a diverse and inclusive approach to learning. Table 5-1 provides the details.

Table 5-1. *Diverse Engagement Approaches*

Access	Build	Internalize
Present Alternatives to Capture Attention	**Present Alternatives to Foster Long-Term Commitment**	**Present Alternatives for Self-control**
Unnoticed information becomes inaccessible in the present and future. Despite teachers' efforts to capture attention, learner preferences vary widely. With evolving interests, skills, and circumstances, diverse engagement methods are crucial to accommodate significant individual differences.	Acquiring skills demands sustained effort. Learners vary in self-regulation abilities influenced by motivation, skills, and the environment. The instructional goal is to enhance self-regulation and self-determination for equitable learning. The environment should offer choices to support learners with diverse motivation and regulation skills.	Fostering learners' intrinsic ability to control emotions and motivations is crucial for effective coping and engagement. While some individuals naturally develop these skills, others may struggle. Unfortunately, some classrooms overlook these skills, making them inaccessible. Explicitly addressing self-regulation enhances the application of UDL principles, recognizing diverse learner needs in managing engagement and affect.

Diverse Representation Approaches: This principle is aligned with the recognition network and addresses the "what" of learning by clarifying what learners are expected to comprehend and engage with. It emphasizes presenting information diversely, offering options for perception, language, symbols, and comprehension. In essence, it facilitates learners' knowledge acquisition.

In line with Universal Design for Learning (UDL) principles, the educational approach centers on accessing, constructing, and internalizing alternatives to enhance perception, language, symbols, and comprehension. Table 5-2 provides the details.

Table 5-2. *Diverse Representation Approaches*

Access	Build	Internalize
Present Alternatives for Perception	**Present Alternatives for Language and Symbols**	**Present Alternatives for Comprehension**
Learning is challenging when information is hard to perceive or requires extraordinary effort. To enhance accessibility, use various modalities (e.g., vision, hearing, touch) and adjustable formats (e.g., enlarging text, amplifying sounds) to ensure that key information is equally clear to all learners. Multiple representations benefit learners with disabilities and enhance accessibility for a broader audience.	Learners vary in their proficiency with different forms of representation, both linguistic and nonlinguistic. While specific vocabulary may be clear to one learner, it could confuse another. Symbols like the equals sign or visual aids like graphs may help some learners but not others. To accommodate these differences, instructional strategies should incorporate alternative representations for improved clarity and comprehensibility, ensuring accessibility for all.	The goal of education is not just to make information accessible but to empower learners to transform it into usable knowledge actively. Constructing usable knowledge relies on information processing skills, such as focusing on specific details and incorporating new information with existing knowledge, categorization, and memorization. Designing and presenting information effectively can provide the necessary scaffolds for all learners to access knowledge.

Diverse Action and Expression Approaches: This principle is aligned with the strategic network and focuses on the "how" of learning by exploring how learners engage in various learning experiences and demonstrate their understanding. It differentiates ways learners express knowledge, providing options for physical action, expression, communication, and executive functions. It comes into play during activities and assessments, allowing learners to practice competencies and apply acquired knowledge.

Complying with Universal Design for Learning (UDL) principles, our pedagogical approach encompasses not only accessing and presenting alternatives for physical action but also building diverse options for expression and communication and internalizing alternatives for executive functions, fostering a comprehensive and inclusive learning environment that accommodates varied abilities and preferences. Table 5-3 provides the details.

Table 5-3. *Diverse Action and Expression Approaches*

Access	Build	Internalize
Present Alternatives for Physical Action	**Present Alternatives for Expression and Communication**	**Present Alternatives for Executive Functions**
Traditional print materials and some interactive educational software present challenges in navigation and interaction, limiting accessibility, especially for learners with physical disabilities, blindness, or dysgraphia. Well-designed curricular materials offer a seamless interface compatible with assistive technologies. This ensures individuals with movement impairments can effectively navigate and express their knowledge, utilizing tools like single switches, voice-activated switches, and expanded keyboards.	Expression mediums aren't universally suitable for all learners or types of communication. Some learners may struggle with certain mediums like a dyslexic student excelling in oral storytelling but facing challenges in written form. Offering alternative modalities is crucial to level the playing field and enable learners to express knowledge effectively and comfortably in the learning environment.	Executive functions associated with the prefrontal cortex enable long-term goal setting, strategic planning, progress monitoring, and strategy modification. Limited by working memory, executive capacity diminishes when managing nonautomatic skills or in the presence of higher-level disabilities. In the UDL framework, scaffolding focuses on expanding executive capacity by reducing lower-level skill demands and enhancing higher-level executive skills. This guideline explores methods for scaffolding executive functions directly, complementing previous guidelines that address lower-level support.

Universal Design for Learning Guidelines

The UDL Guidelines are an essential resource for implementing Universal Design for Learning to optimize teaching and learning based on scientific insights into how humans learn and catering to the diverse needs of learners.

Guidelines to Provide Alternatives to Capture Attention

Guideline 1: Maximize Individual Choices and Independence

In educational settings, offering choices in reaching a learning objective, the context for achievement, tools, or support is often more suitable than providing a choice of the objective itself. Providing such choices fosters self-determination, pride in accomplishment, and a stronger connection to learning. It's essential to recognize that individuals vary in their preference for choices, emphasizing the importance of optimizing the type and level of independence for effective engagement.

- **Provide choices in perceived challenge:**
    - **Example:** In a math assignment, students can choose between different sets of problems based on difficulty levels, allowing them to gauge their challenges.
- **Offer choices in rewards or recognition:**
    - **Example:** Learners can select from a list of potential rewards or recognitions (e.g., verbal praise, extra break time, or a certificate) based on their preferences and motivations.

- **Allow choices in context or content for skills practice:**
  - **Example:** Language arts learners can choose topics for their essays or stories, ensuring engagement by aligning the content with their interests.

- **Provide choices in tools for information gathering or production:**
  - **Example:** In a research project, learners can decide whether to use traditional library resources or digital tools, promoting autonomy in their chosen methods.

- **Allow choices in design elements (color, layout, graphics):**
  - **Example:** Learners working on a presentation can choose the color scheme, layout, and graphics to align with their personal preferences and enhance their engagement.

- **Provide choices in sequence or timing for task completion:**
  - **Example:** For a project with multiple components, learners can decide the order in which they tackle each task, allowing them to manage their time effectively.

- **Involve learners in designing classroom activities:**
  - **Example:** Learners collaborate with the teacher to design a class project, providing input on the format, topics, and activities involved.

- **Include learners in setting personal academic and behavioral goals:**

    - **Example:** Learners work with teachers to establish individual academic achievement and behavior goals, fostering a sense of ownership and responsibility.

Guideline 2: Enhance Relevance, Importance, and Credibility

Learners are captivated by information and activities aligned with their interests and objectives. While realism isn't mandatory, importance and credibility to individual and instructional goals are crucial. Teachers can enhance engagement by emphasizing the practicality of learning and showcasing it through genuine, purposeful activities. Recognizing that not all learners find the same activities equally relevant, it's crucial to offer options that optimize each learner's sense of relevance, importance, and credibility.

- **Vary activities and sources of information:**

    - **Personalized and Contextualized to Learners' Lives:** Students create a project that relates course content to their personal experiences, making the learning more relevant.

    - **Culturally Relevant and Responsive:** Integrate diverse cultural perspectives into literature discussions, ensuring cultural representation in readings and activities.

- **Socially Relevant:** Incorporate current events or community issues into class discussions and projects, fostering awareness and social relevance.

- **Age and Ability Appropriate:** Adapt math problems to different skill levels within the same class, ensuring age and ability appropriateness.

- **Appropriate for Different Groups:** Use literature from various racial, cultural, ethnic, and gender backgrounds, promoting inclusivity in readings and discussions.

- **Design activities with authentic learning outcomes:**

  - Students create a business proposal for a local entrepreneur, applying business concepts to real-world scenarios.

- **Provide tasks for active participation and exploration:**

  - Conduct a geography project where students actively explore and map their local community, promoting hands-on learning.

- **Invite personal response, evaluation, and self-reflection:**

  - After a history unit, students engage in a debate reflecting on the historical events studied, fostering personal responses and critical evaluations.

- **Include activities fostering imagination for problem-solving:**
  - Challenge students to design a sustainable city of the future, encouraging imaginative problem-solving and critical thinking about complex urban issues.

Guideline 3: Reduce Risks and Disturbances

Ensuring a secure learning environment is vital for effective teaching. Teachers should minimize potential risks and disturbances, enabling learners to concentrate on the learning process rather than basic needs or negative experiences. Beyond physical safety, addressing subtler risks and disturbances tailored to individual needs and backgrounds is crucial. For example, language experimentation may be challenging for English Language Learners, and excessive sensory stimulation might disturb specific learners. The optimal instructional environment offers choices that universally mitigate risks and disturbances, fostering a safe space conducive to learning.

- **Create an accepting and supportive classroom climate:**
  - **Example:** Implement peer mentoring programs to foster positive student relationships, creating a supportive atmosphere.
- **Vary the level of novelty or risk:**
  - **Charts, Calendars, and Schedules:** Provide visual schedules to increase predictability, aiding learners in anticipating daily activities.

- **Create Class Routines:** Establish consistent routines to build a sense of security and reduce anxiety.

- **Alerts and Previews:** Use visual or auditory alerts to help learners prepare for changes in activities or schedules.

- **Options for Novelty:** Introduce unexpected elements or surprises within routine activities to enhance engagement.

- **Vary the level of sensory stimulation:**

  - **Background Noise or Visual Stimulation:** Adjust the classroom environment by controlling background noise or visual elements based on learners' sensory needs.

  - **Pace of Work and Breaks:** Vary the pace of lessons and provide breaks to accommodate different attention spans and energy levels.

- **Vary social demands, support, and public display:**

  - **Example:** Offer a range of collaborative and independent learning activities to accommodate different social preferences and comfort levels.

- **Involve all participants in whole class discussions:**

  - **Example:** Use inclusive discussion formats, such as "think-pair-share and all-class response," ensuring active participation from all students.

Guidelines to Provide Alternatives to Foster Long-Term Commitment

Guideline 1: Increase the Prominence of Goals and Objectives

In long-term projects or practices, diverse sources vie for attention. Some learners require assistance to recall the initial goal and its rewards, necessitating periodic reminders to sustain effort and focus amid disturbances.

- **Prompt or require learners to formulate or restate goals explicitly:**

  - **Example:** Before starting a project, ask students to articulate their objectives verbally or in writing, ensuring clarity and commitment to the goal.

- **Display the goal in multiple ways:**

  - **Example:** Create visual aids, charts, or diagrams representing the project goal, appealing to different learning preferences.

- **Promote breakdown of long-term goals into short-term objectives:**

  - **Example:** Break down a complex project into manageable steps, guiding students to focus on short-term objectives contributing to the overall goal.

- **Demonstrate using handheld or computer-based scheduling tools:**

  - **Example:** Introduce students to digital calendars or planners, teaching them to use scheduling tools for organizing tasks and deadlines.

- **Use prompts or scaffolds for visualizing the desired outcome:**

    - **Example:** Provide visual prompts or templates to help students conceptualize the end result of their project, aiding in maintaining focus on the goal.

- **Engage learners in assessment discussions of excellence:**

    - **Example:** Facilitate class discussions about what constitutes excellence in the project, encouraging students to share examples aligned with their cultural backgrounds and interests.

Guideline 2: Modify Requirements and Provisions to Enhance the Challenge

Learners differ not just in skills but also in the challenges that inspire their best efforts. All learners require varied challenges and appropriate resources for successful task completion. To meet demands effectively, learners need flexible resources. Offering a diverse range of challenges and resources allows every learner to find optimally motivating tasks, emphasizing the importance of balancing available resources to meet challenges effectively.

- **Distinguish the degree of difficulty or complexity within core activities:**

    - **Example:** In a math class, offer different sets of problems with varying levels of difficulty, allowing students to choose tasks aligned with their skill level.

- **Provide alternatives in permissible tools and scaffolds:**

    - **Example:** Allow students to use either traditional textbooks or online resources for research projects, providing options based on individual learning preferences.

- **Adjust the levels of flexibility for acceptable performance:**

    - **Example:** In a creative writing assignment, offer flexibility in structure and format, allowing students different degrees of freedom in expressing their ideas.

- **Emphasize process, effort, and improvement in meeting standards:**

    - **Example:** Instead of focusing solely on grades, encourage students to reflect on their learning process, effort, and areas for improvement, promoting a growth mindset over external evaluation and competition.

Guideline 3: Cultivate Teamwork and Social Bonds

In the 21st century, effective communication and collaboration within a learning community are essential for all learners. While varying in ease, these skills are universal goals. Peer mentoring, when well structured, enhances one-on-one support, fostering sustained engagement. Flexible grouping facilitates differentiation, diverse roles, and effective collaboration learning. Providing options for building and utilizing these crucial skills is imperative.

- **Build collaborative learning groups with clear goals, roles, and responsibilities:**
    - **Example:** Assign group projects with specific goals, clearly defined roles, and responsibilities, encouraging collaborative problem-solving.
- **Create school initiatives for promoting positive behavior, incorporating diverse objectives and tailored support systems:**
    - **Example:** Implement a school-wide positive behavior program with tailored objectives and supports, recognizing and reinforcing positive behaviors across diverse student needs.
- **Supply prompts that instruct learners on when and how to seek assistance from peers or teachers:**
    - **Example:** Supply learners with prompts on when and how to seek assistance, fostering a supportive environment where students feel comfortable reaching out for help.
- **Promote and facilitate opportunities for peer interactions and support (e.g., peer tutors):**
    - **Example:** Establish a peer-tutoring program where students support each other academically, creating a culture of collaboration and shared learning.
- **Construct communities of learners engaged in common interests or activities:**
    - **Example:** Form clubs or interest groups where students with shared passions collaborate, fostering a sense of community and shared learning experiences.

- **Create expectations for group work (e.g., rubrics, norms, etc.):**

  - **Example:** Develop clear expectations for group work using rubrics and norms, providing guidelines that promote effective collaboration and mutual understanding of roles and responsibilities.

Guideline 4: Increase Mastery-Oriented Feedback

Effective assessment sustains engagement when feedback is relevant, constructive, accessible, consequential, and timely. Mastery-oriented feedback guides learners toward mastery, focusing on effort and practice rather than fixed performance or compliance notions. This approach highlights the significance of effort over perceived "intelligence" or inherent "ability," fostering successful long-term habits. These distinctions are crucial, especially for learners whose disabilities may have been perceived as permanent constraints.

- **Provide feedback that encourages perseverance and self-awareness:**

  - **Example:** Acknowledge a student's persistent effort during a challenging task, emphasizing their developing efficacy and self-awareness in overcoming obstacles.

- **Provide feedback that emphasizes effort, improvement, and achieving a standard:**

  - **Example:** Praise a student's diligent effort and improvement, highlighting their success in meeting a set standard rather than focusing solely on relative performance.

- **Provide feedback that is frequent, timely, and specific:**

  - **Example:** Regularly provide timely and specific feedback on completed assignments, ensuring students receive consistent guidance and recognition for their efforts.

- **Provide feedback that is substantive and informative:**

  - **Example:** Offer substantive feedback that provides insightful information and guidance, avoiding comparisons and fostering a positive, informative learning environment.

- **Provide feedback that models how to incorporate evaluation into positive strategies:**

  - **Example:** Demonstrate how to analyze errors and wrong answers constructively, guiding students on incorporating evaluation into positive strategies for future success.

Guidelines to Provide Alternatives for Self-Control

Guideline 1: Cultivate Motivational Expectations and Beliefs

Self-control involves learners understanding their intrinsic and extrinsic motivations, setting realistic goals, and fostering positive beliefs in goal attainment. Managing frustration and anxiety during goal pursuit is crucial. Offering multiple options supports learners in staying motivated throughout the process.

- **Provide prompts, reminders, guides, rubrics, and checklists for self-regulatory goals:**

  - **Example:** Supply a checklist for reducing aggressive outbursts in response to frustration, promoting self-
  awareness and regulation.

  - **Example:** Offer reminders to increase on-task orientation amid distractions, assisting learners in managing attention effectively.

  - **Example:** Provide a rubric encouraging the frequent practice of self-reflection and self-reinforcement, fostering a positive learning mindset.

- **Provide coaches, mentors, or agents modeling goal setting:**

  - **Example:** Introduce a mentor who exemplifies setting personally appropriate goals, considering both strengths and weaknesses and guiding learners in effective goal establishment.

- **Support activities encouraging self-reflection and personal goal identification:**

  - **Example:** Design activities that prompt self-reflection, aiding learners in identifying and establishing personal goals aligned with their strengths and weaknesses.

Guideline 2: Support Individual Coping Abilities and Strategies

Modeling self-regulatory skills alone is insufficient for most learners. Sustained apprenticeships, including scaffolding through reminders, models, and checklists, are necessary. These tools aid learners in selecting and trying adaptive strategies to manage emotional responses, whether to external events or internal thoughts. Scaffolds must offer ample alternatives to accommodate individual differences in strategy effectiveness and application independence.

- **Differentiated models, scaffolds, and feedback for managing frustration:**
  - **Example:** Offer various strategies and examples for handling frustration tailored to individual preferences and needs.
- **Differentiated models, scaffolds, and feedback for seeking external emotional support:**
  - **Example:** Consider different learning styles and preferences and provide diverse models and scaffolds for seeking external emotional support.
- **Differentiated models, scaffolds, and feedback for developing internal controls and coping skills:**
  - **Example:** Tailor models, scaffolds, and feedback to help learners develop personalized internal controls and coping skills based on their unique strengths.

- **Differentiated models, scaffolds, and feedback for handling subject-specific phobias and judgments:**
  - **Example:** Address subject-specific challenges by guiding learners to approach difficulties with a growth mindset, focusing on improvement rather than fixed judgments.

- **Use real-life situations or simulations to demonstrate coping skills:**
  - **Example:** Incorporate real-life scenarios or simulations to illustrate effective coping skills in various situations, making the learning experience practical and applicable.

Guideline 3: Foster Self-assessment and Reflective Practices

Learners must develop the capacity for accurate self-regulation by monitoring emotions and reactivity. Metacognition capabilities vary, requiring explicit instruction and modeling. Recognizing progress boosts motivation, while the inability to perceive progress can lead to demotivation. Providing diverse models and scaffolds for self-assessment techniques enables learners to identify and choose optimal strategies for greater independence.

- **Offer devices, aids, or charts for monitoring behavior changes:**
  - **Example:** Provide learners with behavior-tracking devices, visual aids, or charts to facilitate the collection and display of data from their own behavior, aiding in monitoring and recognizing changes.

- **Use activities with feedback and alternative scaffolds for progress understanding:**

  - **Example:** Design activities that offer immediate feedback and present alternative scaffolds such as charts, templates, or feedback displays, ensuring learners can comprehend and track progress in a clear and timely manner.

Guidelines to Provide Alternatives for Perception

Guideline 1: Offer Customizable Information Display Choices

In printed materials, information is fixed, limiting customization. Digital materials, however, offer malleability. Details like call-out boxes can be relocated, enlarged, colored, or removed, enhancing clarity for diverse learners. Digital customization is automatic, but accessibility isn't guaranteed. Collaboration between educators and learners is crucial for tailoring features to specific learning needs, ensuring an optimal match.

- **Flexible Text Size:** Allow users to adjust text size for better readability.

- **Contrast Options:** Provide settings to modify the contrast between the background and text/images.

- **Color Customization:** Allow users to choose different color schemes for information or emphasis.

- **Audio Adjustments:** Enable users to control the volume and rate of speech or sound.

- **Video and Animation Controls:** Allow users to adjust the speed and timing of video, animation, sound, or simulations.

- **Layout Modification:** Provide options for adjusting the layout of visual elements for personalized viewing.

- **Font Selection:** Allow users to choose different fonts for printed materials to enhance readability.

Guideline 2: Offer Substitutes for Auditory Content

While the sound is impactful, especially in conveying emotion, it poses accessibility challenges. Learners with hearing disabilities, those needing extra processing time, or those facing memory difficulties may struggle with information presented solely through sound. Listening is a learned skill, and to ensure universal access, options should be provided for presenting auditory information, including emphasis.

- **Text Equivalents:** Include captions or automated speech-to-text for spoken language.

- **Visual Representations:** Provide visual diagrams, charts, or notations for music or sound.

- **Transcripts:** Offer written transcripts for videos or auditory clips.

- **Sign Language:** Include American Sign Language (ASL) for spoken English.

- **Visual Analogues:** Use visual elements like emoticons, symbols, or images to represent emphasis and prosody.

- **Sensory Equivalents:** Provide visual or tactile (e.g., vibrations) equivalents for sound effects or alerts.

- **Emotional Descriptions:** Include visual and/or emotional descriptions for musical interpretation.

Guideline 3: Offer Substitutes for Visual Content

Visual representations like images, graphics, animations, video, or text effectively convey information, particularly about relationships between objects, actions, numbers, or events. However, they may not be equally accessible to all learners, especially those with visual disabilities or unfamiliarity with certain graphics. Visual information, such as complex visual art, can have multiple meanings and interpretations. It's crucial to provide nonvisual alternatives to ensure universal access.

- **Image Descriptions:** Include text or spoken descriptions for all images, graphics, video, or animations.

- **Tactile Graphics:** Use touch equivalents, such as tactile graphics or objects of reference, for key visuals representing concepts.

- **Physical Objects:** Provide physical objects and spatial models to convey perspective or interaction.

- **Auditory Cues:** Include auditory cues for key concepts and transitions in visual information.

Text is a unique form of visual information, easily transformed into audio for increased accessibility. While text offers permanence, converting it into easily transformable audio retains advantages. Although digital text-to-speech is improving, it still falls short in conveying valuable prosodic information.

- **Accessibility Standards:** Follow accessibility standards (NIMAS, DAISY) when creating digital text. Refer to Appendix B for NIMAS and DAISY standards.

- **Aide or Partner Reading:** Allow for a competent aide, partner, or "intervener" to read text aloud.

- **Text-to-Speech Software:** Provide access to text-to-speech software for auditory support.

Guidelines to Provide Alternatives for Language and Symbols

Guideline 1: Clarify Vocabulary and Symbols

To make information universally accessible, consider learners' diverse backgrounds, languages, and lexical knowledge. Key elements like vocabulary, labels, icons, and symbols should be linked to alternative representations such as embedded glossaries, graphics, charts, or maps. Translate idioms, archaic expressions, culturally exclusive phrases, and slang for clarity and inclusivity.

- **Pre-teaching Vocabulary and Symbols:** Pre-teach vocabulary and symbols in ways that promote connection to learners' experience and prior knowledge, including the following:

  - **Graphic Symbols with Descriptions:** Provide graphic symbols with alternative text descriptions for better comprehension.

- **Highlighting Complexity:** Highlight how complex terms, expressions, or equations are composed of simpler words or symbols.

- **Embedding Support in Text:** Embed support for vocabulary and symbols within the text, using hyperlinks or footnotes to provide definitions, explanations, illustrations, previous coverage, or translations.

- **Support for Unfamiliar References:** Embed support for unfamiliar references within the text, addressing domain-specific notation, lesser-known properties and theorems, idioms, academic language, figurative language, mathematical language, jargon, archaic language, colloquialism, and dialect.

Guideline 2: Clarify Syntax and Structure

Ensure universal access to information by providing alternative representations that clarify the syntactic or structural relationships between elements of meaning. Alternative formats improve understanding for all learners when they encounter unfamiliar syntax in a sentence or encounter a complex structure in a graphical representation.

- **Clarification of Unfamiliar Syntax and Structure:** Clarify unfamiliar syntax or underlying structure in language or math formulas through alternatives that simplify concepts, provide visual aids, or offer step-by-step breakdowns to enhance understanding.

- **Highlight Structural Relations:** Use visual cues or formatting to highlight the structural relations, making them more explicit.

- **Connections to Previous Structures:** Connect to previously learned structures to reinforce understanding and build on prior knowledge.

- **Explicit Relationships:** Emphasize relationships between elements, such as highlighting transition words in an essay or links between ideas in a concept map.

Guideline 3: Assist in Deciphering Text, Mathematical Notation, and Symbols

Ensure universal access to knowledge by providing options that reduce decoding barriers for learners unfamiliar or dysfluent with symbols. Consistent, meaningful exposure to symbols enhances comprehension and use, minimizing cognitive load and promoting information processing for all learners.

- **Text-to-Speech Options:** Allow the use of text-to-speech (TTS) to support learners in auditory comprehension.

- **Mathematical Notation Accessibility:** Use automatic voicing with digital mathematical notation, such as MathML, to enhance accessibility for learners.

- **Voice Accompaniment for Text:** To support auditory learning, provide digital text with an accompanying human voice recording, such as DAISY Talking Books.

- **Multiple Representations:** Allow for flexibility and easy access to multiple representations of notation where appropriate, such as formulas, word problems, and graphs.

- **Clarification of Notation:** Offer clarification of notation through lists of key terms, aiding learners in understanding complex symbols and expressions.

Guideline 4: Facilitate Comprehension Across Languages

Enhance accessibility by providing linguistic alternatives for curricular materials, acknowledging the diverse language backgrounds of learners. Offering alternatives, particularly for crucial information or vocabulary, is essential to ensure understanding, especially for new learners of the dominant language or academic discourse.

- **Multilingual Access:** Make all key information in the dominant language, such as English, also available in first languages, for example, Spanish, for learners with limited English proficiency. Additionally, provide information in American Sign Language (ASL) for learners who are deaf.

- **Vocabulary Support:** Link key vocabulary words to definitions and pronunciations in both dominant and heritage languages. Define domain-specific vocabulary, like "map key" in social studies, using both domain-specific and common terms.

- **Translation Tools:** Provide electronic translation tools or links to multilingual glossaries on the web to support learners in accessing information in their preferred language.

- **Visual Supports:** Embed visual and nonlinguistic supports for vocabulary clarification, such as pictures, videos, and other multimedia elements.

Guideline 5: Demonstrate Using Various Media

Enhance comprehension and accessibility by providing alternatives to text-heavy materials. Illustrations, simulations, images, and interactive graphics can effectively convey concepts and processes, benefiting all learners and making the content accessible to those with text- or language-related disabilities.

- **Multimodal Presentation:** Present key concepts in one form of symbolic representation, such as an expository text or a math equation, along with an alternative form, such as an illustration, dance/movement, diagram, table, model, video, comic strip, storyboard, photograph, animation, or physical/virtual manipulative.

- **Link Text and Representation:** Make explicit links between information provided in texts and any accompanying representation of that information, whether in illustrations, equations, charts, or diagrams, ensuring a comprehensive understanding through multiple modalities.

Guidelines to Provide Alternatives for Comprehension

Guideline 1: Trigger or Provide Prior Knowledge

Information becomes more accessible and easily absorbed when presented in a manner that activates or provides prerequisite knowledge. Barriers arise when learners lack critical prior knowledge or when those who possess it are unaware of its relevance. Offering options that supply or activate relevant prior knowledge or link to prerequisite information helps reduce these barriers and ensures equitable access for all learners.

- **Activate Prior Knowledge:** Anchor instruction by linking to and activating relevant prior knowledge, employing methods such as visual imagery, concept anchoring, or concept mastery routines.

- **Advanced Organizers:** Utilize advanced organizers, including KWL methods and concept maps, to provide learners with a framework for understanding and organizing new information.

- **Pre-teach Prerequisite Concepts:** Pre-teach critical prerequisite concepts through demonstration or models, ensuring learners have the foundational knowledge necessary for upcoming content.

- **Analogies and Metaphors:** Bridge concepts with relevant analogies and metaphors to facilitate understanding by connecting new information to familiar ideas.

- **Cross-curricular Connections:** Make explicit cross-curricular connections, demonstrating the integration of literacy strategies in subjects like social studies and enriching learners' comprehension across disciplines.

Guideline 2: Emphasize Patterns, Crucial Features, Key Concepts, and Relationships

Distinguishing critical information from the unimportant is a key skill of experts. Experts efficiently allocate their time by quickly identifying valuable features and integrating them into existing knowledge. To enhance information accessibility, explicit cues or prompts can assist individuals in focusing on crucial features, ensuring a more effective and targeted assimilation of information.

- **Visual Emphasis:** Highlight or emphasize key elements in text, graphics, diagrams, and formulas to draw attention to critical information.

- **Organizational Tools:** Utilize outlines, graphic organizers, unit organizer routines, concept organizer routines, and concept mastery routines to emphasize key ideas and relationships, aiding learners in structuring information.

- **Exemplify Critical Features:** Employ multiple examples and nonexamples to emphasize critical features, providing learners with varied perspectives for deeper understanding.

- **Cues and Prompts:** Use cues and prompts strategically to draw attention to critical features, guiding learners through important content elements.

- **Integration of Previous Skills:** Highlight previously learned skills that can be applied to solve new problems, fostering connections and transferability of knowledge.

Guideline 3: Facilitate Information Processing and Visualization

Effectively transforming information into usable knowledge involves applying cognitive and metacognitive strategies, such as summarization, categorization, and prioritization. Many learners lack a full repertoire of these strategies. Well-designed materials can offer customized models, scaffolds, and feedback to assist diverse learners in utilizing these strategies effectively.

- **Sequencing Assistance:** Offer explicit prompts for each step sequentially to guide learners through complex tasks.

- **Organizational Choices:** Provide options for organizational methods and approaches, such as tables and algorithms for processing mathematical operations, catering to diverse learning preferences.

- **Interactive Exploration:** Introduce interactive models that facilitate exploration and help learners develop new understandings through hands-on experiences.

- **Scaffolded Information Processing:** Implement graduated scaffolds that support information processing strategies, assisting learners at varying proficiency levels.

- **Flexible Pathways:** Offer multiple entry points to a lesson and optional pathways through content, allowing exploration of big ideas through diverse mediums like dramatic works, arts, literature, film, and media.

- **Information Chunking:** "Chunk" information into smaller, digestible elements to enhance comprehension and aid in memory retention.

- **Progressive Information Release:** Progressively release information, such as using sequential highlighting, to manage cognitive load effectively.

- **Minimize Distractions:** Remove unnecessary distractions unless they are essential to the instructional goal, ensuring a focused learning environment.

Guideline 4: Optimize Transfer and Generalization

All learners require support for memory, generalization, and transfer of learning to new contexts. Varied levels of scaffolding are needed to enhance these abilities, ensuring that information is accessible in diverse situations. Techniques enhancing memorability and guiding learners in employing explicit strategies aid in memory, generalization, and transfer. Multiple representations are crucial, as learning involves interconnected facts, and without such support, information may be learned but prove inaccessible in unfamiliar scenarios.

- **Organization Tools:** Supply checklists, organizers, sticky notes, and electronic reminders to assist learners in managing tasks and information effectively.

- **Memory Aids:** Prompt the use of mnemonic strategies and devices, encouraging techniques like visual imagery, paraphrasing, and the method of loci.

- **Review and Practice:** Incorporate explicit opportunities for review and practice to reinforce learning and enhance retention.

- **Note-Taking Support:** Provide templates, graphic organizers, and concept maps to support effective note-taking during instruction.

- **Knowledge Integration:** Integrate scaffolds connecting new information to prior knowledge using tools like word webs and half-full concept maps.

- **Connecting Ideas:** Embed new ideas in familiar contexts using analogies, metaphors, drama, music, film, or other creative approaches.

- **Generalization Opportunities:** Offer explicit, supported opportunities for learners to generalize their learning to new situations, fostering a deeper understanding.

- **Revisiting Key Concepts**: Provide opportunities over time for learners to revisit key ideas and reinforce linkages between concepts for sustained understanding.

Guidelines to Provide Alternatives for Physical Action

Guideline 1: Diversify Response and Navigation Methods

Learners' ability to navigate physically varies, impacting tasks with motor demands. To mitigate barriers, offer alternative ways for responses, selections, and compositions. Additionally, diverse preferences for navigating through information exist. Instructors should ensure multiple accessible means for navigation and control to provide equitable learning opportunities.

- **Motor Action Alternatives:** Offer alternatives in requirements for rate, timing, speed, and range of motor action needed to interact with instructional materials, physical manipulatives, and technologies.

- **Response and Selection Alternatives:** Provide alternatives for physically responding or indicating selections, offering options beyond traditional pen and pencil marking or mouse control.

- **Interaction Modalities:** Offer alternatives for physically interacting with materials, supporting diverse modes such as hand, voice, single switch, joystick, keyboard, or adapted keyboard for a more inclusive learning experience.

Guideline 2: Enhance Accessibility to Tools and Assistive Technologies

Providing tools is insufficient; effective support for tool usage is essential. Many learners require assistance navigating their environment, and everyone should have the chance to use tools for full classroom participation. Learners with disabilities often rely on assistive technologies, emphasizing the need for curricular designs that don't inadvertently hinder these technologies. Ensuring keyboard commands for mouse actions and maintaining challenges in accessible lessons are crucial considerations.

- **Accessible Keyboard Options:** Ensure access for all learners by providing alternate keyboard commands for mouse actions, allowing keyboard alternatives for enhanced control and independence.

- **Switch and Scanning Options:** Incorporate switch and scanning options to facilitate increased independent access, making learning accessible to a diverse range of students.

- **Alternative Keyboards:** Provide access to alternative keyboards to accommodate individual needs and preferences, enhancing the overall usability of digital interfaces.

- **Customizable Overlays:** Customize overlays for touch screens and keyboards, tailoring the learning environment to individual requirements and promoting a personalized experience.

- **Seamless Software Integration:** Select software that seamlessly integrates with keyboard alternatives and alt keys, ensuring a smooth and inclusive digital learning experience.

Guidelines to Provide Alternatives for Expression and Communication

Guideline 1: Employ Various Media for Communication

Offering alternative media for expression is crucial, especially when specific materials are not essential to the learning goal (e.g., learning oil painting or calligraphy). These alternatives help overcome media-specific barriers for learners with diverse needs and also expand opportunities for all learners to develop a broader range of expression in a media-rich environment. Learning composition, beyond just writing, and understanding the optimal medium for specific content and audience are valuable skills for all learners.

- **Multimodal Composition:** Encourage learners to express themselves through various media, including text, speech, drawing, illustration, comics, storyboards, design, film, music, dance/movement, visual art, sculpture, or video.

- **Manipulatives and Physical Tools:** Incorporate physical manipulatives like blocks, 3D models, or base-ten blocks to enhance hands-on learning experiences.

- **Interactive Web Tools:** Utilize social media and interactive web tools such as discussion forums, chats, web design, annotation tools, storyboards, comic strips, and animation presentations for engaging collaborative learning.

- **Problem-Solving Strategies:** Encourage diverse problem-solving approaches, fostering creativity and critical thinking skills in learners.

Guideline 2: Employ Diverse Tools for Construction and Composition

Schools often prioritize traditional tools over contemporary ones, hindering learners in various ways: (1) failing to prepare them for the future, (2) limiting content and teaching methods, (3) restricting assessment methods, and (4) narrowing the range of successful learners. Current media tools offer a more flexible and accessible toolkit, empowering learners to actively engage in their education and effectively communicate their knowledge. Unless a lesson specifically focuses on learning a particular tool, curricula should allow for various alternatives to match learners' abilities with task demands.

- **Writing Assistance:** Offer support with spellcheckers, grammar checkers, and word prediction software to aid in writing tasks.

- **Speech-to-Text Tools:** Provide text-to-speech software, voice recognition, human dictation, or recording options for diverse communication preferences.

- **Mathematical Tools:** Supply calculators, graphing calculators, geometric sketchpads, or pre-formatted graph paper to facilitate mathematical tasks.

- **Language Support:** Provide sentence starters, sentence strips, story webs, outlining tools, or concept mapping tools to assist with language development.

- **Creative Software:** Utilize computer-aided design (CAD), music notation software, or mathematical notation software for creative expression.

- **Manipulatives and Tools:** To enhance mathematical understanding, offer virtual or concrete mathematics manipulatives, such as base-ten blocks or algebra blocks.

- **Web Applications:** Implement web applications like wikis, animation tools, or presentation software for collaborative and dynamic learning experiences.

Guideline 3: Develop Proficiencies with Progressively Supported Levels for Practice and Performance

Learners should cultivate various proficiencies (e.g., visual, audio, mathematical, reading). Multiple scaffolds are necessary for practice and independence. Curricula should offer varying degrees of freedom, providing ample support for some and more independence for others. Fluency grows through diverse performance opportunities, fostering the synthesis of learning in personal ways. It's crucial to offer options that enhance learners' fluencies.

- **Diverse Models:** Offer differentiated models to emulate, showcasing various approaches, strategies, and skills that lead to similar outcomes.

- **Varied Mentorship:** Provide differentiated mentors, including teachers or tutors who employ diverse approaches to motivate, guide, and provide feedback.

- **Gradual Scaffolding:** Supply scaffolds that can be progressively released as learners gain independence and skills, such as integration into digital reading and writing software.

- **Customized Feedback:** Provide differentiated feedback that is customizable to individual learners, ensuring accessibility.

- **Examples of Novel Solutions:** Present multiple examples illustrating innovative solutions to authentic problems.

Guidelines to Provide Alternatives for Executive Functions

Guideline 1: Facilitate Suitable Goal Setting

Learners may not naturally set effective goals, and merely providing goals for them hinders skill development. The UDL framework advocates for graduated scaffolds to cultivate learners' ability to set their own challenging and realistic goals, fostering skill and strategy development.

- **Estimation Support:** Offer prompts and scaffolds to help learners estimate effort, required resources, and the difficulty level of tasks.

- **Exemplars for Goal Setting:** Provide models or examples showcasing the process and final product of effective goal setting.

- **Guides and Checklists:** Supply guides and checklists as scaffolding tools to assist learners in goal setting.

- **Visible Reminders:** Post goals, objectives, and schedules in prominent locations to serve as constant visual reminders.

Guideline 2: Assist in Planning and Developing Strategies

Effective learners plan strategies to achieve their goals. However, some learners, especially young children in new domains or with executive function challenges, might skip strategic planning. Various options are necessary to support learners in becoming more planful and strategic, such as cognitive prompts, scaffolds to implement strategies, or engaging in decision-making with competent mentors.

- **Reflection Prompts:** Embed prompts encourage learners to "stop and think" before taking action and provide ample space for reflection.

- **Work Showcase Prompts:** Include prompts urging learners to "show and explain their work," facilitating activities like portfolio reviews or art critiques.

- **Checklists and Templates:** Supply checklists and project planning templates to aid in comprehending the problem, prioritizing steps, and creating schedules.

- **Coaching and Mentoring:** Embed coaches or mentors to model think-aloud, guiding learners through the thought process.

- **Goal Breakdown Guides:** Provide guides to help break long-term goals into manageable short-term objectives.

Guideline 3: Support the Management of Information and Resources

Executive function is constrained by the limitations of working memory, especially for learners with disabilities. Remember, working memory acts as a "table's surface" for information, but its capacity is limited. To assist learners who may struggle with organization and preparedness, provide internal scaffolds and external organizational aids, mirroring the strategies used by individuals with strong executive functions.

- **Graphic Organizers:** Offer graphic organizers and templates to assist in data collection and organizing information effectively.

- **Embedded Prompts:** Integrate prompts within the learning material to guide learners in categorizing and systematizing information.

- **Note-Taking Tools:** Supply checklists and guides designed for note-taking, helping learners structure and capture key information.

Guideline 4: Boost the Ability to Monitor Progress

Feedback is crucial for learning, but its effectiveness is improved when it needs more clarity, timeliness, and informativeness. To address this, options should be available to customize feedback, making it more explicit, timely, and accessible. Emphasizing formative feedback enables learners to monitor progress and effectively guide their efforts. This approach is essential to prevent learners from appearing unmotivated or perseverant due to lacking insight into areas needing improvement.

- **Guided Questions:** Ask learners questions that prompt self-monitoring and reflection, such as "What strategies did you use to approach the task?"

- **Visual Progress Displays:** Exhibit progress visually with before-and-after photos, graphs, and charts depicting development over time or maintaining process portfolios.

- **Feedback Identification:** Encourage learners to specify the type of feedback or advice they seek, fostering targeted and meaningful reflections.

- **Self-Reflection Templates:** Utilize templates designed to guide self-reflection on the quality and completeness of their work, providing a structured approach.

- **Diverse Self-Assessment Models:** Present varied models for self-assessment, including role-playing scenarios, video reviews, and opportunities for peer feedback.

- **Assessment Tools:** Implement assessment checklists and scoring rubrics and showcase multiple examples of annotated student work to illustrate performance standards.

Implementation of Universal Design in Learning

Incorporating the seven principles of universal design into educational environments ensures accessibility and inclusivity, fostering a learning atmosphere where every learner can thrive regardless of their individual needs. Table 5-4 discusses the implementation of seven principles of universal design in learning.

Table 5-4. *Implementation of Universal Design in Learning*

Universal Design Principle	Definition Applied to UDL	Example in Learning Design
Equitable Use	This principle emphasizes the creation of accessible and universally appealing instructional designs. The goal is to avoid segregation or stigmatization of any learners, ensuring that the learning experience is open to everyone.	**Diverse Assignment Options:** Offer learners multiple avenues to demonstrate understanding – be it through websites, oral or video presentations, or traditional research papers. **Multi-medium Explanations:** Explain complex concepts using various mediums such as print, audio, diagrams, and written materials with varied reading levels.
Flexibility in Use	This principle ensures that instruction accommodates individual needs, providing a palette of choices for instructors and learners to engage with the materials.	**Varied Instructional Methods:** Foster diverse learning approaches through methods like concept maps, group work, project-based learning, visual outlines in lectures, or storytelling. **Universal Accessibility:** Ensure classroom tools are equally usable for both left- and right-handed learners.

Simple and Intuitive	Simplicity and intuitiveness in instructional design aim to be comprehensible to all learners, regardless of experience, knowledge, language skills, or concentration levels. Unnecessary complexity is avoided.	**Clear Grading Rubrics:** Provide grading rubrics that clearly outline assignment and assessment expectations. **Transparent Course Structure:** Share a straightforward overview of the course's structure.
Perceptible Information	The Perceptible Information principle ensures that key information is equally accessible to all learners, minimizing the need for extraordinary effort or assistance.	**Accessible Instructional Materials:** Select materials accessible via screen readers or reformattable/magnifiable. **Enhanced Readability:** Use font styles, colors, and sizes to promote readability. **Multimodal Information Delivery:** Present the same information through different modalities (vision, hearing, or touch).
Tolerance for Error	Acknowledging variations in individual learning pace and prerequisite skills, Tolerance for Error allows for an inclusive instructional approach.	**Multiple Assignment Drafts:** Permit learners to submit multiple drafts of assignments. **Progress Assessments:** Include practice exercises or tests before summative assessments. **Embracing Growth Mindset:** Use learning portfolios to track progress and encourage a growth mindset.

(continued)

Table 5-4. (*continued*)

Universal Design Principle	Definition Applied to UDL	Example in Learning Design
Low Physical Effort	Low Physical Effort in instruction minimizes nonessential physical efforts to maximize attention to learning.	**Accessible Instructional Materials:** Design materials, including videos, considering varied attention spans. **Adaptive Tools Usage:** Allow the use of word processing or speech-to-text software for writing activities or exams. **Scheduled Breaks:** Provide breaks in longer class sessions.
Size and Space for Approach and Use	Consideration for appropriate size and space ensures that instructional designs accommodate learners regardless of body size, posture, mobility, or communication needs.	**Considering Diverse Needs:** Consider the needs of learners with hearing impairments, language processing differences, cultural distinctions, or other processing difficulties. **Facilitating Engagement:** Use circular seating arrangements when possible for better discussion engagement, especially for learners with hearing impairments or attention deficits.

Universal Design Learning Cycle

Implementing UDL in your courses may seem daunting, leaving you needing guidance on where to begin. Recognizing that UDL is a gradual process, allowing for incremental and iterative implementation is essential. Initiate with small adjustments aligned with UDL principles and progressively refine and expand over time.

Step 1: What Do We Know About the Learner and Context?

The first step involves understanding the individuals and the situation comprehensively. This begins with asking key questions such as "What do we know about the learners?" This inquiry delves into various aspects, including their cultural backgrounds, languages spoken, and personal identities. Additionally, we explore their connections to their environment and the strengths they bring to the learning experience. Understanding their prior knowledge and experiences, as well as recognizing specific needs and preferences, is crucial. The assessment extends to identifying what is currently occupying the learners' attention and their overall well-being and morale.

Simultaneously, we extend our inquiry to the broader context. This encompasses factors such as the time of day and week, the learning environment (whether online, face-to-face, or during a field trip), and whether the activity or setting is familiar or unfamiliar to the learners.

Step 2: What Is the Goal and Purpose?

In the second step, the focus shifts to defining the goal and purpose of our educational endeavors. This involves a series of introspective questions aimed at guiding our intentions:

- **Clarifying the Goal**

    - The question "What is the goal?" centers around establishing a clear objective for the learning experience. This is a foundational step in ensuring that the educational activities are purposeful and aligned with desired outcomes.

- **Understanding the Motivation**

  - The question "Why are we doing this?" delves into the motivations and rationale behind the chosen goal. Understanding the purpose behind the educational initiatives is essential for creating meaningful and relevant learning experiences.

- **Envisioning Success**

  - The inquiry "What could success look like?" involves envisioning the desired outcomes and milestones to characterize a successful educational journey. This step guides the planning process toward measurable and impactful goals.

- **Co-designing with Learners**

  - The question "Can these be co-designed with learners?" emphasizes the possibility of collaboratively shaping goals and purposes with the learners. This inclusive approach ensures that the educational objectives align with the aspirations and perspectives of the learners, fostering a sense of shared ownership and engagement.

Step 3: Identify Possible Barriers to Learning in the Design

The third step focuses on identifying potential barriers to learning within the design. This involves asking critical questions to ensure an inclusive and accessible learning environment.

- **Equity and Discrimination**

  - The question "What in our design could create inequity and discrimination?" prompts a careful examination of the educational design to prevent any elements that might lead to unequal experiences or discrimination.

- **Barriers to Engagement and Motivation**

  - The inquiry extends to potential barriers that may hinder engagement and motivation. This involves considering elements within the design that could impact learners' ability to stay engaged and motivated throughout the learning process.

- **Accessing and Understanding Information**

  - Attention is given to potential barriers related to accessing and understanding information. This inquiry delves into the design's clarity, accessibility, and comprehensibility, ensuring that all learners can readily access and comprehend the provided information.

- **Participation and Expression**

  - Lastly, the assessment considers potential barriers hindering participation in learning and expressing what learners know. This involves examining obstacles that could impede active involvement in the learning process and the ability to articulate understanding.

Step 4: Identify Universal Supports

The fourth step focuses on identifying universal supports that can benefit a broad range of learners. This step involves asking crucial questions to ensure inclusivity and accessibility:

- **Exploring Universal Supports**

  - The question "Which supports, options, and tools could be beneficial for some individuals that we can provide to everyone?" encourages a broad exploration of tools and supports.

- This inquiry aligns with the principles of Universal Design for Learning (UDL), emphasizing the importance of creating an inclusive learning environment that considers diverse needs and preferences.

By addressing this question, educators actively seek to identify tools like assistive technologies and multimodal instructional materials that have the potential to enhance the learning experience for specific individuals, making them universally available. This proactive approach ensures that the educational environment is equipped with resources catering to various learning styles and abilities.

Step 5: Make a Plan Supported by the UDL Guidelines

In the fifth step, we move toward crafting a plan that aligns with Universal Design for Learning (UDL) principles. This involves creating and implementing a comprehensive strategy to support the variability among learners, focusing on allowing learners' unique characteristics to shape the design. The plan addresses key barriers, incorporates useful options, and integrates universal supports, adhering to the UDL guidelines.

As part of this process, it is essential to organize feedback approaches, providing multiple means of engagement to ensure ongoing interaction and participation.

Within the plan, three critical networks are considered, each associated with a specific aspect of learning:

- **Affective Networks (the "WHY" of Learning)**
  - Questions like "Does the lesson provide options that can help all learners regulate their learning, sustain effort and motivation, and engage and interest all learners?" are explored.

- This aligns with the UDL principle of Multiple Means of Engagement, focusing on the emotional and motivational aspects of learning.

- **Recognition Networks (the "WHAT" of Learning)**

  - Questions like "Does the lesson provide options that can help all learners reach higher levels of comprehension and understanding, understand symbols and expressions, and perceive what needs to be learned?" are considered.

  - This aligns with the UDL principle of Multiple Means of Representation, emphasizing diverse ways information can be presented and understood.

- **Strategic Networks (the "HOW" of Learning)**

  - Questions like "Does the lesson provide options to help all learners act strategically, express themselves fluently, and physically respond?" are examined.

  - This aligns with the UDL principle of Multiple Means of Action and Expression, focusing on diverse ways learners can navigate and express their understanding.

Step 6: Teach, Evaluate, and Revise

In the sixth step, the focus shifts to the dynamic cycle of teaching, evaluating, and revising, guided by continuous reflection and responsiveness. This stage involves asking critical questions to refine the educational approach.

- **Assessing Effectiveness**

  - Questions such as "What options, universal supports, and strategies worked well?" are posed.

  - This inquiry aims to identify effective elements in supporting learners, acknowledging the success of specific strategies and universal supports.

- **Feedback Integration**

  - The question "How will the feedback influence my design?" underscores the importance of incorporating feedback into the instructional design.

  - This step aligns with the principles of Universal Design for Learning, emphasizing the iterative nature of the process and the need to adapt based on ongoing feedback.

- **Continuous Improvement**

  - The question "What will I try next time?" underscores the commitment to continuous improvement and adaptation.

  - This aligns with the UDL principle of flexible, ongoing design, promoting a mindset of continuous learning and evolution.

By systematically addressing these questions listed in all six steps of the cycle, instructional designers can plan to tailor the learning experience to accommodate multiple learning styles and abilities, ensuring that all learners can actively participate and succeed in the learning process.

Universal Design for eLearning

WCAG is an abbreviation for Web Content Accessibility Guidelines. It delineates a series of standards and guidelines established by the Web Accessibility Initiative (WAI) under the World Wide Web Consortium (W3C). Its primary objective is to guarantee the accessibility of web content to individuals with various disabilities, encompassing visual, auditory, cognitive, and motor impairments.

POUR Principles

The POUR principles, integral to the Web Content Accessibility Guidelines (WCAG) 2.0, establish a robust framework for web accessibility, focusing on four key aspects: Perceivable, Operable, Understandable, and Robust. These principles are designed to ensure that digital content is inclusive and accessible to users with diverse abilities. Let's explore each principle with illustrative examples:

- **Perceivable**
    - **Guideline:** Present information and user interface components in a manner that users can perceive.
    - **Example:** Text alternatives for nontext content, like images, enable users with visual impairments to comprehend the content through screen readers.

- **Operable**
    - **Guideline:** User interface components and navigation must be operable.
    - **Example:** Keyboard accessibility facilitates navigation for users with mobility impairments who rely on keyboards, ensuring an operable experience.

- **Understandable**

  - **Guideline:** Information and the operation of the user interface must be understandable.

  - **Example:** Consistent navigation labels enhance overall understanding, aiding users in comprehending the structure and organization of the website.

- **Robust**

  - **Guideline:** Content must be robust enough to be reliably interpreted by various user agents, including assistive technologies.

  - **Example:** Valid HTML and CSS ensure compatibility with different browsers and assistive technologies, fostering a consistent and reliable user experience.

Adhering to these POUR principles not only ensures compliance with accessibility standards but also fosters the creation of user-friendly web environments that cater to the diverse needs of all individuals, promoting inclusivity in the digital realm. Table 5-5 discusses the guideline, success criteria, and conformance level for each POUR principle.

Success Criteria: The success criteria of the Web Content Accessibility Guidelines (WCAG) are specific guidelines and requirements that help measure the accessibility of web content. These criteria are organized under four principles: Perceivable, Operable, Understandable, and Robust. Each principle has a set of guidelines, and each guideline has associated success criteria. These success criteria are listed according to the latest version of WCAG 2.2 in operation. Refer to Appendix B for the extensive list of success criteria.

Table 5-5. POUR Principles

POUR Principle	Guideline	Success Criteria	Conformance Level
Perceivable	1.1 Text Alternatives	Provide text alternatives for nontext content.	Level A
	1.2 Time-Based Media	Provide alternatives for time-based media.	Level AA
	1.3 Adaptable	Develop content that maintains its information and structure when presented in various formats.	Level AA
	1.4 Distinguishable	Facilitate improved visibility and auditory access to the content for users.	Level AA
Operable	2.1 Keyboard Accessible	Make all functionality available from a keyboard.	Level A
	2.2 Enough Time	Provide users enough time to read and use content.	Level A
	2.3 Seizures and Physical Reactions	Avoid creating content in a manner that is recognized to induce seizures or physical discomfort.	Level A
	2.4 Navigable	Offer methods to assist users in navigating, discovering content, and establishing their location.	Level AA
	2.5 Input Modalities	Provide similar ways as keyboard and mouse to input the content with switch controls, eye trackers, and voice commands.	Level AA

(continued)

Table 5-5. (*continued*)

POUR Principle	Guideline	Success Criteria	Conformance Level
Understandable	3.1 Readable	Make text content readable and understandable.	Level A
	3.2 Predictable	Ensure that web pages display and function in consistent and expected manners.	Level AA
	3.3 Input Assistance	Help users avoid and correct mistakes.	Level AA
Robust	4.1 Compatible	Enhance compatibility with both existing and future user agents, encompassing assistive technologies.	Level A

Conformance Level: Each success criterion is associated with a level of conformance: Level A, Level AA, and Level AAA. Level A signifies the minimum level of accessibility, Level AA addresses more advanced criteria, and Level AAA signifies the highest level of accessibility.

Evolution of WCAG

The progress of the Web Content Accessibility Guidelines (WCAG) reflects a dynamic journey toward a more inclusive digital landscape. From its early beginnings to the current version, WCAG has continually adapted to the ever-changing digital environment, aiming to set standards prioritizing accessibility for all users. This evolution encompasses technological advancements, user feedback, and a deepening understanding of diverse needs, shaping WCAG into a comprehensive framework that not only meets contemporary challenges but also anticipates the accessibility requirements of the future. Table 5-6 discusses the key milestones and transformations that mark the remarkable journey of WCAG, illustrating its pivotal role in fostering a web that is universally accessible and user-friendly.

Table 5-6. Evolution of WCAG

WCAG Version	Features	Total Guidelines	Success Criteria	Conformance Level	Limitations
WCAG 1.0 (1999)	Foundation for web accessibility and digital content	14	65 checkpoints	A: P1 satisfied AA: P1 and P2 satisfied AAA: P1, P2, and P3 satisfied	• Limited coverage for emerging technologies • Lack of flexibility and adaptability for evolving web content
WCAG 2.0 (2008)	Introduced POUR principles	12	61 success criteria	A: Meet 25 success criteria AA: Meet 25 + 13 success criteria AAA: Meet 25 + 13 + 23 success criteria	• Some criteria were challenging to interpret precisely • Limited guidance on mobile and emerging technologies

WCAG 2.1 (2018)	Addresses mobile and additional cognitive, low vision issues	13	61 + 17 = 78 success criteria	A: Meet 30 success criteria AA: Meet 30 + 20 success criteria AAA: Meet 30 + 20 + 28 success criteria	• Doesn't cover all possible disabilities comprehensively
WCAG 2.2 (2021)	Further enhancements for mobile and cognitive issues	13	61 + 16 + 9 = 86 success criteria	A: Meet 32 success criteria AA: Meet 32 + 24 success criteria AAA: Meet 32 + 24 + 30 success criteria	• May not fully address all emerging technology challenges • Continuous advancements may require future updates

Alignment of POUR Principles to UDL Guidelines

The alignment of the Principles of Universal Design (POUR) with the Universal Design for Learning (UDL) Guidelines marks a significant convergence in creating inclusive and accessible digital environments. This synergy between POUR principles and UDL guidelines represents a holistic approach to design, where the foundations of web accessibility seamlessly intertwine with the broader framework of inclusive educational practices. Table 5-7 explores this alignment and reveals the interconnectedness of these principles and underscores their collective impact in fostering an environment where diverse learners can thrive.

Table 5-7. *Alignment of UDL Guidelines with POUR Principles*

POUR Principles	UDL Guidelines and Checkpoints
Perceivable (P)	**Representation – The "WHAT" of Learning (UDL)**
	- Perception
	- Offer customizable information display choices.
	- Offer substitutes for auditory content.
	- Offer substitutes for visual content.
Operable (O)	**Action and Expression – The "HOW" of Learning (UDL)**
	- Physical Action
	- Diversify response and navigation methods.
	- Enhance accessibility to tools and assistive technologies.
Understandable (U)	**Engagement – The "WHY" of Learning (UDL)**
	- Capturing Attention
	- Maximize individual choices and independence.
	- Enhance appropriateness, importance, and credibility.
	- Reduce risks and disturbances

(continued)

Table 5-7. (*continued*)

POUR Principles	UDL Guidelines and Checkpoints
Robust (R)	**Action and Expression – The "HOW" of Learning (UDL)**
	- Expression and Communication
	- Employ various media for communication
	- Employ diverse tools for construction and composition
	- Build fluencies with graduated levels of support for practice and performance

Extending UDL for Learners with Disabilities

Creating a learning experience for people with disabilities means making content and interactions that everyone can easily access and understand. We want our learning materials to be inclusive and effective for everyone, no matter their abilities. Whether someone faces challenges with seeing, hearing, moving around, or reading and writing (like dyslexia), we aim to ensure that everyone can learn comfortably. We'll explore practical strategies to design content that's not just accessible but also enriching for every learner.

Here are some key considerations for visually impaired learners:

- **Accessible Content**

  - **Text Description:** Provide detailed and accurate text descriptions (alt text) for images, graphics, and other nontext content.

  - **Readable Fonts:** Use clear, legible fonts with sufficient contrast between text and background.

  - **Structured Content:** Organize content with clear headings, lists, and meaningful links.

- **Audio Description**

  - **Narration:** Include audio narration to describe visual elements, actions, and context in videos or presentations.

  - **Separate Audio:** Ensure audio descriptions are provided as separate tracks for flexibility.

- **Screen Reader Compatibility**

  - **Semantic HTML:** Use semantic HTML to structure content and facilitate proper interpretation by screen readers.

  - **Descriptive Links:** Create links with descriptive text that conveys the purpose or destination.

- **Keyboard Navigation**

  - **Accessible Controls:** Design interactive elements that can be easily navigated and activated using a keyboard.

  - **Skip to Content:** Include a "Skip to Content" link to allow users to bypass repetitive navigation menus.

- **Braille Compatibility**

  - **Printable Materials:** Provide materials in formats compatible with braille displays or embossers.

  - **Braille Annotations:** Include braille annotations on physical materials where applicable.

- **Tactile Graphics**

  - **Tactile Diagrams:** Create tactile graphics or diagrams to represent visual information through touch.

  - **Braille Labels:** Include braille labels for tactile graphics or objects.

- **Interactive Learning**

  - **Accessible Interactivity:** Design interactive elements that screen readers can easily interpret.

  - **Keyboard-Friendly Activities:** Ensure that interactive activities are keyboard-friendly.

- **Adaptive Technology Support**

  - **Compatibility:** Ensure compatibility with popular screen readers and other adaptive technologies.

  - **Testing:** Regularly test your content with various assistive technologies to identify and address issues.

- **Flexible Learning Paths**

  - **Customizable Settings:** Learners can customize settings such as text size, color contrast, and audio preferences.

  - **Progress Tracking:** Implement features that enable learners to track their progress.

- **Inclusive Learning Materials**

  - **Multimodal Resources:** Provide content in multiple formats, such as audio, tactile, and braille, to cater to diverse learning preferences.

  - **Variety of Resources:** Offer a variety of learning materials, including podcasts, transcripts, and accessible documents.

Here are some key considerations for designing a learning experience for hearing-impaired learners:

- **Captions and Transcripts**

  - **Video Captions:** Include accurate captions for all video content. Captions should synchronize with the spoken words and include relevant sound information.

  - **Transcripts:** Provide transcripts for audio content, ensuring that all spoken words are documented.

- **Sign Language Interpretation**

  - **Sign Language Videos:** If applicable, provide sign language interpretation for videos. Ensure that sign language videos are clear and easy to understand.

- **Visual Content Emphasis**

  - **Visual Aids:** Emphasize visual elements, such as graphics, images, and diagrams, to complement or convey information independently of audio.

  - **Text Accompaniment:** Provide text-based explanations alongside visual content to ensure understanding.

- **Written Instructions**

  - **Clear Instructions:** Clearly articulate instructions in written form, complementing any verbal instructions.

  - **Visual Guides:** Visual guides, infographics, and step-by-step diagrams enhance understanding.

- **Interactive Transcripts**

  - **Interactive Features:** Implement interactive transcripts highlighting the spoken words as they are played. This helps users follow along with the content.

- **Visual Feedback**

  - **Visual Cues:** Provide visual feedback for interactive elements to compensate for the lack of audio feedback.

  - **Indicators:** Use visual indicators or cues to alert learners to changes in the content.

- **Audio Alternatives**

  - **Text-to-Speech:** Include text-to-speech options for written content to accommodate learners who prefer auditory information.

  - **Audio Descriptions:** If applicable, provide audio descriptions for visual elements to convey essential information.

- **Flexible Learning Formats**

  - **Written Materials:** Offer written materials as an alternative to audio-based instructions or lectures.

  - **Multimodal Resources:** Provide content in various formats, such as text, visuals, and interactive elements.

- **Collaboration Tools**

  - **Chat and Messaging:** Implement chat or messaging tools to facilitate communication among learners and instructors.

  - **Discussion Forums:** Create discussion forums where learners can share thoughts and engage in written discussions.

Here are some considerations to enhance the learning experience for mobility-impaired learners:

- **Accessible Platforms**

  - **Compatibility:** Ensure that the learning platform is compatible with assistive technologies commonly used by individuals with mobility impairments, such as screen readers and voice recognition software.

  - **Keyboard Navigation:** Design the platform to be fully navigable using a keyboard alone, without requiring mouse interactions.

- **Responsive Design**

  - **Mobile Responsiveness:** Optimize the learning platform for mobile devices, considering learners who may use alternative devices with touch interfaces or specialized controls.

- **Flexible Navigation**

  - **Clear Navigation Paths:** Provide clear and intuitive navigation paths with easily clickable buttons and links.

  - **Skip Navigation Links:** Include "skip navigation" links to allow users to bypass repetitive content and directly access the main content.

- **Captioned Videos**

  - **Video Accessibility:** Ensure that all video content includes accurate captions. Captions should synchronize with the spoken words and include relevant sound information.

- **Adjustable Timing**

  - **Timing Considerations:** Avoid time limits on activities and assessments or provide adjustable time limits to accommodate varying mobility speeds.

- **Readable Font and Colors**

  - **Font Size:** Use a readable font size and allow users to adjust the text size based on their preferences.

  - **Color Contrast:** Make sure that there is a sufficient color contrast for text and background elements to aid readability.

- **Voice Commands**

  - **Voice Recognition:** Integrate voice recognition capabilities for learners who may have difficulty with traditional keyboard and mouse interactions.

- **Multimodal Resources**

  - **Diverse Formats:** Provide learning materials in various formats, including text, audio, and visual content, allowing learners to choose the most accessible format for them.

- **Adaptive Learning Paths**

  - **Personalized Learning:** Implement adaptive learning paths that cater to individual learning preferences and abilities.

  - **Progress Tracking:** Allow learners to track their progress and revisit content at their own pace.

- **Collaborative Learning Tools**

  - **Collaboration Platforms:** Utilize collaboration tools, discussion forums, and peer interactions to foster a sense of community among learners.

  - **Accessible Chat:** Ensure that chat features are accessible and support alternative input methods.

- **Accessible Assessments**

  - **Alternative Assessments:** Offer alternative assessment methods that do not rely heavily on physical interactions.

  - **Flexible Evaluation:** Consider various ways to evaluate learners, such as written assignments or online discussions.

- **Remote Learning Support**

  - **Remote Accessibility:** Recognize and accommodate the needs of learners who may be participating in remote learning environments.

  - **Virtual Collaboration:** Facilitate virtual collaboration and engagement opportunities for all learners.

Here are some strategies to enhance the learning experience for individuals with dyslexia:

- **Clear and Readable Text**

  - **Font Choice:** Use sans-serif fonts like Arial or Verdana, as they are often easier to read. Avoid decorative fonts.

  - **Font Size:** Ensure that the text is resizable, allowing users to adjust the font size according to their preferences.

  - **Line Spacing:** Opt for generous line spacing to reduce visual crowding and enhance readability.

- **Readable Colors and Backgrounds**

  - **Color Contrast:** Ensure sufficient contrast between text and background colors for improved readability.

  - **Background Options:** Allow users to customize background colors to reduce visual stress.

- **Multimodal Content**

  - **Audio Support:** Provide audio versions of written content to accommodate different learning preferences.

  - **Visual Elements:** Enhance understanding with visual elements, such as images, diagrams, and videos, to supplement text.

- **Structured Content**

  - **Headings and Subheadings:** Use clear and consistent headings to organize content hierarchically.

  - **Bullet Points and Lists:** Present information in bullet points or lists to improve scan ability.

- **Interactive Learning**

  - **Interactive Formats:** Incorporate interactive formats, quizzes, and activities to engage learners.

  - **Hands-On Exercises:** Include hands-on exercises and practical applications of concepts.

- **Dyslexia-Friendly Fonts and Formats**

  - **Dyslexia-Friendly Fonts:** Consider using fonts specifically designed for individuals with dyslexia, such as OpenDyslexic.

  - **eBooks and Audiobooks:** Offer content in eBook formats with dyslexia-friendly features and provide audiobook options.

- **Text-to-Speech Technology**

  - **Integration:** Allow users to access text-to-speech technology, enabling them to listen to the content.

  - **Highlighting:** Implement synchronized highlighting to aid in tracking the spoken words.

- **Consistent Navigation**

  - **Predictable Layout:** Maintain a consistent layout and navigation structure across the learning platform.

  - **Clear Navigation Paths:** Ensure that users can easily navigate between sections and modules.

- **Assistive Technologies**

  - **Compatibility:** Ensure compatibility with assistive technologies commonly used by dyslexic individuals, such as screen readers and speech recognition tools.

- **Flexible Assessments**

  - **Alternative Assessments:** Offer alternative assessment methods focusing on understanding rather than written expression.

  - **Extended Time:** Provide extended time for assessments to accommodate processing challenges.

- **Guided Reading Tools**

  - **Guided Reading Apps:** Recommend or provide guided reading support tools, such as apps with customizable fonts and background colors.

- **Visual Aids and Mnemonics**

  - **Visual Mnemonics:** Use visual aids, mnemonics, and memory aids to reinforce key concepts.

  - **Color Coding:** Implement color coding for different categories or types of information.

- **Encourage Peer Collaboration**

  - **Peer Support:** Encourage peer collaboration and group activities to foster a supportive learning community.

  - **Discussion Forums:** Provide accessible discussion forums for sharing ideas and insights.

- **User Preferences and Profiles**

  - **Personalization:** Users can set preferences based on their learning needs and save personalized profiles.

  - **Progress Tracking:** Enable learners to track their progress and revisit content at their own pace.

- **Teacher Training and Awareness**

  - **Professional Development:** Train educators and instructional designers on dyslexia awareness and inclusive teaching practices.

  - **Resource Libraries:** Develop resource libraries with guidelines and best practices for creating dyslexia-friendly content.

Pedagogical Approaches for UDL

By seamlessly integrating the principles of the Zone of Proximal Development (ZPD) with Universal Design for Learning (UDL), instructional designers can elevate their pedagogical approaches to new heights, fostering an inclusive, differentiated, personalized, and adaptive learning environment. Through applying this powerful synthesis, instructional designers have the unique opportunity to tailor their teaching strategies to the diverse needs of learners, ensuring that each student navigates their Zone of Proximal Development with targeted support. This holistic approach goes beyond a one-size-fits-all model, acknowledging and embracing the inherent variability in learners. Just as UDL facilitates the creation of instructional materials addressing different learning styles, the integration with ZPD enables instructional designers to pinpoint the precise level of challenge and support necessary for each student's optimal growth. Here's how we can integrate the Zone of Proximal Development with UDL:

- **Individualized Learning Goals**

    - **ZPD Perspective:** ZPD emphasizes the range of tasks that a learner can perform with support but cannot accomplish independently. Identify individualized learning goals based on a student's ZPD to provide the right level of challenge.

    - **UDL Connection:** UDL emphasizes the importance of setting clear, flexible learning goals that cater to diverse learner needs. By considering a student's ZPD within UDL, you can ensure that goals are both challenging and achievable, providing necessary supports.

- **Flexible Methods and Materials**

    - **ZPD Perspective:** ZPD proposes that offering suitable support, commonly referred to as scaffolding, assists learners in bridging the divide between tasks they can independently perform and those they can accomplish with assistance. Teachers should provide guidance and adjust instructional methods accordingly.

    - **UDL Connection:** UDL promotes using flexible methods and materials to meet the needs of learners. By recognizing ZPD, educators can tailor instructional methods and materials to match individual needs, ensuring that learners receive the necessary support to progress.

- **Assessment and Feedback**

    - **ZPD Perspective:** Assessment in the ZPD context involves gauging a learner's ability to complete tasks with varying levels of support. It helps teachers understand the next steps in a student's learning journey.

    - **UDL Connection:** UDL encourages varied forms of assessment and timely feedback. By aligning ZPD with UDL principles, assessments can be designed to measure progress within the ZPD, and feedback can guide adjustments to instruction based on individual needs.

- **Multiple Means of Representation, Engagement, and Expression**

  - **ZPD Perspective:** ZPD primarily focuses on the learner's current developmental level and the potential for growth with support. This aligns with the idea of multiple means of representation and engagement.

  - **UDL Connection:** UDL emphasizes providing multiple means of representation, engagement, and expression to address learner variability. By incorporating the ZPD concept, educators can tailor these multiple means to scaffold learning appropriately, making content accessible to all learners.

- **Collaborative Learning**

  - **ZPD Perspective:** ZPD highlights the importance of social interaction and collaborative learning. Peers, as well as teachers, can provide support to scaffold a learner's understanding.

  - **UDL Connection:** UDL values collaborative learning environments. By acknowledging the Zone of Proximal Development (ZPD), educators can organize collaborative activities to promote effective teamwork among students with varied abilities, ensuring they receive the appropriate level of support and challenge.

Summary

- Universal Design (UD) in education focuses on designing teaching and learning activities to be usable by all individuals, emphasizing accessibility, usability, and inclusivity without the need for adaptation or specialized design.

- The Seven Principles of Universal Design directs the creation of universally accessible environments and products, emphasizing equitable use, flexibility, simplicity, perceptible information, tolerance for error, low physical effort, and size and space considerations for diverse user needs, fostering inclusivity across various aspects of life.

- UDL is a pedagogical approach ensuring equitable opportunities for all students by employing diverse teaching methods that eliminate barriers to learning. It emphasizes flexibility in instructional design to allow personalized adjustments based on individual strengths and needs.

- UDL is rooted in neuroscience research, focusing on three neurological networks: affective, recognition, and strategic.

  - **Diverse Engagement Approaches:** Centers on the "why" of learning, connecting course material with students' personal goals, fostering interest, motivation, and self-regulation through varied engagement methods.

- **Diverse Representation Approaches:** Addresses the "what" of learning by presenting information diversely, offering options for perception, language, symbols, and comprehension, aiming to enhance accessibility and knowledge acquisition.

- **Diverse Action and Expression Approaches:** Focuses on the "how" of learning, providing options for physical action, expression, communication, and executive functions during activities and assessments, creating a comprehensive and inclusive learning environment.

- Implementing UDL involves a gradual, iterative process, starting with small adjustments and progressing to create an inclusive learning environment through a six-step cycle: understanding learners and context, defining goals, identifying barriers, incorporating universal supports, crafting UDL-guided plans, and continuously teaching, evaluating, and revising based on feedback for ongoing improvement.

- WCAG, or Web Content Accessibility Guidelines, developed by the Web Accessibility Initiative (WAI) of the World Wide Web Consortium (W3C), sets standards ensuring web content accessibility for people with disabilities. The POUR principles embedded within WCAG – Perceivable, Operable, Understandable, and Robust – establish a comprehensive framework.

- Integrating ZPD with UDL empowers designers to create inclusive, personalized learning. This synthesis tailors teaching strategies, acknowledging learner variability for optimal growth and adaptive environments.

Let's Brainstorm

These mini-scenarios will make you scratch your head and scribble on your pad.

Mini-scenario 1

Imagine you are tasked with redesigning a public library space to adhere to the Seven Principles of Universal Design. Provide specific design recommendations for each principle, considering diverse users such as those with visual impairments, mobility challenges, and varying technological familiarity. Illustrate how your proposed design enhancements align with the principles and contribute to creating an inclusive library environment accessible to a wide range of patrons.

Mini-scenario 2

As an educator tasked with implementing Universal Design for Learning (UDL) principles, describe how you would design a lesson plan that integrates diverse engagement, representation, and action and expression approaches. Provide specific examples for each approach, considering the unique needs of a diverse group of learners. How would you ensure that your pedagogical approach fosters an inclusive learning environment accommodating varied abilities and preferences?

Mini-scenario 3

Imagine you are tasked with designing a website for a nonprofit organization focused on accessibility and inclusivity. Explain how you would apply the Perceivable, Operable, Understandable, and Robust (POUR) principles from the Web Content Accessibility Guidelines (WCAG) to ensure the website caters to individuals with diverse abilities, including those with visual, auditory, cognitive, and motor impairments. Provide specific examples for each POUR principle to demonstrate your understanding of creating an inclusive and accessible digital environment.

Mini-scenario 4

Imagine you are tasked with implementing Universal Design for Learning (UDL) in your courses. Outline the initial steps you would take in understanding your learners and context (step 1). How would you assess factors such as cultural backgrounds, personal identities, and prior knowledge to ensure a comprehensive understanding of your students? Additionally, explain how you would address broader contextual elements such as the learning environment and the familiarity of activities or settings. Propose specific strategies for initiating this process gradually, in alignment with UDL principles, and discuss the importance of iterative implementation.

Mini-scenario 5

You are a mathematics teacher planning a lesson on fractions. Describe how you would identify and leverage the Zone of Proximal Development (ZPD) for a diverse group of students. Additionally, explain how you would apply Universal Design for Learning (UDL) principles to accommodate different learning preferences within the context of a fraction lesson. Provide specific examples of assessment strategies, learning goals, instructional methods, and collaborative activities that align with both ZPD and UDL to ensure an inclusive and effective learning experience for all students.

PART II

Art of Instructional Design

This part explores the artistic field of instructional design, where structure and creativity combine to create life-changing learning opportunities. The fundamental components of it are instructional design models, which act as project-specific guiding frameworks. We examine their subtleties and relevance, highlighting their function in optimizing the design process and promoting successful learning outcomes.

One of the main tenets of instructional design is creating intelligent learning objectives. Using backward design principles and a variety of taxonomies, we explore the nuances of goal, evaluation, and learning activity alignment. This intentional alignment directs students toward measurable learning objectives while ensuring coherence and relevance.

Similar to drawing the blueprint for a masterpiece, storyboarding develops as a critical step in the instructional design process. We break it down, revealing the key components that make up a strong storyboard. Every element, from interactive features to story arcs, has been carefully chosen to captivate learners and improve understanding.

The core of an effective instructional design is the assessment, feedback, and evaluation of training and educational programs. We examine strategies for developing strong tests, giving prompt feedback, and carrying out in-depth analyses. Instructional designers create learning environments that support growth and continual improvement through iterative refining.

Publishing the learning journey online through learning management systems (LMS) becomes increasingly important as the digital landscape changes. We traverse the landscape of learning management system platforms, investigating various features and adherence to eLearning standards. Through the use of technology, instructional designers increase scalability and accessibility, democratizing access to education for all.

This part essentially sheds light on the artistic quality that permeates instructional design, from conception to execution. By use of methodical planning, imaginative ingenuity, and technological integration, instructional designers craft immersive learning experiences that enable students to flourish in a constantly changing educational environment.

CHAPTER 6

Instructional Design Model

In the expansive realm of instructional design, the effectiveness of learning experiences, courses, and instructional content often hinges on the frameworks and models we employ. With their unique perspectives, instructional design models serve as vital connectors between the artistic and scientific dimensions of instructional design. These models form the backbone of structured learning experiences. Just as architects rely on blueprints to construct buildings, instructional designers draw upon these models to shape and refine the learning journey.

Instructional design models structure and illustrate learning theories and principles, guiding for instructional designers throughout the learning development. These models act as road maps, providing a systematic approach to crafting meaningful learning experiences. From classic methodologies to contemporary approaches, each model brings its unique perspective, offering insights into the art and science of instructional design.

Design models are invaluable tools for optimizing both time and financial resources in instructional design. They provide designers with a clear plan to assess the content required for development or identify existing content assets. This foresight ensures that designers avoid investing time in creating content that may go unused. Additionally, design models empower designers to identify potential gaps in content early in the process,

© Ankita Jiyani Mangtani 2024
A. J. Mangtani, *Instructional Design Unleashed*, Design Thinking,
https://doi.org/10.1007/979-8-8688-0416-8_6

enabling them to address and fill these gaps proactively before entering the development phase. Special consideration must be given to instructional design models for teaching, as any imperfections in the course design will directly affect the learner's ability to achieve a learning objective successfully.

Fundamental Concepts of Instructional Design Models

The four fundamental planning elements identified by experts in the field of instructional design are integral to almost every instructional design (ID) model and are encapsulated in four pivotal questions:

>**Target Audience:** Who is the program developed for? This entails a deep understanding of the characteristics of learners or trainees.

>**Learning Objectives:** What is the desired outcome for learners or trainees? This involves defining specific objectives and outcomes.

>**Instructional Strategies:** How is the subject content or skill best learned? This encompasses the selection of effective instructional methods and strategies.

>**Evaluation Procedures:** How do you determine how much learning is achieved? This involves establishing rigorous evaluation procedures to assess the effectiveness of the instructional design.

It is essential to recognize the interconnectedness of these elements, which could collectively constitute a comprehensive instructional design model. Moreover, additional considerations, such as the learner's context, should be addressed in practice. When integrated with the four fundamental elements, these additional elements form a holistic instructional design model.

Various Instructional Systems Design (ISD) models exist, each differing in the number of steps or phases. However, they can all be categorized into three key activities: defining the instructional goal, designing and developing instructional material, and evaluating the efficacy of instruction. This unified perspective underscores the common thread that runs through diverse ISD models, emphasizing the systematic approach to achieving effective instructional outcomes.

Characteristics of Instructional Design Models

Imagine instructional design as the meticulous planning and execution of a chef preparing a multi-course meal in a restaurant. There are several characteristics that should be present in all instructional design planning efforts.

- **Instructional Design Is Learner Centric:** Just as a chef considers the preferences and dietary needs of diners when crafting a menu, instructional designers customize the learning experience to accommodate learners' unique characteristics and requirements.

- **Instructional Design Is Goal Oriented:** Like a chef who envisions a culinary masterpiece and plans each course with a specific goal, an instructional design model strategically plans learning activities to achieve predetermined educational objectives.

- **Instructional Design Focuses on Meaningful Performance:** Much like a chef doesn't aim for mere consumption but strives for a memorable dining experience with each dish, instructional designers shape learning activities for performances that reflect genuine comprehension and engagement.

- **Instructional Design Assumes Outcomes Can Be Measured Reliably and Validly:** Just as a chef relies on precise measurements to ensure a dish's consistency, instructional designers employ reliable and valid assessment tools to accurately measure and evaluate learning outcomes.

- **Instructional Design Is Empirical, Iterative, and Self-Correcting:** Like a chef refining a recipe based on customer feedback and changing tastes, instructional designers continuously iterate their models based on empirical observations, creating a dynamic and self-correcting learning environment.

- **Instructional Design Is Typically a Team Effort:** Much like a restaurant kitchen where chefs, sous-chefs, and kitchen staff collaborate to deliver a seamless dining experience, instructional design is a team effort where individuals contribute their expertise to create a cohesive and effective learning environment. Each team member plays a vital role, mirroring the teamwork in a successful restaurant.

Benefits of Instructional Design Models

Employing an instructional design model presents a systematic and organized framework for creating, developing, and delivering instructional materials. This approach is instrumental in guaranteeing that all learning content is effective, efficient, and engaging. Several compelling reasons underscore the importance of adopting an instructional design model:

- **Quality Assurance:** An instructional design model ensures the deliberate crafting of learning material with a clear purpose and specific outcomes. Through a systematic approach, potential flaws in the design of learning content can be identified and rectified before integration into training courses, thereby upholding quality standards.

- **Time and Resource Efficiency:** Utilizing an instructional design model allows instructional designers to formulate a comprehensive blueprint for learning materials, encompassing all necessary components, activities, and assessments. This blueprint can be systematically reused for future instructional endeavors, significantly saving time and resources during the content creation.

- **Facilitating Collaboration:** The instructional design model is a structured platform that fosters collaboration among instructional designers, subject matter experts, and stakeholders. This collaborative approach ensures the creation of high-quality instruction tailored to meet the specific needs of both learners and the organization.

- **Enhanced Engagement:** Through adherence to an instructional design model, learning materials are crafted to be engaging and pertinent to the target audience. This systematic process enables instructional designers to discern learners' needs, interests, and preferences, aligning the instruction with these factors to enhance overall engagement.

- **Evaluation and Continuous Improvement:** An integral aspect of the instructional design model involves systematically evaluating the training material's effectiveness and a strategic improvement plan. By collecting data on learners' performance and soliciting feedback, instructional designers can pinpoint areas for enhancement, refine the instruction for subsequent use, and ensure continuous improvement.

Classification of Instructional Design Model

In the diverse realm of instructional design, models are categorized based on their generational evolution, practical applications, and foundational learning theories. This classification (Figure 6-1) provides a comprehensive framework to understand how these models have progressed, where they find practical relevance, and the theoretical underpinnings that shape their design principles.

Classification Based on Generations

Across distinct generations, instructional design models have evolved, reflecting educational philosophy, technology, and learner engagement shifts. From the behaviorism-driven First Generation to the technology-infused Fourth Generation, each era represents a chapter in the narrative of educational design, offering insights into how our understanding of learning has matured over time.

> **First Generation (ISD1):** The first generation of instructional design models, ISD1, emerged in the 1960s. These models were characterized by a linear, step-by-step approach based on the behavioral

paradigm of learning. The ISD1 models were primarily derived from Glaser's action research methodology of the early 1960s.

Second Generation (ISD2): In the 1970s, the second generation of instructional design models, ISD2, addressed the limitations of ISD1 by incorporating systems theory, specifically based on Bertalanffy's general systems theory. While still following a linear, step-by-step approach, ISD2 models expanded the scope of activities in the design process. Tasks such as goal setting, audience analysis, and selecting instructional delivery systems were added. The team collaboration concept, involving an instructional development technician alongside a subject matter expert, was introduced. Dick and Carey's model of systematic design, introduced in 1978, is a well-known example of a second-generation model.

Third Generation (ISD3): The third generation, ISD3, emerged in the 1980s and 1990s. ISD3 models opened the closed-loop system design of ISD2 by introducing iterative procedures in each core phase – analysis, design, development, implementation, and evaluation. This allowed for more flexibility and adaptability, especially in the design of technology-based multimedia systems with increased learner control. The instructional development technician role shifted to instructional design experts, who could manipulate phases based on specific application needs and constraints. ISD3 models paid greater attention to the analysis phase and introduced new evaluation forms at each stage, such as feasibility and maintenance.

Fourth Generation (ISD4): The fourth generation, ISD4, emerged in the 2000s. An iterative, incremental design process and a user-designer approach characterize it. This generation collapses design and development into a single step, emphasizing an iterative series of design decisions. Reigeluth and Nelson's proposal in 1997 suggested cycles of analysis, synthesis, evaluation, and change, collapsing design and development. ISD4 models often integrate elements of rapid prototyping, allowing for greater participation of learners in the overall design process. The focus is on adaptability, flexibility, and user engagement in the design process.

Classification Based on Functional Applications

The instructional design models have diverse landscapes of their functional applications. These functional applications of ID models unfold across three distinctive taxonomies: classroom ID models, product-oriented ID models, and systems-oriented ID models. These taxonomies serve as guiding frameworks, each offering a unique lens through which instructional design principles are strategically deployed to meet specific educational needs.

Classroom ID Models: Classroom ID models cater primarily to professional teachers who perceive their role as instructors. Teachers view ID models as general road maps, offering guidance in a limited scope. These models outline a few functions, serving as a guide for teachers who may not be familiar with systematic instructional development concepts. Developers working with teachers should exercise caution due to potential unfamiliarity with ID concepts.

Product-Oriented ID Models: Product development models assume the creation of instructional products of several hours or days in length, often technically sophisticated. Four key features characterize these models:

- Assumption of the need for an instructional product.

- Emphasis on production rather than selection or modification of existing materials.

- Considerable focus on tryout and revision.

- Assumption that the product must be usable by various instruction managers. Users may have no contact with developers.

Systems-Oriented ID Models: Systems models assume the development of a substantial amount of instruction, such as an entire course or curriculum, with resources allocated to a team of highly trained developers. Key points include the following:

- Assumptions on original production or selection of materials.

- Varied assumptions on technological sophistication.

- High front-end analysis, tryout, and revision.

- Broad dissemination, not involving the development team post-implementation.

- Begin with a data collection phase to assess the feasibility and desirability of instructional development.

- Emphasize analysis of the larger environment before committing to development.

- Assume a larger scope of effort than product development models, with differences primarily in magnitude rather than specific tasks.

Classification Based on Foundational Learning Theories

At the heart of instructional design lies a trio of foundational learning theories: behaviorism, cognitivism, and constructivism. These theories serve as guiding lights, illuminating effective teaching and learning pathways. The ID models align with these theories and help us better understand the pedagogical philosophies that influence instructional design decisions. Whether rooted in observable behaviors, cognitive processes, or the active construction of knowledge, each theory leaves an indelible mark on the design landscape. Refer to Chapter 2 for more details on these theories.

> **Behaviorism:** Behaviorism sees learning as responding to external stimuli, emphasizing observable behaviors shaped through reinforcement. It focuses on measurable outcomes and environmental influences.

> **Cognitivism:** Cognitivism centers on mental processes, highlighting memory, attention, and problem-solving. Learning is an active process of internal information processing, emphasizing understanding cognitive functions.

Constructivism: Constructivism views learning as a social, active process where individuals construct knowledge through experiences, prior knowledge, and social interactions. Learners play an active role in meaning-making and knowledge construction.

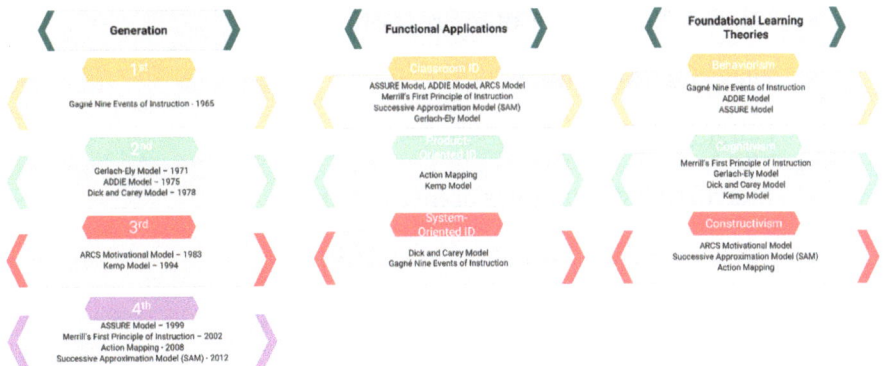

Figure 6-1. *Classification of ID Models*

Instructional Design Models

Over time, numerous instructional design models have been proposed, with only a select few gaining widespread adoption among corporate instructional design professionals. These models define the phases for creating effective learning experiences. According to experts, critical and creative thinking empower instructional designers to determine instruction sequences, content scope, and presentation strategies. While various instructional design models may differ in steps and purposes, many share a common underlying framework – ADDIE. ADDIE – the generic instructional model that underlies diverse instructional design processes – stands for analysis, design, development, implementation, and evaluation. Because of ADDIE's generic nature, it is discussed in detail later in this chapter, and hence, the rest of the models are briefly discussed.

In this section, the following ID models are discussed:

- ADDIE model

- ASSURE model

- ARCS motivational model

- Gagné's Nine Events of Instruction

- Dick and Carey model

- Gerlach-Ely model

- Merrill's First Principles of Instruction

- Successive Approximation Model (SAM)

- Action mapping

- Kemp model

ADDIE Model

Analysis

The first step of the ADDIE model, the Analysis phase (Figure 6-2), lays the foundation for effective instructional design, and a well-structured Analysis phase enhances resource utilization and paves the way for successful training programs. It's crucial to look closely at four main things: figuring out what you want to teach (instructional goals), breaking down the steps needed to teach it (instructional analysis), finding out what your learners already know (learner analysis), and deciding what you want them to be able to do by the end (learning objectives).

Think about it like baking a cake. First, you must decide if you want a vanilla or chocolate cake – that's your instructional goal. Then, when you plan out each step, it's like writing down the recipe for the cake, ensuring you don't miss anything important. Baking is not merely about combining flour, sugar, and eggs; it involves understanding the nuances of ingredient proportions, mixing techniques, and baking temperatures that's the instructional analysis.

The learner Analysis phase is akin to assessing individuals' prior knowledge involved in the cake-baking endeavor. Whether it's a one-on-one conversation with a friend eager to learn baking or conducting surveys among a group of aspiring bakers, understanding their existing knowledge helps tailor the instructional content. For instance, if a learner is already adept at frosting techniques, there's no need to delve into those details, saving valuable time and effort.

The analogy extends to emphasize the importance of learning objectives in the context of baking a cake. Learning objectives are the recipe for success, framed as skills, attitudes, or knowledge. Crafting specific objectives, such as mastering the art of creating a moist sponge, skillfully frosting layers, and understanding the optimal baking time, ensures a focused and productive learning experience.

Design

The second step of the ADDIE model, the Design phase, emphasizes three key elements: assessments, course format selection, and instructional strategy creation. While designing assessments before content creation may seem unconventional, it is important to understand how learners' skills, attitudes, or knowledge will be tested.

While designing good assessment questions, utilize the insights gained from the Analysis phase and consider four critical areas: instructional goals, learners, context, and the assessment itself. Ensure the assessment question's context mirrors real-world scenarios and is aligned with

- Specific learning or performance objectives framed in learning objectives

- Learners' proficiency levels gained during learner analysis

The choice of course format is defined as the medium for presenting content to learners. Consider factors like where the course will be taught in a traditional classroom setting, an online synchronous or asynchronous class, a correspondence course through a self-paced workbook, or a blended course combining different methods. It's advised to align the format with the assessment method. For instance, if learners will be tested on the skills of pastry chefs, a baking and confectionery section with the required equipment is recommended for instruction.

Once assessments and course format are established, the instructional strategy is crafted, encompassing lectures, readings, discussions, projects, worksheets, assessments, and activities. Dick and Carey outline five essential learning components: pre-instructional activities, content presentation, learner participation, assessment, and follow-through activities. For pre-instructional activities, motivation and illustration of course objectives are crucial. Content presentation should be concise and relevant and include practical examples. Learner participation involves practice tasks and feedback, ensuring active engagement. Assessments should include final evaluations, practice assessments, and attitude assessments. Finally, follow-through activities focus on reviewing the entire course strategy to help learners internalize and apply instruction post-class.

Development

The third step of the ADDIE model, called the Development phase, is pivotal in creating effective course materials and ensuring a seamless instructional experience. Using a practical analogy of baking a cake, the key steps involved in the Development phase are as follows:

After analysis and design, instructional designers craft a sample of materials like a preliminary cake recipe. This sample, including a portion of the recipe and instructions, is shared with the client for initial feedback.

Regular collaboration with the client is essential. Share progress in the sample, seeking ideas and feedback. This approach prevents investing time in extensive materials that may face disapproval or faults later.

Upon approval of the sample, develop the course materials. Utilize the instructional strategy from the Design phase and consider client feedback. The aim is to create comprehensive, effective materials aligned with the strategy.

Conduct a thorough review session with the client after developing course materials. Gather final comments or suggestions to ensure alignment with expectations and instructional goals.

The final step involves a real-time rehearsal of the course. In our example, demonstrate the instructions and recipe to a friend, treating them as the learner. Timing is crucial, considering any client-set constraints. Prepare a feedback assessment for areas of improvement.

Implementation

The fourth step of the ADDIE model, called the Implementation phase, comprises three integral parts: training the instructor, preparing the learners, and arranging the learning space.

In many scenarios, the instructional designer transitions into the role of the instructor. However, there are cases where a different trainer takes the reins. Using the example of baking a chocolate cake, it's essential to train the instructor thoroughly after completing analysis, design, and development. This involves imparting knowledge on course objectives and planned activities and familiarizing them with all media and assessments to be used. For professional contexts, where multiple instructors may be involved, developers often train numerous instructors across diverse areas. In cases that involve conducting a synchronous class online, an instructor needs to be trained on the LMS, attendance system, and software for conducting classes.

The preparation of learners is crucial for effective instruction. Higher education entails ensuring learners have the necessary tools and knowledge to participate. This includes verifying prerequisites, conducting orientations, and providing training on relevant software. In baking example, it is vital to ensure learners know class details and the equipment and ingredients they need, such as an oven, flour, and eggs. In cases that involve conducting a synchronous and asynchronous class online, a learner needs to be trained on the LMS, accessing the class recordings, discussion forums, and resource materials.

Beyond merely providing sufficient seating and equipment, arranging the learning space involves meticulous planning. In the example, if the instructor plans to show a video on basic tips and techniques to bake a cake, ensuring the availability of necessary technology is paramount. Testing multimedia equipment beforehand prevents technical glitches during instruction. Instructors should also handle practicalities like printing handouts, having a ready whiteboard with markers, electrical connections, and equipment, and ensuring a smooth, distraction-free learning environment.

The traditional offline classroom should have a whiteboard, marker, duster, projector, projector screen, mic, speaker print handout for safety and device handling, and course materials handouts.

The online classroom settings should have high-speed Internet, virtual team meeting tools, discussion forums, and login credentials for learners using LMS.

Evaluation

The fifth step of the ADDIE model, called the Evaluation phase, comprises two critical components: formative evaluation and summative evaluation.

Formative Evaluation

- **One-to-One Evaluation**

    - Tailor evaluations to match the age and abilities of learners, ensuring clarity, impact, and feasibility.

    - Use a series of pre-planned questions to assess effectiveness systematically.

    - Continuously refine instructional materials based on feedback.

- **Small Group Evaluation**

  - Test changes made from one-to-one assessments in a group setting.

  - Evaluate clarity, impact, and feasibility, adding an attitude questionnaire for a comprehensive assessment.

  - Gather feedback on the instructional activities, instructor, and materials.

- **Field Trial**

  - Conduct a real-time rehearsal of all instructional activities in a setting similar to the instructional environment.

  - Assess clarity, impact, and feasibility in real time.

  - Implement changes based on field trial feedback.

Summative Evaluation

- **Reaction**

  - Document learner reactions through statements on clarity, relevance, and instructor effectiveness.

  - Use a Likert scale to measure agreement.

- **Learning**

  - Administer post-tests to evaluate knowledge acquisition, using multiple-choice questions for understanding rules and skills assessment for baking techniques.

  - Assess attitudes through questionnaires similar to reaction surveys.

- **Behavior**

  - Measure the transfer of knowledge, skills, and attitudes from training to performance.

  - Evaluate application in real-world scenarios, such as observing learners utilizing baking skills in real-world scenarios.

- **Results**

  - Assess the broader impact of training on various aspects, including confidence, enjoyment, and performance.

  - In professional settings, evaluate effects on profits, productivity, morale, and job satisfaction.

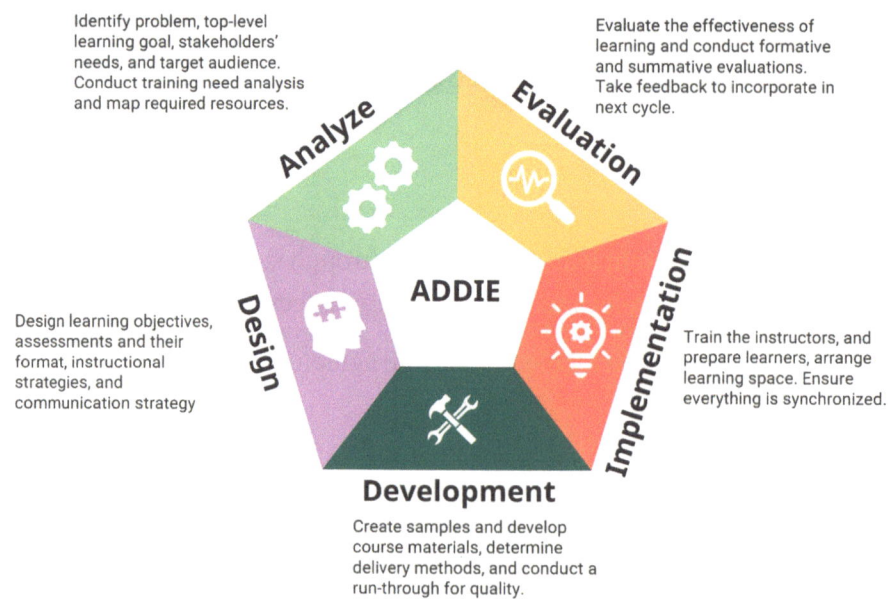

Figure 6-2. *ADDIE Model*

ASSURE Model

The ASSURE model, while stemming from the foundations of the ADDIE model (Figure 6-3), distinguishes itself as a framework tailored to leverage technology and media for enhanced learner engagement. As an evolved version of ADDIE, the six-step ASSURE model aligns with the overarching design phases of its predecessor while emphasizing the integration of media to achieve specific learning outcomes. The model adopts an individualized approach, recognizing the uniqueness of each learners' learning style. Suited for both novice and experienced educators, ASSURE serves as a guiding road map, encouraging strategic thinking and the gradual development of instructional expertise over time. The six components of ASSURE are described in Figure 6-3.

Figure 6-3. *ASSURE Model*

ARCS Motivational Model

With the surge in online education, there's a renewed interest in John Keller's Instructional Model of Motivation, which addresses the challenge of motivating learners in virtual settings (Figure 6-4). Grounded in expectancy theory, it addresses the challenge of motivating virtual learners by linking activities to personal needs and positive expectancy for success. Motivational design, central to the model, encompasses Attention, Relevance, Confidence, and Satisfaction, ensuring sustained learner engagement. Originally designed for classrooms and professional settings, the ARCS model adapts to evolving learner needs. Keller's research on human motivation underpins this practical guide for instructional designers and educators, facilitating the capture of attention, enhancing relevance, boosting confidence, and fostering satisfaction in the online learning journey. Aligning the following targeted strategies given in Table 6-1 with the ARCS model helps create an eLearning environment that engages learners effectively and promotes sustained satisfaction throughout the educational journey.

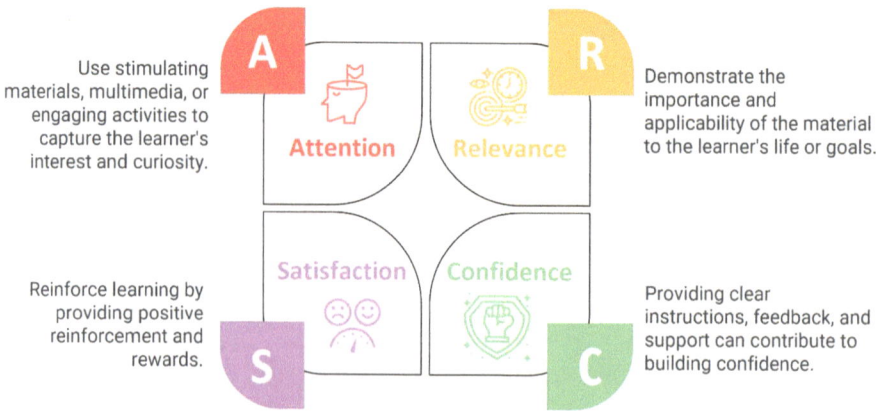

Figure 6-4. *ARCS Motivational Model*

Table 6-1. *Design Strategies for the ARCS Motivational Model*

ARCS Model	Specific Design Strategies
Attention	Perceptual Arousal: Integrate multimedia, relatable examples, and humor.
	Inquiry Arousal: Pose questions, encourage brainstorming, and incorporate role play.
	Variability: Blend web-based content, videos, and virtual field trips.
Relevance	Goal Orientation: Clearly articulate learning goals, emphasizing practical benefits.
	Motive Matching: Identify individual learner needs, offering options to accommodate.
	Familiarity: Connect content to real-life experiences, fostering relevance.
Confidence	Learning Requirements: Clearly outline expectations and evaluation criteria.
	Success Opportunities: Offer choices and meaningful experiences for successful learning.
	Personal Control: Link outcomes to effort and abilities, enhancing a sense of control.
Satisfaction	Intrinsic Reinforcement: Encourage and support learners' enjoyment of the learning process.
	Extrinsic Reinforcement: Provide positive feedback and motivational reinforcement.
	Equity: Maintain consistency within assignments and across the learning framework.

Gagné's Nine Events of Instruction

Robert Gagné's Nine Events of Instruction is a comprehensive model rooted in the information processing model (Figure 6-5). It focuses on cognitive events during adult learning and emphasizes the importance of internal (learner's knowledge) and external (instruction stimuli) conditions.

Gagné's nine-step instructional design process ensures flexibility, making it widely used, especially in online training. The Conditions of Learning, outlined in Gagné's book, identify mental prerequisites for effective learning, aligning with the information processing model.

The Events of Instruction is a strategic framework for trainers and instructional designers. Each event corresponds to a specific instructional stage, facilitating engagement and knowledge retention. These nine events are divided into three segments, and each corresponds to an information-processing stage, as given in Table 6-2.

Table 6-2. Cognitive Process Involved in Gagné's Nine Events of Instruction

Segments	Event	Learner Cognitive Process
Preparation	Gain Attention	Reception (stimuli activate brain receptors)
	Inform the Learner of the Objective	Expectancy (helps create learning expectations)
	Stimulate Recall of Prior Information	Retrieval (retrieves known information from memory)
Instruction with Practice	Present Information	Selective perception (process newly displayed information)
	Provide Guidance	Semantic encoding (makes the connection between memory and new information)
	Elicit Performance	Responding (integrates encoding)
	Provide Feedback	Reinforcement (confirms correctness)
Assessment and Retention	Assess Performance	Retrieval (retrieves acquired information and reinforces understanding)
	Enhance Retention and Transfer	Generalization (acquired information cements and applied in future)

Figure 6-5. *Gagné's Nine Events of Instruction*

Dick and Carey Model

Walter Dick and Lou Carey develop the Dick and Carey instructional design model (Figure 6-6). This model employs a systems approach, viewing each part of the instructional design process as interconnected components. It offers a comprehensive nine-step process for planning and designing effective learning initiatives. This approach encompasses all five stages of the ADDIE model, providing additional depth and structure. The model considers components such as the instructor, learners, materials, instructional activities, delivery system, and learning environments as interconnected units, working collaboratively to achieve predetermined learning outcomes. The success of the systems approach lies in the careful linkage between each component shown in Table 6-3, particularly the relationship between instructional strategy and desired learning outcomes. The nine basic steps, excluding summative evaluation, form a cohesive set of procedures within this systematic framework.

Table 6-3. *Linkage Between Steps of the Dick and Carey Model*

Input	Step	Actions Within Each Step	Output
Identify the problem that can be resolved with instruction	Identify Instructional Goal	Define the need by explicitly establishing the instructional goal according to the identified need. Provide a general description of learners, the performance context, and the tools available.	Educational goal and an overview of the learners, performance context, and tools.
Output from step 1	Conduct Instructional Analysis	Identify the sequential actions individuals undertake while achieving the goal.	Visual representation of the key steps necessary to achieve the instructional goal, graphical depiction of the skills, knowledge, and attitudes needed for each key step, and specification of entry behaviors required to initiate instruction.

(continued)

Table 6-3. (*continued*)

Input	Step	Actions Within Each Step	Output
Output from step 2	Identify Entry Behaviors	Carry out an analysis of learners to determine their existing knowledge, skills, preferences, and attitudes. Conduct a context analysis of the performance setting and the learning environment.	Comprehensive explanations of learners, the performance environment, and the learning environment encompass details about resources and tools.
Output from steps 2 and 3	Write Performance Objectives	Combine information on the skills to be acquired, the characteristics of the target group, the learning context, and the performance context. Use this synthesis to formulate learning outcome statements customized for the content domain, the characteristics of the learners, and the learning and performance contexts.	Establish performance objectives for each step in the model, specifying the skills to be acquired, the conditions under which they will be performed, and the criteria for successful performance.

(*continued*)

Table 6-3. (*continued*)

Input	Step	Actions Within Each Step	Output
Output from step 4	Develop Assessment Instructions	Examine performance objectives to determine the appropriate format for the test item or task, such as objective, live performance, or product creation, aligning it with the desired behavior or action. Apply rules of item formatting to generate test items, directions, and rubrics as needed.	Sufficient criterion-referenced test items for each performance objective to develop various tests required for the instruction, encompassing entry behaviors tests, pre-tests, practice tests, and post-tests.
Output from steps 1 to 5	Develop Instructional Strategy	Combine materials generated from steps 1 through 5 to ascertain a suitable instructional strategy. Integrate the following elements into the strategy, considering current theories of learning, findings from learning research, characteristics of the instructional delivery medium, content, learner characteristics, and contexts' characterIstics (e.g., resources).	Explanations for the following components: • Pre-instructional activities • Presentation of information • Practice and feedback • Testing • Follow-through activities

(*continued*)

Table 6-3. (*continued*)

Input	Step	Actions Within Each Step	Output
Output from step 6	Develop and Select Instructional Materials	Adhere to the instructional strategy to generate the prescribed instruction, encompassing tasks such as writing, selecting, organizing, packaging, etc.	Suitable instructional materials, including • Instructor's manual • Student's manual • Student instruction • Projection (e.g., PowerPoint) • Video • Web-based instruction
Output from step 7	Design and Conduct Formative Evaluation	Execute and assess materials and tests with individuals from the target group through one-to-one evaluations. Similarly, implement and evaluate materials and tests with a small group from the target population through small group evaluations. Lastly, implement and evaluate materials and tests in the intended learning setting, often called a field trial.	Report of learner-based data on feasibility considering learners, resources, and the setting. Evaluate the clarity of vocabulary, descriptions, examples, illustrations, sequence, size of chunks, and learner attitudes. Gather learner-based data from entry behavior tests, pre-tests, post-tests, attitude questionnaires, and/or interviews.

(*continued*)

Table 6-3. (*continued*)

Input	Step	Actions Within Each Step	Output
Output from step 8	Revise Instructions and Update Continuously	Summarize and analyze data from one-to-one, small group, and field trials. Revise instructional materials and procedures based on learner-based data tables that highlight areas of strength and identify problems in the materials or procedures.	Instructional content was adjusted after each one-to-one trial. Subsequent revisions were made after a small group trial. Further refinements were implemented following a field trial.
Instruction of high quality, suitable for the domain, learners, and setting, which has undergone formative evaluation and subsequent revision	Design and Conduct Summative Evaluation	Carry out a comprehensive expert judgment evaluation encompassing congruence analysis, content analysis, design analysis, and feasibility analysis. Additionally, perform a field trial evaluation, including outcomes analysis, to assess the impact on learners, the job, and the organization, and conduct management analysis.	Demonstrate the materials' potential for meeting the organization's needs. Provide evidence of their effectiveness with the target learners in the specified setting.

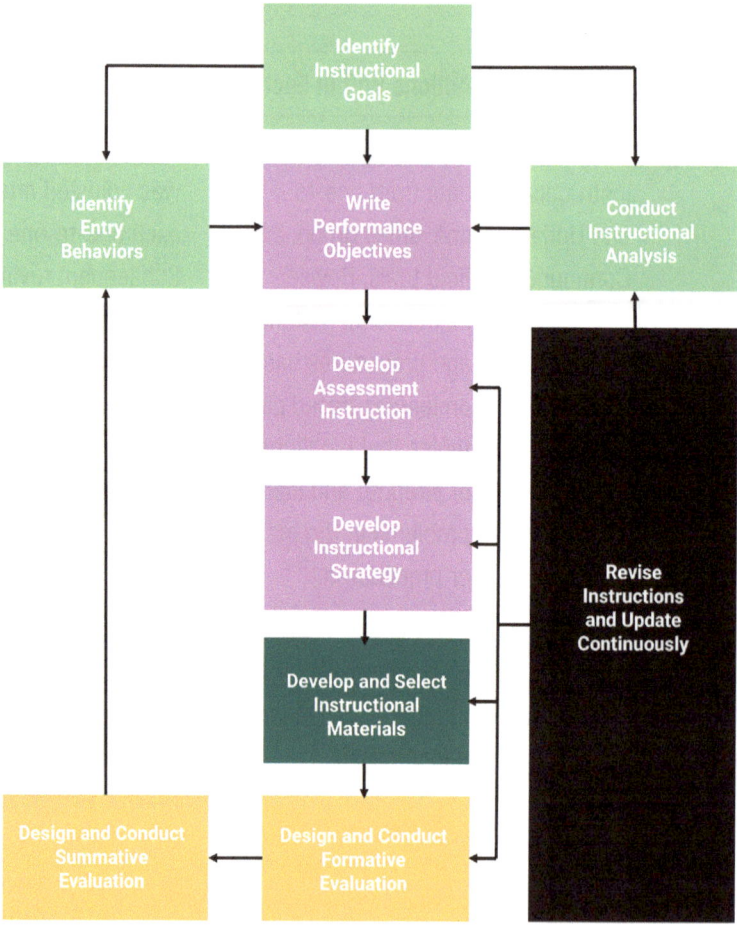

Figure 6-6. *Dick and Carey Model*

Gerlach-Ely Model

The Gerlach and Ely instructional model, developed by Vernon S. Gerlach and Donald P. Ely in 1971, embraces foundational teaching and learning principles (Figure 6-7). Suitable for both K-12 and higher education, this model focuses on systematic planning, clearly defining teaching goals and methods for achieving desired learning outcomes.

Comprising ten elements, the Gerlach and Ely model integrates linear activities (content/objectives, prerequisites, planning, evaluation, post-analysis) and concurrent activities (strategy, groupings, timing, space, resources). Notably, it is an authoritarian model most fitting for instructional planning when learning objectives and instructional content are predetermined. Its distinctive feature involves synchronous completion of content selection and specification of objectives, making it particularly applicable in K-12 settings.

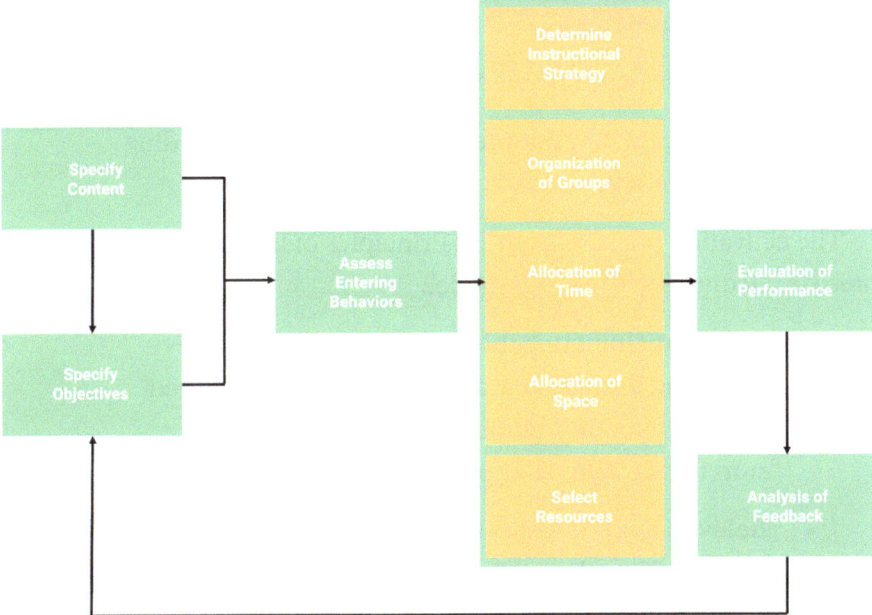

Figure 6-7. *Gerlach-Ely Model*

Merrill's First Principles of Instruction

Merrill's emphasis on effective instruction is crucial in online education, where the simplicity of course launch can neglect thoughtful learning experience design. His dedication to distilling common elements in instructional design theories led to Merrill's First Principles of Instruction, offering essential guidelines for creating engaging and effective learning experiences. Merrill's principles uniquely adopt a design-oriented approach, providing practical guidelines to establish a foundation for new knowledge, deviating from a traditional focus on knowledge acquisition (Figure 6-8). These principles emphasize effective and efficient instruction, highlighting that learning is promoted when the learners

- Participate in resolving real-world issues

- Activate existing knowledge as a basis for acquiring new information

- Apply the demonstrated new knowledge

- Incorporate new knowledge into their understanding or perspective

The activation phase in Merrill's Principles of Instruction comprises three key elements of the ARCS model:

- **Attention:** Initiating the learning process involves capturing the learner's attention, which is achieved through attention-grabbing techniques such as storytelling, humor, or posing challenging questions.

- **Relevance:** Following attention, the next step is to establish the relevance of the learning material by connecting it to the learner's goals, interests, or real-life experiences.

- **Confidence:** The final step is to bolster the learner's confidence in acquiring and applying new information. This is accomplished by providing clear instructions, offering constructive feedback, and creating a supportive learning environment.

While aligned with Gagné's Nine Events of Instruction, Merrill's principles specifically focus on problem-solving, distinguishing them from Gagné's model. Merrill's activation phase aligns with Gagné's initial three events. Merrill's "demonstration" phase in his principles corresponds to Gagné's fourth and fifth events. The "application" phase in Merrill's model aligns with Gagné's sixth, seventh, and eighth events. The "integration" phase in Merrill's model resonates with Gagné's ninth event.

PROBLEM-CENTERED

Show task – Provide a worked example.
Task level – Ensure learners are engaged.
Problem progression – Start with a basic problem, then progress with complex ones.

INTEGRATION

Watch me – provide an opportunity to demonstrate and share learning.
Reflection - include reflection activities to recognize progress.
Creation – Encourage learners to transfer their learning.

ACTIVATION

Previous experience – Tap into existing knowledge.
New experience – Build upon new experience upon existing one.
Structure - Begin with a basic problem, then build the complexity to scaffold learning.

DEMONSTRATION

Demonstrate consistency – Provide relevant examples and demonstrate the content.
Learner Guidance – Provide multiple representations of ideas, concepts, and perspectives.
Relevant media – Ensure media supports effective learning.

APPLICATION

Practice consistency – Align practice activities.
Diminish coaching – Gradually remove the learning support.
Varied problems – Provide opportunities to apply learning to different contexts.

Figure 6-8. *Merrill's First Principles of Instruction*

Successive Approximation Model (SAM)

The Successive Approximation Model (SAM), developed by Michael Allen, is a streamlined alternative to the ADDIE model, emphasizing early feedback and iterative development (Figure 6-9). SAM utilizes a recursive process, deviating from a linear approach, and is particularly popular in the tech space.

The SAM model comprises three main components: Preparation, Iterative Design, and Iterative Development. The core principle of SAM is its iterative nature, highlighting that each step is intended to be revisited and repeated throughout the course development process.

In the Preparation phase, crucial project information is gathered, its scope varying based on the project or course. The "Savvy Start" concludes this phase, promoting activities like brainstorming, sketching, and prototyping, with the involvement of stakeholders such as colleagues, advisors, and learners.

Transitioning to the Iterative Design phase, the goal is to design and prototype the material for stakeholder evaluation. This phase prioritizes assessing a tangible product over a conceptual one, facilitating comprehensive review and testing.

In the final Iterative Development phase, the fully developed prototype is implemented. After its use, the material undergoes evaluation and can be revisited through development and implementation phases as necessary. SAM's emphasis on iteration and early feedback establishes it as a practical model for course development.

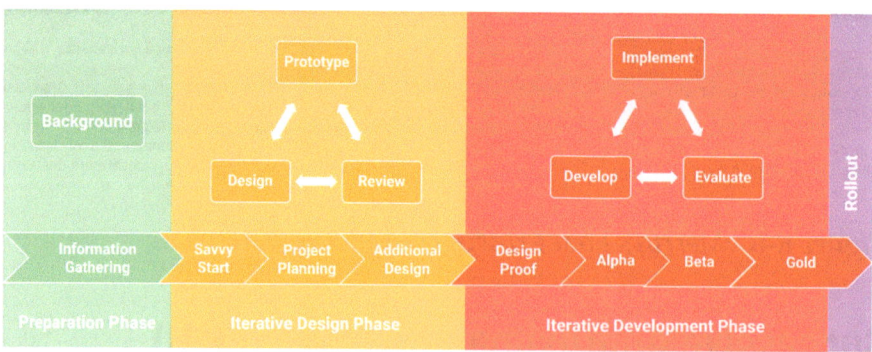

Figure 6-9. *Successive Approximation Model (SAM)*

Action Mapping

Action mapping, developed by Cathy Moore, is a strategic approach to instructional design that shifts the focus from knowledge acquisition to performance improvement. It begins by identifying measurable improvements desired in the organization due to training, emphasizing what people need to do on the job rather than what they need to know (Figure 6-10).

372

Key Steps in Action Mapping

- **Identify the Business Goal:** Define the measurable improvement sought in the organization.

- **Determine On-the-Job Actions:** List specific behaviors people must exhibit to achieve the desired change.

- **Design Practice Activities:** Develop activities that allow individuals to practice each required behavior.

- **Identify Supporting Knowledge:** Identify what individuals need to know to complete the activities successfully.

Practical Steps for Creating an Action Map

- **Kickoff Meeting:** Meet with clients and stakeholders to identify the core problem or goal.

- **Define Necessary Actions:** Identify actions required to achieve the goal and understand why they are not currently performed.

- **Create the Action Map:** Place the business goal in the center and add necessary actions, collaborating with subject matter experts and clients.

- **Prioritize Actions:** Determine which actions contribute most to the goal and which are commonly performed incorrectly.

- **Address Root Cause:** Understand why the problem has arisen in the first place.

- **Create a Prototype:** Develop a prototype, gather feedback, and ensure its effectiveness.

- **Practice Activities:** Work with subject matter experts to practice activities.

- **Finalize Project Plan:** Complete the project plan and obtain approval.

Action mapping offers a practical and focused approach to designing learning solutions, aligning them closely with business objectives and emphasizing real-world behaviors over theoretical knowledge.

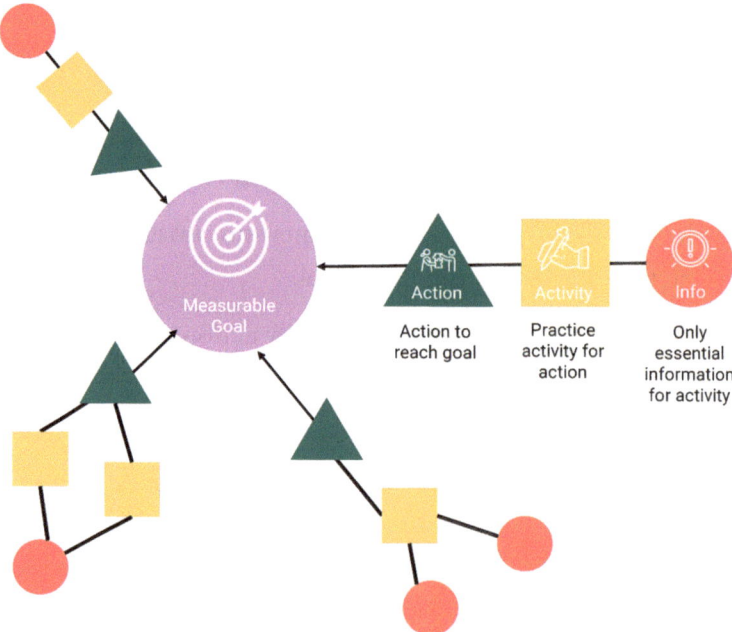

Figure 6-10. *Cathy Moore's Action Mapping Model*

Kemp Model

The Morrison, Ross, and Kemp instructional design model presents a nonlinear approach, distinct from the linear structure of the Dick and Carey model. Experts describe it as an open iterative circle, starting from the inner center oval with process patterns, moving to the second outer circle with evaluation forms, and concluding with the outermost circle encompassing planning, implementation, project management, and support services (Figure 6-11). Unlike linear models, this circular design implies a continuous cycle with no defined starting point.

Inside the innermost circle, nine elements form a logical, clockwise pattern, with the instructional problem at the 12 o'clock position emphasizing a nonsequential, nonlinear presentation. The absence of lines and arrows between elements reinforces this characteristic. Notably, the first oval underscores the importance of revision and formative evaluation at each design stage.

The outer oval introduces unique features to the model, incorporating planning, project management, and support services, making it comprehensive. Kemp's model emphasizes continual planning and appropriate project management throughout the design process.

The Morrison, Ross, and Kemp model revolves around four key components: the learner, objectives, methods, and evaluation. When integrated with an instructional problem or learning activity, these components result in a complete instructional design plan. While the model doesn't prescribe a specific starting point, a logical order of instructional activities validates its practical application, involving learner characteristics, objectives, content, sequence, instructional strategies, delivery, evaluation, and resources.

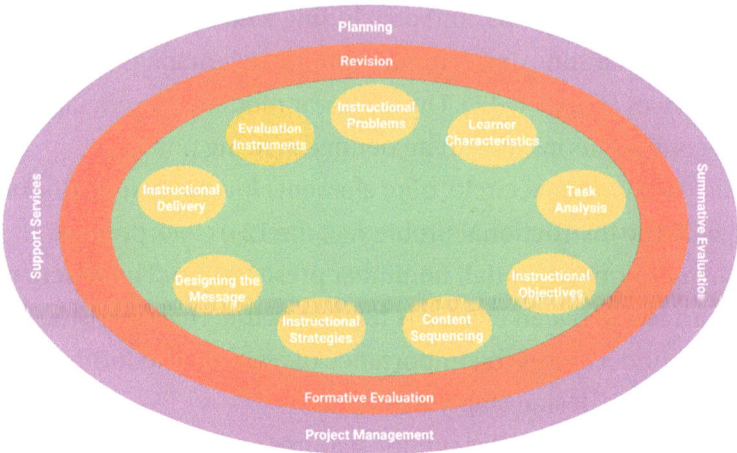

Figure 6-11. *Kemp Model*

Selection of ID Model

Choosing the appropriate instructional design strategy is critical to developing effective learning experiences. The selection of a strategy involves careful consideration of various factors. These considerations encompass the following:

- **Audience**

    The audience plays a pivotal role in determining the instructional design strategy. Identify the characteristics of your target audience, such as their age, educational background, prior knowledge, and learning preferences. For instance, distinguishing between pedagogy for children and andragogy for adults is crucial. Adult learners, being self-directed, have distinct learning needs compared to children.

- **Subject**

  Different subjects demand varied instructional
  approaches. Subjects like Mathematics may require
  extensive practice, while others like Music may
  emphasize listening skills. In a corporate context,
  the strategy must align with course categories, such
  as soft skills, technical skills, business/leadership
  development, or selling skills.

- **Learning Environment**

  Learning can occur in diverse environments, from
  corporate settings to online platforms. Consideration
  should be given to factors such as whether learning
  occurs at workstations, during travel, on various
  devices, collaboratively, or through self-directed
  methods.

- **Align with Learning Theories**

  Explore the learning theories that underpin
  different instructional design models. For example,
  constructivist theories may align with models
  emphasizing collaboration and active learning,
  while behaviorist theories may be reflected in more
  structured, step-by-step approaches.

- **Iterative or Agile Approaches**

  Determine whether your project would benefit from
  an iterative or agile approach. Models like SAM
  (Successive Approximation Model) are iterative and
  allow ongoing feedback and adjustments.

- **Consider Multimedia and Technology Integration**

  Choose a model that supports effective integration
  if your instructional content involves multimedia or
  technology. Some models may be better suited for
  incorporating interactive elements and multimedia.

- **Accessibility and Inclusivity**

  Ensure that the chosen instructional design model
  facilitates the creation of accessible content for
  all learners. Consider inclusivity and diverse
  learning needs.

- **Parameters/Constraints**

  The success of an eLearning initiative hinges on the
  ability to develop courses within the parameters set
  by the client organization. Parameters may include
  time, cost, and technical specifications. When these
  parameters become constraints, instructional designers
  must adhere to specific strategies that align with the
  given limitations.

SWOT Analysis of ID Models

Each model offers a distinctive approach, focusing on various facets of
the instructional design process. Table 6-4 provides a comprehensive
SWOT analysis that aims to provide valuable insights into the strengths,
weaknesses, opportunities, and threats of prominent instructional design
models, fostering a nuanced understanding for seasoned professionals
and those embarking on exploring these methodologies.

Table 6-4. SWOT Analysis of ID Models

ID Model	Strength	Weakness	Opportunity	Threat
ADDIE Model	ADDIE exhibits flexibility, ease of modification, and compatibility with models like SAM, ensuring adaptability and versatility in instructional design.	ADDIE's potential drawbacks include a linear approach, excessive detail impeding creativity, and its time-consuming nature, posing challenges in dynamic or time-sensitive educational settings.	ADDIE has the potential to capitalize on emerging technologies for dynamic learning experiences, foster enhanced collaboration among stakeholders, and continuously improve its processes based on feedback and advancements in instructional design practices.	ADDIE faces challenges such as potential resistance to its linear approach, competitive pressure from models emphasizing quicker cycles, and the risk of technological constraints hindering adaptation to evolving educational landscapes.
ASSURE Model	ASSURE's learner-centric approach, coupled with technology utilization and aligning programs with learner needs, enhances learning through technology and ensures clear, well-stated objectives.	ASSURE is perceived as academically focused with a narrow scope and potential time constraints.	ASSURE has the opportunity to expand its application beyond academic settings and modify its approach to better meet workplace needs.	The model faces the threat of being time-consuming, which may pose challenges in fast-paced educational environments or when efficiency is a critical factor.

(continued)

379

Table 6-4. (*continued*)

ID Model	Strength	Weakness	Opportunity	Threat
ARCS Motivational Model	ARCS keeps learners engaged throughout the learning experience, encourages active participation, is flexible across various learning settings, maintains relevance over time, and is easy to apply.	ARCS may necessitate a significant rethinking of the facilitator's instructional approach and can pose challenges when dealing with learners of different motivation levels in a group.	ARCS can be further integrated into diverse learning experiences, including training, professional development, and various educational settings, leveraging its flexibility and adaptability.	The model may face threats in situations where facilitators resist or find it challenging to rethink their instructional approaches, potentially limiting the effective implementation of the ARCS model.

Gagné's Nine Events of Instruction	Gagné's model is systematic, easy to follow, and adaptable to various learning situations, instilling confidence in learners.	The model may foster learner dependency on guided information, potentially hindering problem-solving skills and requiring significant time in the development phase.	Gagné's model has the opportunity to incorporate more interactive and exploratory elements to enhance problem-solving skills and engagement.	The risk lies in the potential resistance to adaptability and evolves beyond its systematic structure, limiting its effectiveness in dynamic learning environments.
Dick and Carey Model	Dick and Carey's flexibility makes it applicable across various subject areas in education and the business world, emphasizing goal-oriented design that considers learners' prior knowledge and needs.	The model may be time-consuming, requiring significant effort in the design process, and lacks consideration for variables such as changes in student composition from year to year.	Dick and Carey have the opportunity to enhance adaptability by addressing time constraints and incorporating strategies to account for variables, expanding its applicability.	The model may face threats in situations where time for instructional design is limited or when significant variations in student demographics challenge the adaptability of the design from one iteration to another.

(continued)

381

Table 6-4. (*continued*)

ID Model	Strength	Weakness	Opportunity	Threat
Gerlach-Ely Model	The Gerlach and Ely model increases efficiency in instructional design, streamlining the process for educators and allowing for effective program assessment.	Tendency toward teacher-directed approaches and misalignment with ADDIE, missing key components like design task inventory and performance objectives.	The model has the opportunity to enhance adaptability and alignment with contemporary educational practices, addressing limitations and incorporating emerging instructional strategies.	The model may face threats if its teacher-directed nature and deviations from established models limit its applicability and acceptance in diverse educational settings.
Merrill's First Principles of Instruction	Merrill's principles, rooted in research and cognitive psychology, provide a dependable foundation for instructional design.	The principles might not encompass all facets of instructional design, possibly overlooking crucial considerations.	There is potential to adapt and integrate these principles with emerging educational technologies, fostering more dynamic learning experiences.	Inadequate implementation of the principles poses a risk of suboptimal learning outcomes.

	Strengths	Weaknesses	Opportunities	Threats
Successive Approximation Model (SAM)	SAM facilitates quick improvements and flexibility through iterative prototyping, allowing for creative exploration and reducing costs and timeframes.	The iterative nature of SAM can lead to repetition, increased possibilities for mistakes, and the potential for wasted resources.	SAM offers the opportunity for early feedback by putting the course in front of real people, enabling rapid assessment and guiding future steps.	The broad testing of various ideas in SAM may make it challenging to determine the optimal outcome, potentially requiring significant analysis to focus the team on the best approach.
Action Mapping	Action mapping is a collaborative tool for securing stakeholder buy-in and effectively managing project scope in instructional design.	The model may face challenges due to potential stakeholder and SME resistance to active participation in the workshop process.	Action mapping has the opportunity to enhance stakeholder engagement and overcome resistance, fostering a more inclusive and participatory design approach.	The model may face threats if stakeholder resistance persists, hindering the effectiveness of collaborative design efforts and impacting project outcomes.

(continued)

Table 6-4. (*continued*)

ID Model	Strength	Weakness	Opportunity	Threat
Kemp Model	Kemp is designed for flexible delivery modes, employing a holistic approach suitable for diverse learners and subjects, incorporating various instructional methods.	The model's unstructured nature, resource intensity, and the need for expertise in implementation can be perceived as drawbacks.	Kemp has the opportunity to adapt to emerging instructional technologies and diverse learning environments, enhancing its versatility.	Its perceived lack of a step-by-step process and the demand for significant resources may pose challenges when a more structured or resource-efficient approach is preferred.

Summary

- Identifying the target audience, learning objectives, instructional strategies, and evaluation procedures forms the basis of instructional design, emphasizing a holistic approach to designing effective learning experiences.

- Instructional design is learner centric and goal oriented and focuses on meaningful performance, relying on measurable outcomes, iterative processes, and collaborative teamwork, akin to a chef crafting a culinary masterpiece.

- Utilizing an instructional design model ensures quality assurance and time and resource efficiency, facilitates collaboration, enhances learner engagement, and enables systematic evaluation and continuous improvement, providing an organized framework for creating effective and efficient instructional materials.

- Models are categorized based on generational evolution, practical applications, and foundational learning theories, providing a comprehensive framework for their progression, relevance, and theoretical underpinnings.

- The various ID models are derived from the generic structure of the ADDIE model, and selecting the ID model for any project depends on various factors such as audience, subject, learning environment, alignment with learning theories, iterative or agile approaches, multimedia integration, budget, time, team expertise, accessibility, and inclusivity.

Let's Brainstorm

These mini-scenarios will make you scratch your head and scribble on your pad.

Mini-scenario 1

You have been appointed as the head of the instructional design department at a prestigious educational institution. Your responsibility is to categorize and recommend instructional design models based on three key criteria: foundational theory, functional application, and generations. The institution is keen on implementing various models to cater to various subjects, learning environments, and evolving technological landscapes. How would you approach this task, considering the foundational theories that underpin each model, their practical applications across disciplines, and their alignment with different generations of instructional design methodologies? Provide insights into how your classification would optimize instructional design practices for the institution's varied educational offerings.

Mini-scenario 2

As an instructional designer working on a project to develop an online course for a diverse audience, you've been asked to ensure flexibility, easy modification, and compatibility with other design models. How would you leverage the strengths of the ADDIE model to meet these requirements? Additionally, what potential weaknesses might you need to address in the context of this project?

Mini-scenario 3

You are tasked with designing a training program to improve customer service skills for a retail organization. Stakeholder and SME engagement is crucial for success. How would you employ the action mapping model by Cathy Moore to collaboratively gather insights, secure buy-in, and manage the scope of the training program? What potential challenges might you anticipate regarding stakeholder and SME participation, and how would you address them within the action mapping?

Mini-scenario 4

You lead an instructional design team responsible for creating an organization's comprehensive cybersecurity training program. The training must be effective and responsive to rapidly evolving threats and accommodate end-user feedback throughout the development process. How would you decide between adopting the ADDIE and Successive Approximation Model (SAM)? Discuss how the flexibility of SAM could allow for quick adjustments in response to emerging threats, while ADDIE's systematic approach might ensure a thorough and structured development process. Consider each model's advantages and potential challenges in the context of this cybersecurity training initiative.

Mini-scenario 5

You are tasked with developing a training program for a multinational corporation aiming to standardize employee onboarding across various departments and regions. The program must be adaptable to diverse learning needs yet maintain a cohesive structure to ensure consistency in content delivery. Considering this scenario, discuss how you would decide between the Kemp model and the Dick and Carey model. Examine how Kemp's flexibility could cater to diverse needs, while Dick and Carey's systematic approach might ensure a standardized onboarding process. Consider each model's advantages and potential challenges in achieving the organization's goal of harmonizing onboarding practices globally.

CHAPTER 7

Objective Taxonomy

In the educational landscape, learning objectives are guiding beacons for learners and instructors, offering a road map for what is expected and where attention should be focused. For learners, well-crafted learning objectives provide clarity and communicate the essential elements of their learning journey. Understanding what is important and expected not only aids in efficient study methods but also cultivates the crucial skill of transfer – applying knowledge flexibly across various contexts, indicative of deep learning.

Learning objectives also play a pivotal role in developing learners' metacognitive skills – the ability to reflect on and direct their thinking. In self-directed learning, assessing objectives, evaluating knowledge and skills, planning approaches, monitoring progress, and adjusting strategies are integral components.

Instructors, too, benefit significantly from clearly defined learning objectives. These objectives aid in selecting appropriate instructional materials, organizing course content, and assessing the relevance and suitability of various resources. Furthermore, well-articulated learning objectives streamline the creation of assessments that align seamlessly with the learning activities and materials within the course. This alignment, where objectives, assessments, and learning activities work harmoniously toward learning goals, enhances the educational experience.

© Ankita Jiyani Mangtani 2024
A. J. Mangtani, *Instructional Design Unleashed*, Design Thinking,
https://doi.org/10.1007/979-8-8688-0416-8_7

Domains of Learning

In 1956, Benjamin Bloom introduced the Taxonomy of Learning Domains, a three-part model categorizing learning into distinct domains. These domains include the cognitive, affective, and psychomotor domains, each addressing different aspects of the learning process.

Cognitive Domain: The cognitive domain encompasses learning skills primarily associated with mental processes. These processes involve a hierarchical structure of skills, ranging from basic knowledge acquisition to higher-order skills such as synthesis and evaluation. Bloom's original taxonomy outlined six levels of cognitive complexity: knowledge, comprehension, application, analysis, synthesis, and evaluation. While initially focused on describing levels of attainment rather than process skills, the 2001 version of Bloom's Taxonomy incorporates added features to better guide educators in constructing optimal learning experiences. Skill clusters within the cognitive domain organize essential learning skills for each process, providing educators with a comprehensive framework.

Affective Domain: Contrary to the common perception of learning as purely intellectual, the affective domain recognizes that learning extends beyond cognitive functions. It involves the acquisition of attitudes, behaviors, and physical skills. This domain delves into feelings, emotions, and attitudes, acknowledging the multifaceted nature of the learning experience.

Psychomotor Domain: The psychomotor domain pertains to specific physical functions, reflex actions, and interpretive movements. Notably, unlike the cognitive and affective taxonomies outlined in 1956 and 1964, the Psychomotor domain was not thoroughly described until the 1970s. This domain underscores the importance of physical skills and practical application in learning.

Bloom's Taxonomy of Learning Domains has become an invaluable tool for educators. It provides a structured framework to understand and address diverse aspects of the learning experience. Educators can design comprehensive and effective learning experiences that encompass intellectual, emotional, and practical dimensions by categorizing learning into cognitive, affective, and psychomotor domains. Table 7-1 highlights the aspects of all three domains of learning.

Table 7-1. *Aspects of Domains of Learning*

Aspects	Cognitive Domain	Affective Domain	Psychomotor Domain
Nature of Learning	Intellectual processes (knowledge, analysis, etc.)	Development of attitudes, values, and emotions	Physical skills, reflex actions, interpretive movements
Examples	Solving math problems, memorizing facts	Developing empathy, appreciating diversity	Playing a musical instrument, typing, and scientific experiments
Mode of Learning	Lectures, readings, discussions, problem-solving	Experiential learning, group activities, discussions	Simulations, demonstrations, hands-on practice

(*continued*)

Table 7-1. (*continued*)

Aspects	Cognitive Domain	Affective Domain	Psychomotor Domain
Assessment Methods	Tests, quizzes, essays	Self-reflection, peer assessments, projects	Practical exams, demonstrations, observations
Development Over Time	Progression from basic knowledge to higher-order skills	Evolution of attitudes and values over experiences	Sequential development from basic to complex skills
Integration with Other Domains	Interaction with affective and psychomotor domains	Influence of attitudes and emotions on thinking and actions	Interconnected with cognitive and affective domains

Robert Mager's Learning Objectives

Learning objectives serve as critical markers of achievement, defining the significant and essential learning outcomes learners are expected to reliably demonstrate after completing an educational activity. A well-crafted learning objective concisely describes the knowledge, skills, and attitudes learners are poised to gain.

The three key components integral to a learning objective are as follows:

- **Behavior:** Describes the specific action or skill the learner will perform

    - **Example:** "...analyze a given dataset to identify patterns and trends."

- **Condition:** Specifies the tools, situations, settings, or constraints under which the behavior will occur

  - **Example:** "...using Microsoft Excel, learners will calculate and interpret financial ratios."

- **Standards:** Establishes the criteria for acceptable learner performance, considering factors like accuracy, productivity, time, or level of proficiency.

  - **Example:** "...with at least 90% accuracy, learners will troubleshoot common software issues."

A practical learning objective incorporates all three components:

Example: "...utilizing Python programming, the learner will write code to implement a sorting algorithm and explain its functionality."

It's important to note that while including all components is ideal, there are instances where conditions or criteria may not be necessary for every learning objective.

Aligning Learning Objectives

Implementing backward design methodology begins with defining the learning objectives for a lesson, module, or course – outlining what learners are expected to learn and accomplish. This process then unfolds "backward," leading to the creation of assessments that substantiate learners have mastered the outlined objectives. Finally, instructors devise learning activities and instructional materials that align seamlessly with and bolster the achievement of these objectives.

The three sequential steps of backward design (Figure 7-1) involve the following:

- **Identify Desired Results:** Clarify the knowledge, skills, or abilities learners should have acquired by completing a module, lesson, or semester.

393

- **Determine Acceptable Evidence:** Specify the types of documentation necessary for evaluating learners' progress toward proficiency in the learning objectives and course goals.

- **Plan Learning Activities and Instructional Materials:** Strategically plan engagement activities and direct instruction to support learners in achieving the learning objectives and course goals, progressively moving toward self-directed learning.

The alignment of the following three components is crucial for the effectiveness of the course:

- **Learning Objectives:** Clearly articulate the knowledge and skills learners must acquire by the course's conclusion.

- **Assessments:** Serve as evaluative tools, allowing instructors to gauge how learners meet the established learning objectives.

- **Learning Activities and Instructional Materials:** Carefully chosen to cultivate learner learning, steering them toward attaining the stated objectives.

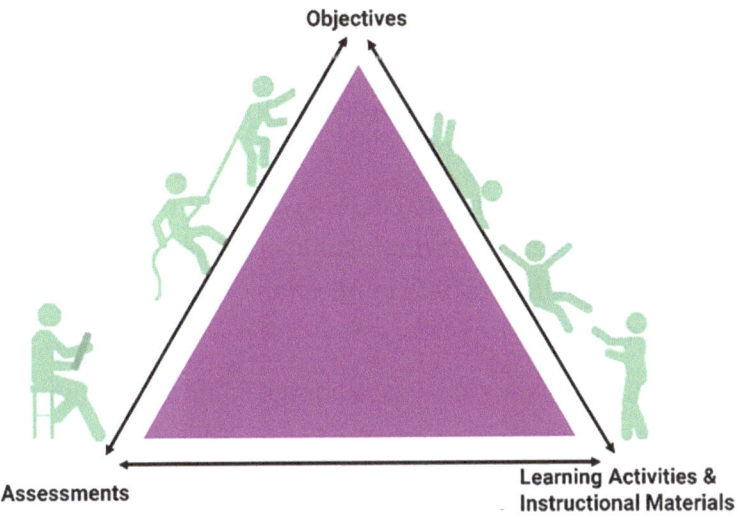

Figure 7-1. *Backward Learning Approach*

Learning Domains in Objective Crafting

Learning is an intricate, ongoing process that shapes the brain's architecture through information absorption, processing, connection, cataloging, and application. Learning is often categorized into three domains to navigate this continuous growth: cognitive, affective, and psychomotor. These domains encapsulate various levels of learning, progressing from foundational surface-level understanding to more intricate, profound comprehension.

When crafting learning objectives, a crucial consideration is which domain(s) is pertinent to the designed learning experience. This decision hinges on factors such as the nature of the experience, the developmental levels of participating learners, and the duration and intensity of the educational encounter.

> **Cognitive Domain:** Learning objectives in this domain aim to impact different levels of cognitive complexity, aligning with the intellectual demands of the content.

Affective Domain: Assessing the affective domain is intricate due to diverse belief systems, but the process through which learners cultivate attitudes can be evaluated.

Psychomotor Domain: Learning objectives in this domain evaluate a learner's ability to perform tasks, focusing on the practical application of acquired knowledge.

These learning domains are widely adopted among educators for several key reasons:

- Assist educators in tailoring their teaching strategies to the complexity of the subject matter. By categorizing objectives based on cognitive processes and learning levels, educators can create lesson plans and learning activities that align with the intellectual demands of the content, ensuring a more nuanced and effective instructional approach.

- Provide a structured framework for educators to organize and sequence learning objectives, ensuring a coherent and logical progression of skills and knowledge acquisition. This systematic approach aids in crafting well-rounded courses that cater to the various needs of learners.

Moreover, the utility of learning domains extends to the assessment phase of education. Educators rely on these domains to create assessments that align with the stated learning objectives, enabling them to measure and evaluate learner outcomes effectively. This ensures that assessments are valid and reflective of the intended educational goals.

Crafting Effective Learning Objectives

The ABCD approach of writing objectives is an excellent way to configure instructional objectives. In this method, "A" is for the audience, "B" is for behavior, "C" is for conditions, and "D" is for the degree of mastery needed.

ABC Structured Examples

Upon successful completion of this learning segment, the specified audience (A – aspiring bakers) will demonstrate the following behaviors (B) under defined conditions (C):

A. Audience

- Culinary arts learners specializing in the art of bread baking

B. Behavior

- Prepare and bake artisanal bread from scratch, exhibiting proficiency in mixing, kneading, and shaping techniques for various types of bread, including sourdough and baguettes.

- Apply advanced baking skills to achieve desired textures and crusts, demonstrating an understanding of fermentation, proofing, and oven management.

C. Conditions

- Without using pre-mixed ingredients, learners will independently source and measure raw materials, emphasizing hands-on experience in dough preparation.

Learners will work with professional-grade ovens and equipment within a fully equipped commercial kitchen, simulating real-world bakery conditions to enhance their practical skills.

Upon completing this learning segment, culinary arts learners will prepare and bake artisanal bread from scratch without using pre-mixed ingredients.

ABCD Structured Examples

Upon successful completion of this learning segment, the specified audience (A – aspiring bakers) will demonstrate the following behaviors (B) under defined conditions (C) and achieve a specified degree of mastery (D):

D. Degree of Mastery

- Successfully produce artisanal bread that meets established quality standards, as evaluated by taste, texture, and appearance.

- Demonstrate efficient time management and organizational skills in executing the bread-baking process, from ingredient preparation to final product, achieving a high level of proficiency in artisan bread craftsmanship.

Upon completing this learning segment, culinary arts learners will prepare and bake artisanal bread from scratch without using pre-mixed ingredients and meet established quality standards, as evaluated by taste, texture, and appearance.

For writing objectives using the ABCD method, follow these steps to create meaningful and measurable ones:

Step 1: Identify Knowledge, Skills, and Changes in Attitude

Begin by pinpointing the essential knowledge, skills, or changes in attitude that you intend learners to acquire by the conclusion of the course, module, or learning activity. This foundational step sets the direction for the subsequent objectives.

Step 2: Align with Course Objectives or Learning Goals

Review your overarching course objectives or broader learning goals. Align your weekly or modular-level learning objectives with these higher-level outcomes. Ensuring consistency and cohesion between these levels enhances the overall effectiveness of your educational framework.

Step 3: Choose Action Verbs

Select action verbs that precisely articulate observable and measurable outcomes or behaviors aligned with the identified knowledge or skills. Utilize learning taxonomies as valuable tools to assist in identifying appropriate action verbs. These taxonomies help create specific and measurable learning objectives, providing a more comprehensive educational strategy.

Dale's Cone of Experience

In accordance with Dale's research findings, the efficacy of instructional methods follows a hierarchical structure. The least effective method at the top of the hierarchy entails learning through verbal symbols, specifically by listening to spoken words. Conversely, the most effective methods, positioned at the bottom, involve direct and purposeful learning experiences, such as hands-on activities or field experiences. These direct experiences are deemed to mirror real-life situations, fostering a deeper understanding closely.

Dale's cone serves as a graphical representation of the average retention rates associated with various teaching methods (Figure 7-2). As one progresses down the cone, the likelihood of enhanced learning and increased information retention rises. The implication is that when selecting instructional methods, the involvement of learners in the learning process is paramount for reinforcing knowledge retention.

Notably, the cone underscores the effectiveness of "action-learning" techniques, revealing up to 90% retention rates. Dale emphasizes the importance of catering to perceptual learning styles, which are sensory based. Maximizing sensory channels in interactions with educational resources enhances the probability of accommodating diverse learning preferences. Instructors should design activities that leverage real-life experiences to optimize learning.

Table 7-2 summarizes the learning outcomes of each level of the cone of experience.

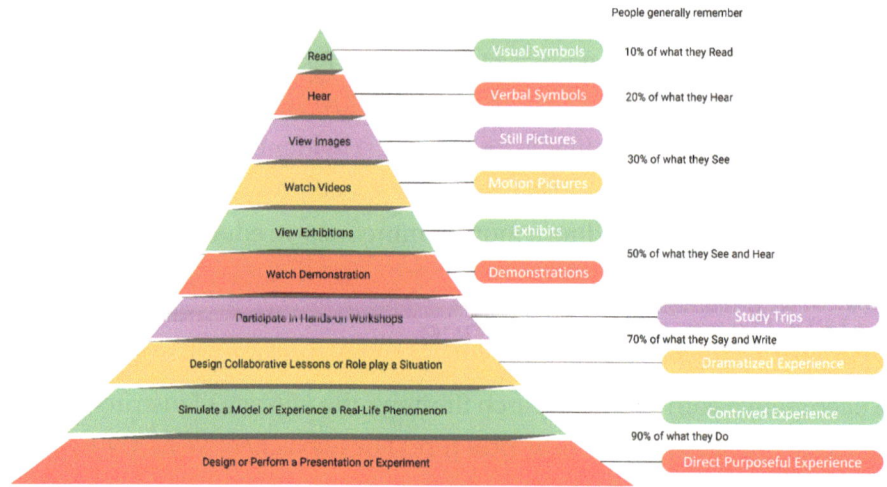

Figure 7-2. *Dale's Cone of Experience*

Table 7-2. *Outcomes of Each Level of the Cone of Experience*

Cone of Experience	Learners Will Be Able To
Read	Define
Hear	Describe
	Explain
	List
View images	Apply
Watch videos	Demonstrate
View exhibitions	Practice
Watch demonstrations	
Participate in hands-on workshops	Analyze
Design collaborative lessons or role play a situation	Create
Stimulate a model or experience a real-life phenomenon	Design
Design or perform a presentation or experiment	Evaluate

Bloom's Taxonomy of the Cognitive Domain

To design effective learning experiences, it is essential to select appropriate action words carefully. The initial group of words falls within the cognitive domain, which concentrates on cultivating mental skills and expanding the learner's knowledge base (Figure 7-3). This domain comprises six distinct categories: knowledge, comprehension, application, analysis, synthesis, and evaluation. These categories represent a continuum of increasing cognitive complexity – from knowledge to evaluation.

Commencing with knowledge, the focus is assessing the learner's ability to recall data or information. Moving forward, comprehension scrutinizes whether the learner can demonstrate an understanding of the acquired knowledge by articulating it in their own words, such as summarizing a theory.

Application follows, evaluating the learner's capability to apply acquired knowledge in novel situations. This aspect emphasizes the practical utilization of theoretical concepts in real-life scenarios.

The analysis category is designed to differentiate between facts and opinions. Synthesis, on the other hand, highlights the learner's capacity to amalgamate disparate elements or concepts to construct a coherent pattern or structure, thereby establishing a new meaning.

Lastly, the evaluation category gauges the learner's proficiency in forming judgments regarding the significance of concepts.

The "action words" explain how people think and deal with knowledge. A learning objective statement has a verb (an action) and an object (usually a noun).

- The verb usually shows [actions connected with] the intended cognitive process.

- The object normally describes the knowledge learners are supposed to get or put together.

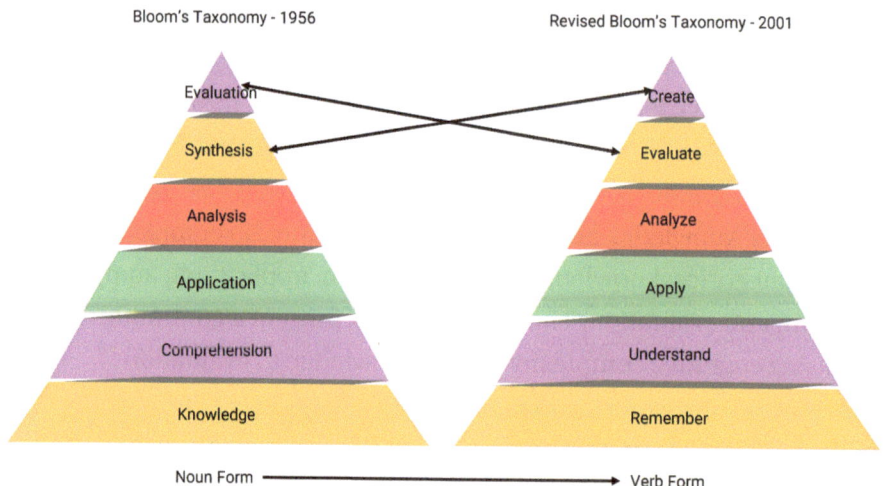

Figure 7-3. *Comparison of Original and Revised Bloom's Taxonomy*

According to Anderson and Krathwohl, the two dimensions of the revised taxonomy are the knowledge dimension and the cognitive process dimension, which represent the type of knowledge to be learned and the cognitive process to be used to acquire that type of knowledge, respectively.

The cognitive process dimension (Figure 7-4) delineates a spectrum of ascending cognitive complexity, from the basic act of remembering to the more advanced stage of creating. Anderson and Krathwohl elaborate on this dimension by identifying 19 distinct cognitive processes that provide additional clarity within the six categories.

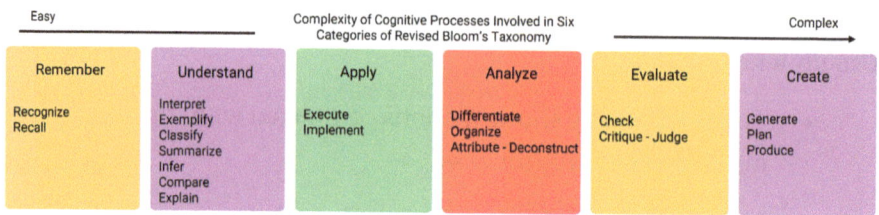

Figure 7-4. *Complexity of Cognitive Process*

In this updated framework, knowledge forms the foundation for six cognitive processes. The authors introduced a distinct taxonomy to categorize the various types of knowledge applied in cognition.

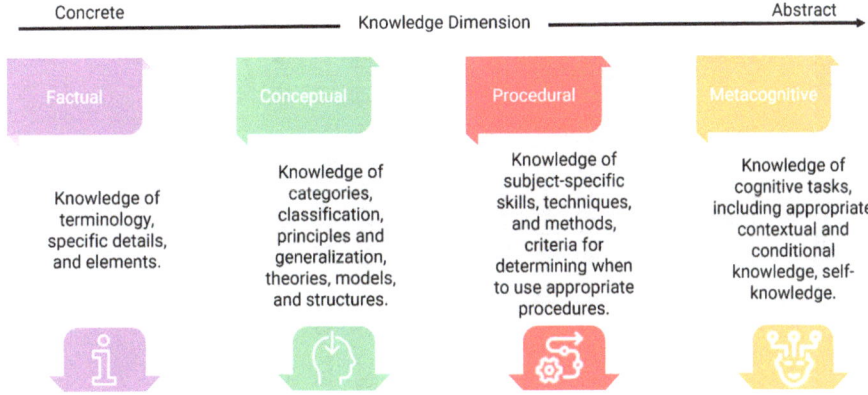

Figure 7-5. *Knowledge Dimensions*

Notably, the primary distinctions do not reside in mere listings, changes from nouns to verbs, or the renaming of components, nor do they involve repositioning the last two categories. The significant differences are evident in the practical and comprehensive additions, illustrating how the taxonomy intersects with and influences different types and levels of knowledge – factual, conceptual, procedural, and metacognitive (Figure 7-5). This integration is represented in Table 7-3 to understand how learning co-occurs at both knowledge and cognitive process levels.

Table 7-3. *Integration of Knowledge Dimensions with Bloom's Taxonomy*

Knowledge Dimension/ Cognitive Process Dimension	Factual	Conceptual	Procedural	Metacognitive
Remember Access pertinent information stored in long-term memory.	List the chemical formula of water.	Recognize the principles of supply and demand.	Recall the steps to assemble a primary circuit.	Identify techniques for improving critical thinking skills.
Understand Create understanding from instructional messages, encompassing verbal, written, and visual communication.	Outline key characteristics of a recently discovered species.	Categorize programming languages based on their primary functions.	Explain the steps to troubleshoot a computer network.	Anticipate personal reactions to a challenging ethical dilemma.
Apply Execute or apply a procedure in a specific situation.	Address commonly raised concerns about product features.	Offer guidance to beginners on effective marketing strategies.	Conduct pH tests on various soil samples.	Employ strategies tailored to one's individual strengths.
Analyze Deconstruct the material into fundamental components and ascertain the interconnections and overall structure or purpose of each part.	Determine the most exhaustive compilation of project tasks.	Differentiate characteristics of classical and contemporary literature.	Incorporate adherence to regulatory standards.	Examine and break down personal biases.

Evaluate Formulate assessments guided by established criteria and standards.	Verify coherence across various reference materials.	Assess the significance of research findings.	Assume a stance on the effectiveness of data collection methods.	Contemplate one's development and achievements.
Create Assemble elements to create a cohesive entity; rearrange them into a novel pattern or structure.	Develop a detailed record of monthly expenses.	Establish a committee of subject matter experts.	Architect a streamlined system for customer query resolution.	Build a reflective journal to track personal learning insights.

405

Implementation of Bloom's Taxonomy

First Method: Gershon recommends a practical strategy for implementing Bloom's Taxonomy by dividing it into distinct sections, making it more manageable for educators and instructional designers. In the cognitive domain, he categorizes levels into remember and understand, apply and analyze, and evaluate and create. Following is an example of how to implement this method.

Description: Learners explore the nuanced concept of symbolism in literature. They delve into using symbols as literary devices to convey deeper meanings, emotions, or themes in various forms of written expression, such as novels, poems, and short stories. Through engaging discussions and analyses, learners understand how authors employ symbolism to enhance the richness and complexity of their works, fostering a deeper appreciation for the artistry and interpretation involved in literary expression.

Remember and Understand (Segment One)

- Individually, learners list symbols commonly found in literature.

- Summarize their understanding of how symbolism functions in writing.

- Read an excerpt with symbols, rephrasing it to convey personal interpretation.

- Share identified symbols and interpretations with a partner.

Apply and Analyze (Segment Two)

- Use knowledge to explain how symbolism enhances a specific literary work.

- Interpret a poem using their understanding of symbolic elements.

406

- Analyze a short story's use of symbolism, explaining its impact.

- Examine provided sources on literary symbolism and extract key insights.

Evaluate and Create (Segment Three)

- Evaluate the strengths and weaknesses of different symbolic interpretations.

- Write a report on a literary work, focusing on the effectiveness of its symbolism.

- Craft a short creative piece integrating symbolism.

- Develop a solution to a literary challenge, considering symbolic elements.

Second Method: As an alternative, Crowe suggests dividing Bloom's Taxonomy into lower-order cognitive skills (LOCS) and higher-order cognitive skills (HOCS). LOCS encompasses the first three levels (remember, understand, and apply), while HOCS includes the subsequent three levels (analyze, evaluate, and create). This approach is particularly useful in aligning assessments with learning activities, especially in college-level Biology classes. Following is an example of how to implement this method.

Description: In this biology lesson, learners delve into the fundamental biological process of cellular respiration. They explore the intricate series of metabolic reactions occurring within cells to produce energy through adenosine triphosphate (ATP). Throughout the lesson, learners examine cellular respiration's key stages and principles, understanding its crucial role in sustaining life and providing the energy necessary for various cellular activities. Through practical examples and analyses, learners gain insights into the efficiency, significance, and adaptations related to cellular respiration in diverse organisms.

Lower-Order Cognitive Skills (LOCS)

- Recall and list the stages of cellular respiration.

- Summarize the main principles behind each stage in writing.

- Read through information on cellular respiration, rephrasing it for clarity.

- Share their knowledge about cellular respiration with a partner.

- Apply knowledge to explain the importance of cellular respiration in energy production.

Higher-Order Cognitive Skills (HOCS)

- Analyze a hypothetical cellular respiration scenario and predict outcomes.

- Evaluate the efficiency of different organisms' cellular respiration processes.

- Analyze and interpret various sources discussing cellular respiration.

- Evaluate the favorable and unfavorable aspects of different energy production pathways.

- Write a comprehensive report on cellular respiration, emphasizing its significance.

- Create a model representing cellular respiration processes creatively.

- Propose a solution to enhance cellular respiration efficiency in specific conditions.

Krathwohl Taxonomy of Affective Domain

Many individuals commonly associate learning with intellectual or cognitive functions. However, it extends beyond mental processes to encompass attitudes, behaviors, and physical skills. The affective domain delves into our feelings, emotions, and attitudes, forming a hierarchical structure ranging from simpler to more complex sentiments. This hierarchical arrangement is underpinned by the principle of internalization, signifying the transformation of one's affect toward something from a general awareness to a point where it consistently guides and controls behavior. As complexity increases, individuals become more engaged, committed, and internally motivated.

The inception of the affective domain dates back to 1964, delineating skills and behaviors aligned with attitudes and values. Progression through its levels leads to self-reliance and internal motivation. Although learning objectives aligned with the affective domain may initially seem challenging to articulate and assess, they often reflect outcomes closely linked to profound thinking and lifelong learning.

The affective domain comprises five levels, progressing from the lowest to the highest (Figure 7-6):

- **Receiving:** Becoming aware of an idea opposing existing attitudes through exposure, listening, or reading.

- **Responding:** Taking steps, albeit reluctantly, toward the new idea, such as talking about it or adjusting one's structure.

- **Valuing:** Advocating for new ideas and expressing support in arguments or debates.

- **Organization**: Examining and balancing existing and new ideas to facilitate integration.

- **Characterization:** The final step involves internalizing the new attitude, value, or belief, transforming the individual into a fervent proponent or advocate of the new idea.

Figure 7-6. *Krathwohl Taxonomy of Affective Domain*

Implementation of Krathwohl Taxonomy

Description: In a high school environmental science class, the teacher aims to cultivate a sense of environmental responsibility and appreciation for nature among learners. The specific objective is to have learners recognize the importance of sustainable and long-lasting practices in their daily lives.

Implementation of Five Levels

- **Receiving:** Begin with an engaging documentary on environmental issues, ensuring learners are exposed to the realities of ecological challenges.

- **Responding:** Facilitate class discussions where learners express their initial thoughts and emotions regarding environmental problems, encouraging open dialogue.

- **Valuing:** Design group activities where learners collaboratively brainstorm and prioritize sustainable actions, fostering a shared sense of responsibility.

- **Organizing:** Challenge learners to create personal action plans, organizing and prioritizing eco-friendly practices they commit to adopting.

- **Characterizing:** Conclude the unit with reflective assignments, asking learners to analyze how their new values manifest in their behavior and choices beyond the classroom.

Outcome: Learners internalize a deep appreciation for the environment, developing a value system that guides their actions and promotes sustainable behaviors in their daily lives.

Psychomotor Domain

In the traditional context, the psychomotor domain and its associated objectives are concerned with the tangible encoding of information. This involves engaging in movements and activities employing gross and fine muscles to express or interpret information and concepts effectively. Within the psychomotor domain, two distinct types of movements are recognized: conscious and subconscious.

Conscious physical movement encompasses the coordination of movements and motor skills, with the evaluation of development based on factors such as speed, precision, distance, procedural proficiency, and technical execution. This aspect requires intentional and calculated efforts to refine and enhance one's motor skills and coordination.

On the other hand, subconscious movements within the psychomotor domain refer to innate autonomic responses or reflexes. These natural bodily reactions occur without conscious thought, emphasizing the automatic and instinctive nature of certain motor responses.

This taxonomy is highly valuable for skill development, demanding dedicated and focused practice to elevate proficiency levels. By honing conscious and subconscious movements, individuals can achieve a comprehensive mastery of motor skills, contributing to their overall competence and effectiveness in various activities.

Simpson's Taxonomy of Psychomotor Domain

Elizabeth Simpson's psychomotor taxonomy (1972) builds upon the works of Bloom, Krathwohl, and others. It centers on motor skill utilization and coordination, emphasizing skill mastery progression from observation to invention. Harrow's and Simpson's psychomotor domains prove especially beneficial for children, young individuals, and adults aiming to develop skills beyond their comfort zones. Dave's psychomotor domain stands out for its simplicity and applicability in corporate development.

Simpson's psychomotor taxonomy encapsulates seven progressive levels, allowing educators and instructional designers to guide individuals through the developmental stages of acquiring and mastering motor skills effectively (Figure 7-7).

- **Perception**
    - **Definition:** Applying sensory information to motor activity.
    - **Example:** A mechanic adjusts the pressure on a wrench based on the feel to tighten a bolt securely.

- **Set**

  - **Definition:** Demonstrating readiness to act.

  - **Example:** A basketball player displays eagerness and focus before attempting a free throw.

- **Guided Response**

  - **Definition:** Imitating a displayed behavior or using trial and error.

  - **Example:** A guitarist follows a tutorial video to learn a new chord progression.

- **Mechanism**

  - **Definition:** Converting learned responses into habitual actions with proficiency and confidence.

  - **Example:** A cyclist smoothly shifts gears without hesitation after practicing for an extended period.

- **Complex Overt Response**

  - **Definition:** Skillfully performing complex patterns of actions.

  - **Example:** A pianist flawlessly plays an intricate musical piece, seamlessly transitioning between challenging passages.

- **Adaptation**

  - **Definition:** Modifying learned skills to meet special events.

  - **Example:** A chef creatively adjusts a recipe based on the availability of ingredients without compromising flavor.

413

- **Origination**

  - **Definition:** Creating new movement patterns for a
    specific situation.

  - **Example:** A martial artist develops a unique
    and effective technique tailored for countering a
    specific opponent's style.

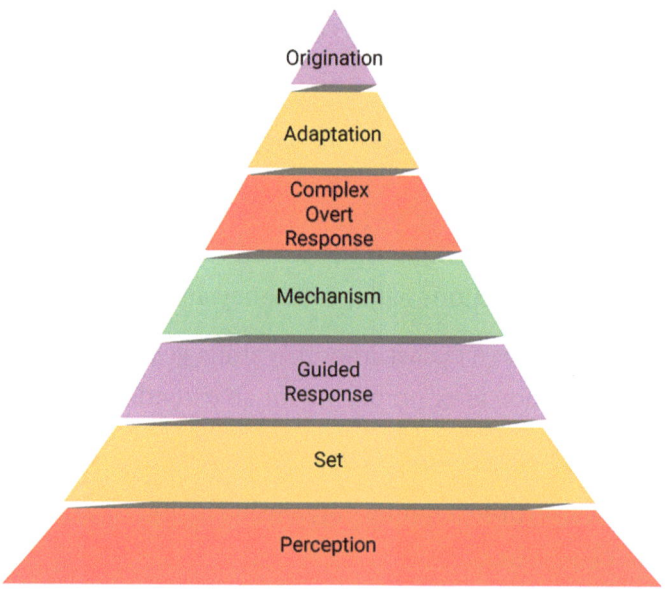

Figure 7-7. *Simpson's Taxonomy of Psychomotor Domain*

Implementation of Simpson's Psychomotor Taxonomy

Description: Building carpenter skills and developing proficiency in
constructing a wooden bookshelf.

- **Perception:** Identify the characteristics of each wood
  type, understand tool functions, and recognize safety
  protocols.

- **Set:** Establish mental and physical readiness by organizing tools, selecting the wood, and preparing the workspace.

- **Guided Response:** Imitate the demonstrated techniques, such as measuring and cutting wood, under the supervision of an experienced carpenter.

- **Mechanism:** Convert learned responses into habitual actions by repetitively assembling basic structures, gaining confidence and proficiency.

- **Complex Overt Response:** Execute complex actions, including precise measurements, intricate joinery, and finishing details, demonstrating high proficiency.

- **Adaptation:** Modify techniques and approaches based on unforeseen issues, such as adjusting measurements or troubleshooting structural problems.

- **Origination:** Create a customized bookshelf by introducing innovative design elements, showcasing an ability to originate new movement patterns for a specific carpentry project.

Dave's Taxonomy of Psychomotor Domain

Dave's taxonomy of the psychomotor domain, developed by R. H. Dave in 1975, is a classification system that outlines a hierarchy of physical skills and abilities. Unlike Bloom's cognitive domain, which focuses on thinking and intellectual skills, Dave's psychomotor domain is centered around physical actions, movements, and manual skills. The taxonomy is structured linearly, representing the stages of skill development from simple to complex. It emphasizes acquiring physical proficiency through practice, repetition, and refinement of motor skills. Dave's taxonomy is often used in educational settings, particularly in areas that require hands-on learning, such as vocational training and technical education. Due to its pragmatic orientation, it is recognized as generally easier to apply in corporate development environments.

Dave's taxonomy encompasses various stages of skill development (Figure 7-8):

- **Imitation**

    - **Definition:** Observing and replicating someone else's actions.

    - **Example:** Learning to play a musical instrument by watching online tutorials and replicating the finger movements demonstrated by an expert musician.

- **Manipulation**

    - **Definition:** Guided by instruction to perform a specific skill.

    - **Example:** Following a cooking recipe with step-by-step instructions to prepare a specific dish, guided by explicit directions and visual aids.

- **Precision**

    - **Definition:** Achieving accuracy, proportion, and exactness in skill performance without relying on the original source.

    - **Example:** A skilled archer consistently hits the bullseye on a target after extensive practice, achieving accuracy and exactness in arrow placement.

- **Articulation**

    - **Definition:** Combining two or more skills in a sequenced and consistent manner.

    - **Example:** A skilled dancer seamlessly combining various dance movements, choreographed in a routine that flows smoothly from one step to the next.

- **Naturalization**

    - **Definition:** Combining two or more skills sequentially and consistently performed effortlessly and automatically.

    - **Example:** A professional typist effortlessly typing on a keyboard, combining accuracy, speed, and fluidity without conscious thought or effort.

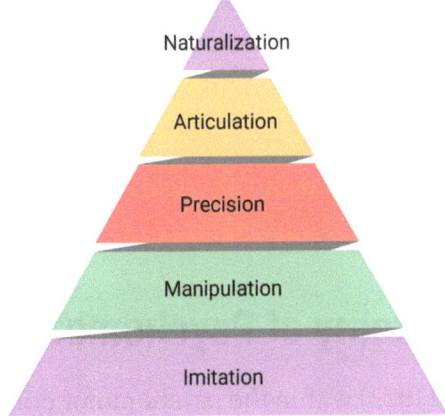

Figure 7-8. *Dave's Taxonomy of Psychomotor Domain*

Implementation of Dave's Psychomotor Taxonomy

Description: Developing culinary skills and mastering the technique of julienning vegetables.

- **Imitation:** Observe and replicate the chef's precise knife movements and hand placements.

- **Manipulation:** Follow step-by-step instructions and guidance to practice the julienne technique under the instructor's supervision.

- **Precision:** Refine the technique to achieve consistent and precise cuts without relying on external guidance, ensuring uniformity in the size and shape of the julienned vegetables.

- **Articulation:** Combine the julienne technique seamlessly with other culinary skills, such as sautéing and plating, to create a well-coordinated and visually appealing dish.

- **Naturalization:** Execute the julienne technique effortlessly and automatically, demonstrating mastery with minimal mental and physical exertion, even in a fast-paced culinary setting.

Harrow's Taxonomy of Psychomotor Domain

Harrow's taxonomy (1972) within the psychomotor domain is designed to foster the development of physical fitness, dexterity, agility, and precise body control, aiming for a high level of expertise. The taxonomy is meticulously organized based on the degree of coordination, encompassing both involuntary responses and acquired capabilities. Commencing with simple reflexes, it progresses through a continuum of increasingly sophisticated and expressive movements that demand refined coordination and precision.

418

Harrow (1972) presents her model without tying it to a specific, overarching criterion but rather seeks a critical order. Mastery at each lower level is imperative for the subsequent ascent in the hierarchy of movements. Particularly beneficial for educators in physical education, Harrow's taxonomy offers valuable insights, with Level 3 being notably relevant for preschool and elementary school teachers. This level includes a comprehensive battery for assessing learners' perceptual abilities, diagnosing challenges, and proposing tailored remedial exercises.

The taxonomy's foundational levels begin with simple reflexes, while the apex comprises intricate neuromuscular coordination, reflecting the spectrum of movement complexities. The following are the six levels of Harrow's taxonomy (Figure 7-9):

- **Reflex Movements**

    - **Definition:** Reflex movements encompass actions that occur instinctively, without the need for learning, in response to specific stimuli. Notable examples of reflex movements include flexion, extension, stretch, and postural adjustments. These involuntary responses form the foundation for more complex motor skills.

    - **Example:** Reflex movements, such as the automatic withdrawal of a hand from a hot surface or the automatic adjustment of posture to maintain balance, illustrate the innate nature of these responses.

- **Basic Fundamental Movements**

    - **Definition:** These movements represent inherent patterns resulting from the amalgamation of reflex movements. These serve as the building blocks for intricate and skilled motor activities. Fundamental movements include walking, running, pushing, twisting, gripping, grasping, and manipulating.

419

- **Example:** The coordinated combination of reflex movements, like postural adjustments and limb extensions, contributes to fundamental movements such as walking or the manipulation of objects, forming the essential groundwork for more sophisticated motor skills.

- **Perceptual Skills**

  - **Definition:** These skills pertain to interpreting diverse stimuli, enabling individuals to make adjustments within their environment. This category encompasses visual, auditory, kinesthetic, or tactile discrimination, suggesting a fusion of cognitive and psychomotor behavior. Examples include coordinated movements like jumping rope, punting, or catching, where perceptual understanding is essential.

  - **Example:** Engaging in activities that demand precise spatial awareness, such as catching a ball or executing a complex dance routine, showcases the integration of perceptual skills into coordinated physical movements.

- **Physical Activities**

  - **Definition:** These activities demand endurance, strength, vigor, and agility, contributing to developing a robust and efficiently functioning body. Examples encompass activities requiring strenuous effort for extended periods, muscular exertion, wide-ranging motion at the hip joints, and quick, precise movements.

- **Example:** Endurance sports, weightlifting, and activities necessitating rapid and precise hip movements exemplify physical activities that foster overall bodily strength, endurance, and agility.

- **Skilled Movements**

  - **Definition:** These movements arise from the mastery of complex tasks, demonstrating a high degree of efficiency in execution. These movements are evident in various domains, such as sports, recreation, and dance, reflecting a refined level of proficiency.

  - **Example:** Accomplished performances in sports, intricate dance routines, and skilled recreational activities exemplify the culmination of learning and practice, resulting in highly efficient and aesthetically pleasing skilled movements.

- **Nondiscursive Communication**

  - **Definition:** This involves conveying messages through bodily movements, from facial expressions to sophisticated choreography. Examples encompass body postures, gestures, and facial expressions, especially pronounced in expertly executed dance movements and choreographies.

  - **Example:** In dance, the fluency of nondiscursive communication is evident as skilled performers use body postures, gestures, and facial expressions to convey emotions and narratives without verbal communication.

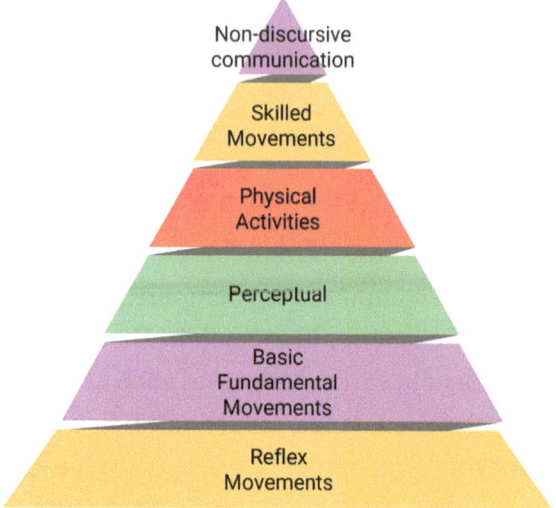

Figure 7-9. *Harrow's Taxonomy of Psychomotor Domain*

Implementation of Harrow's Psychomotor Taxonomy

Description: Advancing yoga proficiency and mastery in a complex yoga posture, such as the Crane Pose.

- **Reflex Movements:** Participants instinctively react to the instructor's guidance, attempting to lift their feet off the mat and balance on their hands.

- **Basic Fundamental Movement:** Learners engage in simplified preparatory exercises, focusing on building strength in the arms and core and mastering the basic mechanics required for the Crane Pose.

- **Perceptual:** Practitioners refine the Crane Pose by adjusting hand placement and distributing body weight based on instructor cues and their perceptual awareness.

- **Physical Activities:** Participants engage in a series of physically demanding poses, enhancing endurance, strength, and agility as prerequisites for sustaining the Crane Pose.

- **Skilled Movements:** Learners gradually refine the Crane Pose, achieving a level of efficiency in the execution, emphasizing precision, balance, and controlled movements.

- **Nondiscursive Communication:** Practitioners embody the Crane Pose with fluidity and grace, utilizing body language as a form of nondiscursive communication to convey strength, balance, and mastery in the yoga practice.

Using Taxonomies

Creating meaningful learning objectives involves strategically using educational taxonomies like Bloom's, Krathwohl's, Simpson's, Dave's, and Harrow's. These frameworks guide precise objective formulation, ensuring comprehensive learning experiences. In parallel, differentiated instruction thrives on adapting to individual needs, with taxonomies offering a road map to tailor lessons, fostering an inclusive and engaging educational environment.

Craft Learning Objectives

Instructional designers and educators can utilize the discussed taxonomies to articulate precise learning objectives, describing what learners should achieve post-educational activity completion. A well-structured objective details knowledge, skills, or attitudes, ensuring observability and measurability.

To create impactful learning objectives using these taxonomies, the following are the essential steps:

- **Select Domain:** Choose the cognitive, affective, or psychomotor domain that aligns with the nature of the learning objective you aim to design.

- **Determine Complexity Level:** Clearly define the complexity level at which you want learners to showcase their learning. Each taxonomy provides a structured framework for identifying the expected depth of understanding or proficiency.

- **Choose Action Verb:** Align the learning objective with an action verb corresponding to the chosen domain and complexity level. This step ensures that the objective is both focused and measurable, clarifying what learners should be able to do. Refer to Appendix C for action verbs for all taxonomies.

Differentiated Instructions

Differentiated instruction is a pedagogical approach that tailors lessons to accommodate the varied needs, interests, strengths, and abilities of learners in the classroom. Gershon's methodology divides lesson objectives into three distinct outcomes:

- All learners will be able to...

- Most learners will be able to...

- Some learners will be able to...

Strategically aligning these outcomes with varying levels of each taxonomy ensures differentiation in cognitive, affective, and psychomotor domains. Table 7-4 presents the example from each taxonomy.

Table 7-4. Differentiation Instructions

Taxonomy/ Outcomes	Bloom's Taxonomy (Cognitive)	Krathwohl's Taxonomy (Affective)	Simpson's Taxonomy (Psychomotor)	Dave's Taxonomy (Psychomotor)	Harrow's Taxonomy (Psychomotor)
All	**Understand** All learners will comprehend the foundational concepts.	**Receive** All participants will receive information about cultural diversity and its significance in fostering a global perspective.	**Perceive** All learners will perceive the key elements and techniques of basic procedural skills, such as understanding the components of a swimming stroke or culinary technique.	**Imitate** All learners will imitate the basic elements of procedural skills, such as following culinary techniques.	**Reflex Movement** All learners will exhibit reflex movements, such as automatic adjustments ir running form or immediate reactions in woodworking to maintain balance.

(*continued*)

425

Table 7-4. (*continued*)

Taxonomy/ Outcomes	Bloom's Taxonomy (Cognitive)	Krathwohl's Taxonomy (Affective)	Simpson's Taxonomy (Psychomotor)	Dave's Taxonomy (Psychomotor)	Harrow's Taxonomy (Psychomotor)
Most	**Apply** Most learners will apply acquired knowledge to solve problems.	**Respond** Most individuals will respond constructively to feedback, adjusting their behavior accordingly.	**Guided Response** Most learners will engage in guided responses, actively following instructions and feedback to pioneer innovative dance moves.	**Articulation** Most participants will articulate their skills by combining various elements and showcasing advanced culinary techniques in a coordinated and expressive manner.	**Perceptual** Most learners will focus on perceptual skills, demonstrating advanced awareness and discrimination in activities like adapting running techniques to different terrains or adjusting woodworking methods for various materials.

Some	Evaluate	Characterization	Origination	Naturalization	Nondiscursive Communication
	Some learners will critically evaluate information with reference to evidence, examples, and reasons.	Some employees will characterize themselves as adaptable, embracing change enthusiastically in the workplace.	Some participants will originate new approaches and techniques, pioneering original dance moves	Some individuals will reach the stage of naturalization by executing complex and nuanced movements effortlessly, whether in advanced athletic performance or intricate woodworking.	Some learners will engage in nondiscursive movements, efficiently using body postures or gestures in dance routines without the need for verbal communication.

427

SMART Learning Objectives

Effective learning objectives require addressing key questions: Who will perform the task, how much or how well, of what, and by when? Start by identifying content areas for participants to learn. Select measurable and observable action verbs that precisely describe desired learner performance. Opt for higher-order verbs, avoiding vague terms like "understand" or "know." Ensure each objective stands alone, avoiding the combination of separate actions or topics. Clearly specify the conditions under which the action will occur, commencing statements clearly indicating expected post-activity participant capabilities. An example of a good way to begin the statement is: "By the end of the course/module/section/topic/learning activity, learners/participants/employees should be/will be able to ...."

Employ the SMART method to craft effective learning objectives (Figure 7-10). This method serves as a quick and straightforward checklist while constructing useful objectives.

Figure 7-10. *SMART Learning Objectives*

- **Specific:** What action to perform after learning and by whom? Use simple, direct language to tell learners exactly what to learn and perform after training. Avoid vagueness or confusion.

- **Measurable:** How do you measure the success of learning? Set objectives that can be observed and measured objectively. It should quantify the amount of change expected in learners after learning. Avoid using vague terms like "know" or "understand." Ensure any observer can agree on whether the objective is met.

- **Attainable:** Can the learners achieve the objective with available resources in a given time frame? Ensure learners have the knowledge and resources to complete the objective. Avoid setting tasks that are too challenging or too easy.

- **Relevant:** Are the objectives aligned with the instructional methods/learning activities and assessment? Teach valuable skills that matter to learners. Ensure the training is meaningful and applicable to their needs, especially in job-related scenarios.

- **Time Based:** When will the objective be achieved, and when can it be applied successfully? Set objectives that learners will use soon – tomorrow or next week – not something they won't need until much later. Avoid training on things they'll forget before applying them on the job.

Example

SMART Learning Objective Template

By ___(time)___, the ___(audience)___ will ___(verb=performance)___ as measured by ___(assessment/measure + standard/criteria)___.

Before

Develop a marketing strategy.

After

By the end of the quarter, marketing team members will create and present a comprehensive digital marketing strategy, as evaluated through the successful implementation of targeted advertising campaigns and measurable audience engagement metrics.

Figure 7-11. *Example: SMART Objectives*

Explanation

The initial objective, "Develop a marketing strategy," lacks specificity and measurability. In contrast, the SMART version provides clarity by outlining the desired outcome – a comprehensive digital marketing strategy. It sets a specific time frame (end of the quarter). It includes measurable criteria, such as the successful implementation of targeted advertising campaigns and measurable audience engagement metrics, enhancing accountability and assessment capabilities (Figure 7-11).

Crafting objectives without clear, measurable standards can create confusion about what to study or the expected performance. SMART objectives offer clarity; guide educators in defining, teaching, assessing, and providing meaningful feedback; and shape the learning session's scope, instructional method, and assessment quality.

Summary

- There are three learning domains: cognitive domain focuses on intellectual processes and critical thinking; affective domain focuses on emotions, attitudes, and values; and psychomotor domain focuses on physical skills and coordination. All three domains collectively foster comprehensive learning outcomes.

- Dale's cone posits a hierarchy of instructional methods, ranging from least effective (verbal symbols) to most effective (hands-on experiences). Direct, purposeful learning, such as action-learning techniques, enhances retention rates, emphasizing the importance of engaging learners in real-life experiences.

- Crafting effective learning objectives involves the ABCD method: audience, behavior, conditions, and degree of mastery. Educators create measurable outcomes that enhance overall educational effectiveness by identifying knowledge, aligning with higher-level goals, choosing precise action verbs, and ensuring consistency.

- Bloom's Taxonomy is a hierarchical model categorizing cognitive skills from basic to complex levels, aiding educators in designing effective learning objectives.

- Revised Bloom's Taxonomy is the updated version that enhances the original taxonomy, focusing on verbs and clarifying language.

- The affective domain in learning, outlined by Krathwohl's taxonomy, encompasses attitudes and behaviors, progressing through five levels.

- The psychomotor domain focuses on developing motor skills, coordination, and physical abilities. It emphasizes integrating cognitive and physical processes for effective learning and performance.

- Simpson's taxonomy proposes a hierarchical structure for psychomotor skills, categorizing them into simple to complex levels. It is divided into seven levels: perception, set, guided response, mechanism, complex overt response, adaptation, and origination.

- Dave's taxonomy stresses the importance of practice and refinement in achieving competence in psychomotor skills. It focuses on the acquisition of skills through a process of imitation, manipulation, precision, articulation, and naturalization.

- Harrow's taxonomy defines stages of motor skill development, emphasizing the progression from reflex movements to skilled activities, and includes reflex movements, basic fundamental movements, perceptual skills, physical activities, and skilled movements.

- Craft SMART objectives by clearly defining specific, measurable, achievable, relevant, and timely goals, ensuring practicality and alignment with learner needs.

Let's Brainstorm

These mini-scenarios will make you scratch your head and scribble on your pad.

Mini-scenario 1

Imagine you are a biology teacher designing a unit on photosynthesis for high school learners. Considering the cognitive domain, how would you formulate a SMART learning objective for your learners related to understanding the process of photosynthesis?

Mini-scenario 2

As a school counselor, you aim to enhance learners' empathy and understanding of diverse cultures. How would you frame a SMART learning objective, based on Krathwohl's affective domain, to foster a more inclusive and culturally sensitive school environment?

Mini-scenario 3

As a martial arts instructor, how would you design a SMART learning objective for your learners to demonstrate the precise execution of complex karate kata, incorporating advanced techniques and stances, within a specified time frame during their upcoming grading session?

Mini-scenario 4

In a woodworking class, you want learners to refine their skills in crafting intricate joints. According to Dave's taxonomy, how would you frame a SMART learning objective for learners to manipulate tools and materials with precision, ensuring the creation of flawlessly articulated joints in their final project within a set time frame?

Mini-scenario 5

As a teacher working with young children with special abilities, your objective is to help them develop foundational movement skills. To create a differentiated SMART instruction using Harrow's taxonomy, design individualized activities that cater to each child's unique strengths and needs. For instance, set a goal for a child with enhanced spatial awareness to balance on one foot while incorporating personalized adaptations, ensuring achievement within an appropriate time frame based on their capabilities and progress.

CHAPTER 8

Storyboarding Fundamentals

Now equipped with an understanding of the science behind instructional design, the role of instructional design (ID) models, and the crafting of SMART objectives to structure courses effectively, the next crucial step is consolidating these elements into a cohesive plan – a storyboard. The storyboard serves as a visual road map, combining the instructional design principles and objectives in a tangible format. Its historical roots trace back to the early days of filmmaking, with visionaries like Walt Disney playing a pivotal role in its popularization, particularly in animation. As the decades unfolded, storyboarding expanded its influence beyond film to encompass diverse industries, such as advertising, television, video games, and graphic design. The digital era further transformed the process, introducing computer software for more efficient and flexible storyboarding. Today, storyboarding is ubiquitous in education and business and is used for planning presentations, instructional materials, and multimedia projects. Its evolution reflects a dynamic synergy between traditional techniques and modern technological advancements, making it an indispensable component of the creative process across various fields.

© Ankita Jiyani Mangtani 2024
A. J. Mangtani, *Instructional Design Unleashed*, Design Thinking,
https://doi.org/10.1007/979-8-8688-0416-8_8

What Is a Storyboard?

Storyboarding in instructional design (ID) is the systematic process of mapping out the essential elements of a course following the guidelines of a specific instructional design model. Drawing a parallel to a film director's meticulous planning of each shot, angle, and detail through storyboards, the ID storyboard pre-visualizes the course's content. Instead of scenes, it delineates the structure, logical sequence of content, visual design, and interactivity of the proposed course. Embedded within storyboards are sketches, detailed descriptions, and notes encapsulating the course's content, learning objectives, and instructional strategies. Functioning as a blueprint for your learning initiative, it assists you and your instructional design team in mapping the learning journey, pinpointing potential gaps or redundancies, and ensuring comprehensive coverage of all learning objectives.

Anatomy of Storyboard

Mastering storyboard creation is a pivotal skill in instructional design, constituting an indispensable component of the overall instructional design process. By providing a comprehensive blueprint, including the following elements (Figure 8-1), instructional designers empower the development team to bring the envisioned learning experience to life.

- **Header/Introduction:** Incorporating project name, course name, lesson name, duration, and other crucial information related to the project, course, or lesson.

- **Slide/Screen Title and Number:** Adding subtopics of the project, course, lesson, and any other pertinent information to the slide/screen title and the number of slide/screen for the streamlined organization of storyboards.

- **On-Screen Text:** Presenting the full onscreen text, including learner instructions such as "Click NEXT to learn more." This transparency allows subject matter experts to envision the content precisely.

- **Graphic/Media Resources:** Including images or animations and their precise screen placement. To enhance clarity, consider using thumbnail sketches as visual aids instead of verbal descriptions.

- **Narration/Audio Script/ Voice-Over:** Integrating narration corresponding to each screen, accompanied by audio notes to provide additional context for each slide/screen of the storyboard.

- **Navigation/Interaction Instructions:** Detailing all possible learner actions and interactions with the screen, encompassing interactive elements like quizzes and diverse responses and explicitly outlining how learners progress from one page to the next, utilizing programming logic forms, such as if/then statements, whenever applicable.

- **Development/Programmer Notes/Comments:** Any specific information about a particular screen/ slide or interactivity to be mentioned in detail using the terminologies and style guides you use with the development or programmer team.

- **Reviewer Notes/Comments:** Any detail or additional explanation added or required to be added to the storyboard should be planned and communicated with other stakeholders under this section of the storyboard.

- **Accessibility Notes:** Provide details to make the content accessible to various learners by providing alt text for the graphic elements, a color contrast ratio, keyboard navigation, color palette, font size, and font face used according to branding guidelines.

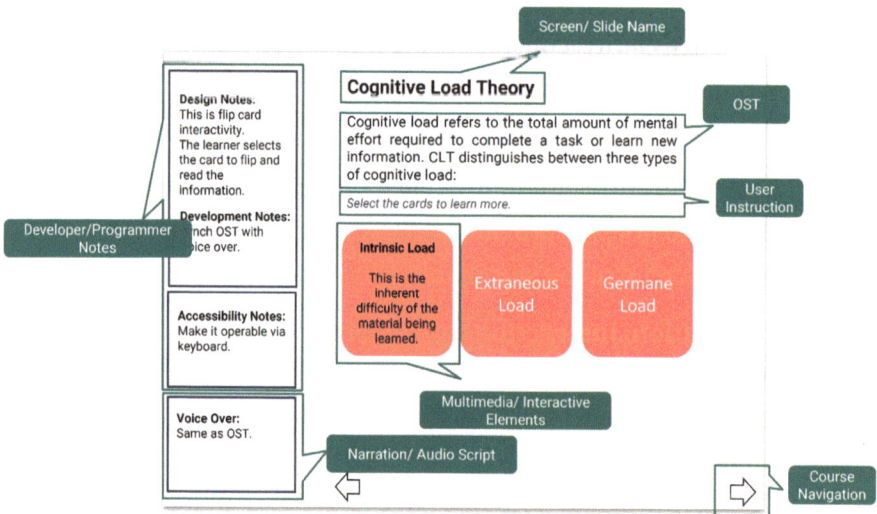

Figure 8-1. *Anatomy of Storyboard*

Benefits of Storyboard

Storyboarding is a crucial aspect of instructional design, offering stakeholders and subject matter experts a preview of the course's flow and content presentation. This visual blueprint facilitates the following:

- Effective communication between stakeholders and reviewers

- Provide opportunities to propose edits and adjustments before the development phase and keep track of changes within the storyboard

- Time-saving and averting trial-and-error scenarios

- Ensure adherence to research-backed best practices,
  simplifies production and post-production of
  the course

Types of Storyboards

The flexibility of instructional design storyboards lies in their adaptability
to diverse learning needs and preferences. Storyboards can be tailored
to accommodate various instructional strategies, delivery formats, and
multimedia elements. Choose a storyboard format that fits your project's
needs, considering visual elements, reviewer familiarity, technical
literacy, and interaction complexity. Tailor your storyboard to the specific
requirements of your project to ensure an effective and engaging learning
experience.

Micro Storyboard

Before crafting the final storyboard, it proves beneficial to formulate a
comprehensive course outline as a visual road map. This process helps
delineate the content flow, pinpoint optimal locations for interactive
elements, and establish the overall course layout (Figure 8-2).

- Commence the outline with a straightforward
  approach using notecards

- Begin by assigning a notecard for each primary topic
  and arranging them chronologically

- Subsequently, generate notecards for individual
  subtopics and place them beneath their corresponding
  main topics

- Conclude by incorporating notecards for the course introduction and conclusion. This phase allows for strategic decisions regarding quizzes, simulations, or other interactive elements

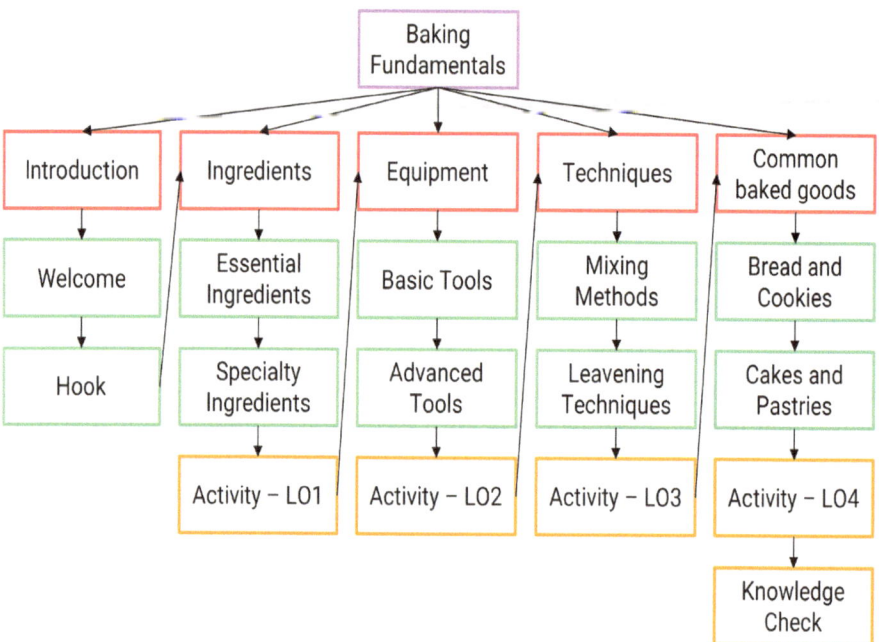

Figure 8-2. *Example of Micro Storyboard*

One notable advantage of the micro storyboard is its accessibility. Devoid of technical jargon and designed to be comprehensible without a background in instructional design (ID), it offers a universally applicable approach. This one-size-fits-all framework caters to SMEs, whether they possess extensive eLearning knowledge or are relatively new.

This condensed format has been carefully crafted to align seamlessly with the preferences of this audience. The first row serves as a quick reference point, highlighting the central theme of each corresponding slide sequence, which is then detailed in the green boxes. Simple and intuitive arrows guide SMEs through each course section, maintaining a transparent and efficient flow.

Each section culminates in a strategically placed formative assessment labeled as an "activity." These activities include one-question quizzes, matching games, or drag-and-drop scenarios. Significantly, each formative assessment is intricately tied to a specific learning objective (LO). This approach ensures that learners can gauge their understanding of the content by measuring their proficiency in achieving each learning objective.

Word-Based Storyboard

This storyboard style is centered around textual content and is typically generated through word processing tools such as Microsoft Word. It involves detailed written descriptions for every slide or scene, encompassing text, visuals, and any programming notes that may be relevant (Figure 8-3). This particular format is well suited for situations where the primary emphasis lies on the written content, with less focus on incorporating elaborate visual elements or interactive components.

Program/Course/Module Name: Baking Fundaments			
Section Name: Introduction			
Page No. - 01		Lesson/Topic No.	01
Page Title - Welcome		Lesson/Topic Name	Introduction about Baking

On-Screen Visuals/Text

Welcome to the world of baking, where the magic of transforming simple ingredients into delightful treats has a rich history. In this segment, we'll explore the historical significance and evolution of baking, uncovering the roots that have shaped the diverse art of baking we know today.

On-Screen Text	Animation Description
Introduction to Baking	Bring the element to the screen with a piece of music.

Graphics Description

Cover Image of the course: Baking.jpg

Interactivities

<A description of the Interactivities>
Navigation button "Launch Module"

Notes to Developers:
<Specific instructions to developers, if any>
The on-screen text will appear in synch with the VO.

Figure 8-3. *Example of Word-Based Storyboard*

PowerPoint-Based Storyboard

A PowerPoint-based storyboard visually represents your eLearning course, where each slide corresponds to a specific scene or screen. This format is particularly advantageous when you aim to visualize the course layout, incorporating graphics and planning animations (Figure 8-4). Additionally, it proves valuable for presenting the storyboard to stakeholders, facilitating discussions, and obtaining constructive feedback on the visual elements and course structure.

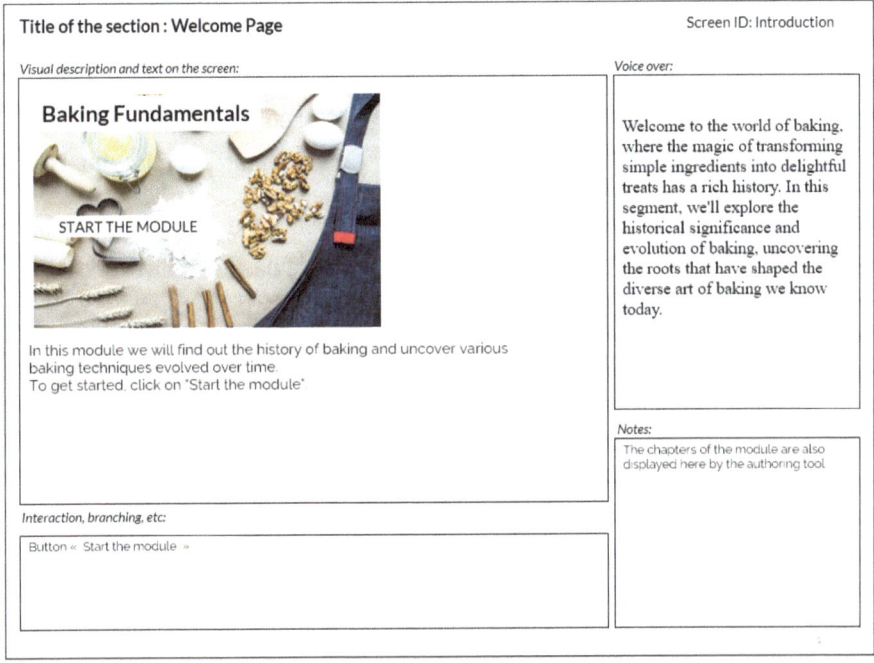

Figure 8-4. *Example of PowerPoint-Based Storyboard*

Prototype-Based Storyboard

A prototype-based storyboard transcends traditional static representations by providing an interactive experience. Utilizing software tools such as Adobe XD or Axure RP, this format enables the creation of dynamic and engaging prototypes. Reviewers can actively experience the design, interact with various elements, and gain a more realistic preview of the final course. Prototypes prove incredibly beneficial when aiming to showcase intricate interactions or assess the course's usability in a simulated environment before moving into the development phase.

The differences between all four types of storyboards are detailed in Table 8-1.

Table 8-1. *Differences Between Types of Storyboards*

Parameter	Micro Storyboard	Word-Based Storyboard	PowerPoint-Based Storyboard	Prototype-Based Storyboard
Format Type	Textual and concise	Textual, descriptive	Visual and slide based	Interactive and dynamic
Representation Style	Minimalist, focused on essential details	Narrative, detailed	Visual, focusing on layout, graphics, and animations	Interactive, allowing for user engagement and testing
Interactivity	Absent	Limited	Limited (may include animations)	Highly interactive, simulating user experience
Tool Usage	Created manually or using simple tools	Typically made using word processing software	Utilizes Microsoft PowerPoint or similar tools	Developed using prototyping software (e.g., Adobe XD)
Audience	Suitable for quick communication with SMEs	Suitable for detailed communication with stakeholders	Suitable for presenting to stakeholders and obtaining feedback	Suitable for demonstrating complex interactions and usability testing

(continued)

Table 8-1. (*continued*)

Parameter	Micro Storyboard	Word-Based Storyboard	PowerPoint-Based Storyboard	Prototype-Based Storyboard
Usability Testing	Not designed for detailed usability testing	Limited usability testing capabilities	Limited usability testing capabilities	Well suited for usability testing and user feedback
Visual Elements	Emphasis on essential visual cues and content	Primarily text based, little visuals	Focus on visual layout, graphics, and animations	Rich visual representation with interactive elements
Collaboration	Quick collaboration with SMEs or team members	Collaboration requires manual document sharing and updates	Collaboration may involve sharing PowerPoint files for feedback	Collaboration facilitated through prototype sharing and testing
Flexibility	Quick to create and modify	Can be time-consuming to update and modify	Easily editable, with changes reflected visually on slides	Allows for iterative changes and adjustments easily

Choosing Storyboard Format

Choosing the right storyboard format depends on various factors, including the project requirements, audience, level of detail needed, and the development team's and stakeholders' preferences. It is a common practice for instructional designers to employ a blend of these formats throughout various phases of the project to ensure a comprehensive and tailored approach. Refer to Appendix D for different storyboard formats and templates.

- **Project Complexity:** A micro or Word-based storyboard may suffice for simple projects, while complex projects with intricate interactions benefit from prototype-based storyboards.

- **Stakeholder Engagement:** If stakeholder engagement and feedback are crucial, PowerPoint or prototype-based storyboards may be preferable due to their visual and interactive nature.

- **Usability Testing:** If early usability testing is a priority, a prototype-based storyboard allows for simulating user interactions and refining the design before development.

- **Timeline and Resources:** Micro storyboards are quick to create, while prototype-based storyboards may require more time. Consider the project timeline and available resources.

Creating Effective Storyboards

Crafting compelling storyboards is the cornerstone of seamless content development. Storyboarding lays the foundation for a well-structured and engaging narrative for eLearning modules, film projects, or presentations. Figure 8-5 explores critical insights and strategies to create impactful storyboards that captivate and communicate precisely.

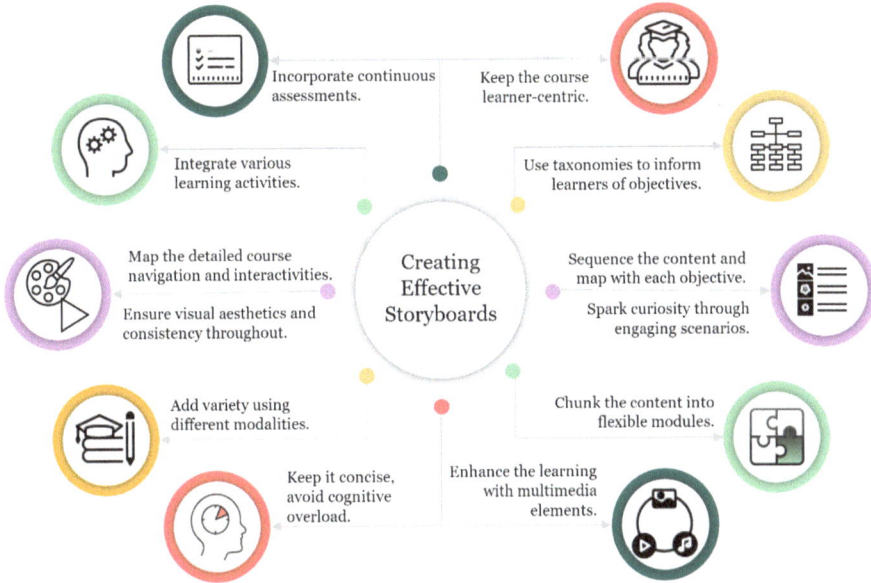

Figure 8-5. *Different Aspects of Creating Effective Storyboards*

Storyboarding Tools

The choice of storyboarding tools depends on the project's needs, the required collaboration level, and the preference for design elements or multimedia integration in the storyboarding phase. Four tools – Storyboard That, MakeStoryBoard, Canva, and Storyboarder – are scrutinized across different parameters to determine the optimal fit for project needs.

The analysis is typically organized and presented in three distinct tables. Table 8-2 addresses specifications of the tools. Table 8-3 summarizes the evaluation of the tools based on different parameters. Table 8-4 presents the SWOT analysis of the tools. The specifications table delves into the technical details of each tool, the evaluation table shifts focus to usability, and the SWOT analysis provides a strategic perspective, collectively serving as a comprehensive guide for stakeholders to make informed decisions based on a thorough understanding of each tool's strengths, weaknesses, opportunities, and threats.

> **Storyboard That:** Storyboard That is a browser-based Storyboard Creator tailored for crafting storyboards and graphic organizers. Featuring a drag-and-drop interface, it empowers users to design scenes using a customizable art library encompassing characters, scenes, and props. The platform offers six layout options to enrich visual storytelling.

> **MakeStoryBoard:** MakeStoryBoard is a visualization solution dedicated to helping users create dynamic storyboards. Adding images, sketches, and descriptions to communicate ideas effectively fosters collaboration by enabling team members to work together. Project-specific access permissions allow users to maintain privacy and control over editing, commenting, and feedback.

> **Canva:** Canva stands as a widely embraced online graphic design tool, providing a user-friendly and versatile platform for designing visual materials. It allows image importation, offers templates for banners and logos, and provides access to premium stock images and templates through paid options. Known for its simplicity and extensive design capabilities.

Storyboarder: Storyboarder, a free, open source desktop application, facilitates rapid story visualization. Users can swiftly create and showcase animatics, expressing story ideas without producing a full-fledged movie.

Table 8-2. *Storyboarding Tools Specifications*

Name of Tool	Type of Application	Subscription	Collaboration	Version Control	Output Format
Storyboard That	Browser based	Paid	Yes	No	PDF and images
MakeStoryboard	Browser based	Paid	Yes	Yes	PDF, video, and images
Canva	Browser based and desktop	Paid	Yes	No	PDF, video, and images
Storyboarder	Desktop application	Free	No	No	Zip, videos, PDFs, and images

Table 8-3. *Storyboarding Tools Evaluation*

Name of Tool	Key Features	Pros	Cons	Use Cases
Storyboard That	• Drag-and-drop • Extensive art library • Layout option	• User-friendly interface • Impressive performance • Flexibility of characters' expressions	• Limited editing capabilities • Dated visual styles • Lack of integration with other tools • Navigation challenge	• Education • Film and media production • Quick scenario creation
MakeStoryboard	• Dynamic visualization • Collaborative capabilities • Limited free plan	• Access control • Versioning and backup • PDF download • Data security	• PDF download may take a long time in some cases	• Software development • Education market campaign

Canva	• Extensive templates • Collaboration • Sharing on social media platforms directly • Download created storyboard	• Design library • Ready-made templates • Various layouts and orientation	• No search function for photos • Limited text editing options • Elements alignment issues	• Brand development and outreach • Book proposals • Education
Storyboarder	• Sleek interface • Six different drawing tools	• Free and easy to use for beginners • Export boards to another app/software	• Software lagging problem • Dated visuals • Time taking	• Media production • Education • Animated videos

Table 8-4. *Storyboarding Tools SWOT Analysis*

Name of Tool	Strength	Weakness	Opportunity	Threat
Storyboard That	Aligns with Cognitive Load Theory.	Limited 3D perspective options. Limited mobile optimization.	Integration with Learning Management Systems (LMS).	Evolving eLearning industry standards impacting tool relevance.
MakeStory board	Access controls for data security.	Limited resizing options for elements could be improved.	Integration with eLearning authoring tools. Enhanced compatibility with SCORM standards.	Increased competition from more feature-rich eLearning tools.
Canva	A variety of device-compatible templates for diverse creative projects.	Despite its extensive library, users may encounter limitations in fully customizing certain elements, restricting complete creative freedom for advanced designers.	Addressing user requests for additional eLearning features. Collaboration features improvement for eLearning teams.	Potential changes in eLearning design trends impacting Canva.

(*continued*)

Table 8-4. (*continued*)

Name of Tool	Strength	Weakness	Opportunity	Threat
Storyboarder	Auto-saving drawings enhance user convenience, preventing potential data loss during the creative process.	Being a desktop application may limit the tool's versatility, as users may prefer cloud-based solutions that offer cross-device compatibility.	Introducing collaborative features, such as real-time editing or commenting, could open up opportunities for Storyboarder's use in team-based creative projects.	Rapid technological changes may render certain features or visuals outdated, impacting the tool's relevance.

Summary

- A storyboard serves as a blueprint for your learning initiative; it assists instructional designers in mapping the learning journey, identifying potential gaps or redundancies, and ensuring comprehensive coverage of all learning objectives.

- The core components of a storyboard include screen name, onscreen text, graphics, user and navigation instructions, multimedia and interactive elements, developer or programmer notes, narration, or audio script.

- Storyboarding is a vital component of instructional design, providing stakeholders and subject matter experts with a preview of the course's flow and content presentation, fostering effective communication, allowing for proposed edits, minimizing trial and error, and ensuring adherence to research-backed best practices for streamlined course production and post-production.

- There are four types of storyboards: micro storyboard, Word-based storyboard, PowerPoint-based storyboard, and prototype-based storyboard. The format selection depends on factors such as project complexity, stakeholder engagement, usability testing, timeline, and resources.

Let's Brainstorm

These mini-scenarios will make you scratch your head and scribble on your pad.

Mini-scenario 1

An instructional designer recently completed a storyboard for an eLearning module by simply incorporating the content provided by the client without any additional modifications. However, it failed to meet the expected outcomes upon launching the module. Explore potential reasons for the failure and propose steps the instructional designer could take to address and prevent similar issues in the future.

Mini-scenario 2

As an instructional designer working on a project, you find yourself in a situation where the subject matter expert (SME) is extremely busy and needs help to review the Word-based storyboard you've created.

Recognizing the time constraints, what strategies would you employ to effectively communicate with the SME and ensure the approval of the storyboard? Additionally, which specific format for the storyboard would you choose to convey the content flow in a way that allows the SME to grasp the essence quickly and provide timely feedback on its accuracy?

Mini-scenario 3

While creating a prototype-based storyboard for a software training module, you encounter a situation where the key stakeholders have varying levels of technical expertise. How would you design your prototype to cater to technical and nontechnical stakeholders, ensuring that the essential instructional elements are communicated effectively?

Mini-scenario 4

As an instructional designer, you find that the developer or programmer working on your eLearning module project frequently reaches out to you for clarifications on the storyboard you've created. Analyze the situation and identify potential reasons for their inquiries. What elements in your storyboard might need more clarity that necessitates frequent communication with the developer? Propose strategies you would implement to address these issues and enhance the communication process with the development team.

Mini-scenario 5

Discuss your approach as an instructional designer for a project where the content is visually rich, and there's a need to sequence and chunk the information carefully to prevent cognitive overload. Elaborate on the storyboard format you would choose and provide reasons for your decision. How does this format help select, organize, and present visual-heavy content in a way that optimizes learning and avoids cognitive overload for the audience?

CHAPTER 9

Assessment, Feedback, and Evaluation Approaches

The primary goal of education and professional development is to ensure that courses are effective and practical. Designing such courses requires thorough data collection and analysis using assessment, feedback, and evaluation methods.

Throughout the life cycle of a course, spanning from initial ideation to final delivery, stakeholders spanning from learners to trainers contribute invaluable insights. Employing a 360-degree feedback approach allows educators and instructional designers to gather diverse perspectives on course performance and learner outcomes. This feedback is essential for continuously improving and aligning future courses with audience needs.

The assessment methods provide a structured framework for evaluating learner progress against predetermined objectives. By analyzing the attainment of learning goals, instructional designers can tailor interventions to address specific areas of improvement and optimize the learning experience.

© Ankita Jiyani Mangtani 2024
A. J. Mangtani, *Instructional Design Unleashed*, Design Thinking,
https://doi.org/10.1007/979-8-8688-0416-8_9

There needs to be more than the insights from the assessment to improve course effectiveness. Effective evaluation methods offer a deeper understanding of overall course effectiveness, identifying strengths, weaknesses, and areas for refinement. By applying established evaluation models, instructional designers can acquire actionable insights to inform strategic decision-making and drive continuous improvement initiatives.

This chapter unveils the symbiotic relationship between assessment, feedback, and evaluation methods in pursuing course excellence. By employing a comprehensive approach that integrates these elements seamlessly into the course development and delivery process, educational institutions and training providers can unlock the untapped potential to foster a culture of ongoing learning and growth.

Continuous Improvement Approach

The cyclic continuous improvement approach comprises four elements: assessments, feedback, evaluation, and iterative improvement (Figure 9-1).

Assessment
This phase captures insights into student learning by following various assessment methods at different stages of the program

Feedback
This phase aims to collate the feedback from different stakeholders involved in the learning process, i.e., learners, instructors, industry professionals, etc.

Iterative Improvement
This phase monitors the program's performance, effectiveness, and implements the identified improvements and adjustments to program.

Evaluation
This phase evaluates the program's overall effectiveness by following models such as Kirkpatrick's Training Evaluation.

Continuous Improvement Approach

Figure 9-1. *Continuous Improvement Approach*

The cyclic nature indicates that iteratively implementing the feedback gathered from assessments and evaluation methods improves the program experience. The later section decodes all the elements.

What Is Assessment?

Assessment of learners' learning is a comprehensive process beyond mere grading. It is essential to collect data to measure the extent and scope of learner learning in a course. It's about understanding the progress of learner learning and identifying their strengths and weaknesses.

Assessment measures learning and influences learner motivation and study behaviors based on the types of tasks presented. As John Biggs, a prominent figure in higher education, notes, how learners perceive assessment significantly impacts their learning approach. Therefore, it's crucial to carefully consider the assessment methods to ensure they align with desired learning outcomes.

Effective assessment methods seamlessly integrate grading, learning objectives, and learner motivation, providing valuable insights into what learners have learned, how proficiently they've learned it, and areas where they may have encountered difficulties.

What to Assess?

In Chapter 7, we explored the importance of crafting effective and SMART learning objectives, and these objectives help envision the knowledge, skills, and attitude you expect learners to possess upon completion.

The following inquiries offer valuable guidance in defining learning objectives and selecting suitable assessment strategies:

- What specific content knowledge should learners demonstrate, and at what proficiency level?

- What cognitive abilities or critical thinking skills do I aim for learners to cultivate and exhibit through course assessments?

- Are there particular professional competencies or attitudes I anticipate learners acquiring during the course?

459

Remember, the validity of a test or assignment relies on its ability to accurately measure the intended learning outcomes, making alignment between assessment tasks and desired learning essential.

Purpose of Assessment

As instructional designers, recognizing assessments' diverse roles in quantifying learner progress and informing instructional strategies is essential. Assessments serve three primary purposes, listed in Table 9-1.

Table 9-1. *Purpose of Assessment*

Assessment of Learning	Assessment for Learning	Assessment as Learning
It helps evaluate whether learners are meeting grade-level standards and achieving learning objectives.	It provides real-time insights into learner understanding as they learn, allowing educators to adapt their instructional strategies and lesson plans accordingly.	It helps actively engage learners in learning, fostering critical thinking skills, problem-solving abilities, and self-directed learning.
Common examples include • Exams • Portfolios • Final projects • Standardized tests	Common examples include • Formative assessment • Diagnostic assessment	Common examples include • Self-assessment • Peer assessment • Ipsative assessments

Types of Assessment

Let's understand six different types of assessment with the help of analogy.

> **Diagnostic Assessments:** Picture yourself as a chef preparing a new recipe. Before you start cooking, thoroughly examine all the ingredients to determine

their quality and freshness. This initial examination helps you identify the problems or areas for improvement before you start cooking.

Formative Assessments: As you cook, you taste the dish at various stages to adjust the seasoning and flavors. Each taste test informs your next steps, allows you to make ongoing improvements, and ensures the final dish is delicious, presentable, and well balanced.

Summative Assessments: After you've finished cooking, you serve the dish to the diners. Their feedback provides a final judgment on the overall quality and success of the dish. This final assessment determines whether the dish meets expectations and satisfies the diners.

Ipsative Assessments: If you were to taste the dish you made today and compare it to a dish you made last week, you'd be engaging in ipsative assessment. Analyzing your current performance to your past performance helps you to keep an eye on your progress and growth over time.

Norm-Referenced Assessments: Now, if you're participating in a cooking competition. Judges compare your dish to those made by other participants. They evaluate the dishes using a set of predefined criteria and compare them to others. This comparison resembles norm-referenced assessments, where you are compared to your peers based on your performance.

Criterion-Referenced Assessments: Finally, envision where you are following a recipe step-by-step from a cooking book. At the end, you compare your prepared dish to the picture of the dish in the book with the description of how it should taste. This comparison is akin to criterion-referenced assessments, where performance is evaluated against specific criteria or standards.

Table 9-2 presents more details on each type of assessment.

Table 9-2. *Types of Assessment*

Assessment Types	Purpose	Examples	Results of test	When to conduct
Diagnostic assessments	Identifying the gaps in existing knowledge	• Pre-test • Survey or questionnaire • Learner interview • Checklist • Graphic organizers	Helps to customize lesson plans to learners' existing knowledge	Before starting a new concept or skill, learn the objective
Formative assessments	Fine-tuning instructional strategies based on results of tests	• One-minute write-up • Summary • Class discussions • Group projects • Entry and exit tickets • Short, regular quizzes	Helps determine the learner's overall and individual progress	Throughout the unit, section, or learning objective

(*continued*)

Table 9-2. (*continued*)

Assessment Types	Purpose	Examples	Results of test	When to conduct
Summative assessments	Determining what learners have learned and if it aligns with their goals	• Traditional test • Essays • Research papers • Presentations • Recording podcast • Creating infographic	Analyzing the results of diagnostic and summative tests gives a clear picture of learner progress	At the end of the unit, section, or learning objective
Ipsative assessments	Measuring learner's performance based on their previous achievements	• Self-assessment surveys • Portfolios • Two-stage testing	The results help learner's personal development, education planning, career development, and goal setting	At various stages: • At the beginning of the learning program • Periodically, throughout the program • At the end of the program

(*continued*)

Table 9-2. (*continued*)

Assessment Types	Purpose	Examples	Results of test	When to conduct
Norm-referenced assessments	Measuring learner achievement to determine language ability, grade readiness, physical development, and institute admission decision	• IQ tests • Physical assessments • Standardized admission test	Evaluate the effectiveness of the program and instructional strategies by analyzing the performance of the norm group	At various stages: • At the beginning of the learning program • Periodically, throughout the program • At the end of the program
Criterion-referenced assessments	Measuring learners' mastery of specific learning objectives or criteria	• Standardized test • Rubrics • Performance tasks • Checklists • End-of-unit or end-of-course assessment	Helps determine the strengths, weaknesses, and proficiency level achieved with the predetermined criteria or standards	At the end of the unit, learning objective, or final test

Creating Effective Assessment

Creating effective assessments is an art. Suppose you want learners to become great soccer players. You spend your efforts and time teaching them dribbling skills, passing techniques, and teamwork strategies. But when it comes to the big game, they're only tested on how fast they can run. As a result, they never get to show off their actual play, leaving the game disappointed and confused.

Similarly, if you want learners to become critical thinkers but only test them on memorizing dates and names, they'll miss the chance to use their skills meaningfully. It's like practicing for a soccer game but playing basketball instead – the skills don't match up. Effective teaching and testing should be like playing the right sport with the proper rules so learners can shine and show what they've learned.

Table 9-3 lists the effective test or assessment for each taxonomy level.

Table 9-3. *Creating Effective Assessment*

		Bloom's Taxonomy of Cognitive Domain			
Remember	**Understand**	**Apply**	**Analyze**	**Evaluate**	**Create**
• Fill in the blanks	• Diagrams	• One-minute write-up	• Case studies	• Debates	• Grant proposal
• MCQ/MRQ	• Infographics	• Presentation	• Research paper	• Discussions	• Outline
• True and false	• One-minute write-up	• Portfolio	• Review paper	• Presentation	• alternative solutions
• Match	• Presentation	• Discussion forum	• Analysis paper	• Provide alternative solutions	• Research proposal
• Quizzes	• Provide working examples	• Short answers		• Report	
	• Quizzes				
	• Short answers				
	• Essays				

Krathwohl's Taxonomy of Affective Domain

Receive	Respond	Value	Organization	Characterization
• MCQ/MRQ	• Critical questioning	• Debate	• Focus groups	• Critical reflection
• Fill in the blanks	• Feedback and peer evaluation	• Rating scale	• Solve novice problems	• Group projects
• List	• One-minute paper	• Reflection paper	• Analyze and contrast	• Self-evaluation
• Match	• Role-play	• Report on extra-curricular activities		
• Memory tests	• Written assessments	• Ungraded paper		
• One-minute write-up		• Self-report		
• Qualitative interviews				

What Is Feedback?

There are three different lenses to perceive feedback: the learner, the educator, and how teaching and learning work together. Feedback helps learners identify and improve their mistakes before they snowball into more significant problems.

Learners get feedback from tests, assignments, and class activities. Good feedback focuses on what learners understand and need to work on. The feedback process includes setting expectations, tracking progress, recognizing strengths, and suggesting ways to improve.

To effectively learn from feedback, educators must establish trust. They need to know the expectations; listen to feedback from learners, colleagues, principals, and parents; make necessary changes; and assess whether those changes help.

When giving feedback on learning, two key aspects must be considered: understanding the content thoroughly and mastering the tools or study skills necessary for effective learning. Constructive feedback requires a deep understanding of learners' characteristics, including their learning styles, preferred study techniques, and any limitations they may face.

Purpose of Feedback

The feedback enhances learning outcomes and cultivates a sense of autonomy and self-efficacy among learners. The intent of giving feedback to learners is to let them

- Recognize and acknowledge specific strengths demonstrated during and after learning

- Focus on actionable steps for progress with clear guidance on areas for improvement

- Inspire and motivate them to actively engage with their assessments and strive for continuous improvement

- Adapt and adjust their learning strategies and approaches

- Foster the development of self-regulation skills, empowering them to monitor, evaluate, and adjust their learning approaches and styles effectively

Types of Feedback

Let's understand different types of feedback with the help of an analogy.

Formative vs. Summative Feedback

Formative feedback is like tending to a garden throughout the growing season. Like a gardener who regularly monitors soil moisture, adjusts fertilizer levels, and prunes away any overgrowth or disease. This continuous care and attention help the plants thrive and develop strong roots, just as formative feedback nurtures learners' growth and learning throughout the educational program.

Summative feedback is akin to harvesting the fruits of your labor in the garden at the end of the growing season. After months of nurturing and tending to the plants, a gardener gathers the ripe fruits and assesses their quality. This final evaluation provides a comprehensive overview of the success of the gardening efforts, similar to a summative assessment providing an overall review of the learners' learning outcomes after an educational program.

Individual vs. Generic Feedback

Individual feedback is like customizing care for each plant in the garden. Just as each plant has unique needs based on its species, size, and location, each learner has unique talents, strengths, and weaknesses. Individual feedback addresses specific strengths and areas for improvement tailored to each learner's learning style and progress.

Generic feedback is like a one-size-fits-all approach to fertilizing, spreading a general fertilizer over the entire garden. While it provides a basic level of nourishment, it may overlook the diverse needs and requirements of individual species and varieties. Similarly, it may not meet the nuanced needs of each learner. Generic feedback may miss opportunities to effectively inspire, motivate, and support their learning journey.

Automatic vs. Manual Feedback

Automatic feedback is like using an automated watering system in the garden. This system delivers water to the plants at preset intervals, ensuring consistent hydration without manual intervention. Similarly, automatic feedback tools provide immediate responses to learners' work, such as automated quizzes or grading software.

Manual feedback is like hand-watering each plant in the garden. The gardener requires direct involvement and attention to assess the soil moisture, adjust watering levels, and provide personalized care. Manual feedback offers a more personalized and detailed evaluation of learner progress.

Oral vs. Written Feedback

Oral feedback is like a gardening mentor walking alongside you in the garden, offering real-time advice and tips as you tend to the plants. This immediate exchange allows for on-the-spot clarification and discussion, as oral feedback facilitates direct interaction and dialogue between teacher and learner.

Written feedback is like leaving detailed instructions and observations on plant care for specific species in the garden journal. Similarly, providing written feedback offers a lasting record of guidance and suggestions that learners can revisit over time, submitting a structured and comprehensive feedback form for ongoing learning and improvement.

Characteristics of Effective Feedback

Educators should exhibit heightened sensitivity when delivering timely feedback and be attuned to subtle shifts in learners' behaviors to provide tailored guidance and support. Table 9-4 decodes the mnemonic MICROSENSOR to understand the characteristics of effective feedback.

Table 9-4. *Characteristics of Effective Feedback*

Mnemonic Letter	Meaning	Explanation
M	Meaningful	Meaningful and manageable feedback allows the learner to take actionable steps for improvement without overwhelming them.
I	Insightful	Insightful and informative feedback helps learners identify what they did well and what they need to work on.
C	Constructive	Constructive feedback encourages the learners' development.
R	Relevant	Relevant feedback recognizes the specific behavior, outcome, or result.
O	Open	Open and honest feedback encourages candid conversations between educators and learners.
S	Specific	Specific feedback offers clear reasons and explanations for the learner's performance and opportunities for growth.
E	Educational	Educational feedback helps learners focus on their performance details and how they can excel in the future.
N	Neutral	Neutral and nonjudgmental feedback avoids bias or personal opinions.
S	Supported	Feedback supported with evidence or reasons allows the learner to rectify the mistakes.
O	Observational	Feedback observations should be recorded to assist in recalling feedback.
R	Reflective	Reflective feedback helps the learner to link the input to their performance.

What Is Evaluation?

Evaluating an educational program requires assessing its effectiveness in achieving its intended objectives and providing value to both the learners and the organization. To approach the evaluation process, one must determine the following:

- **Alignment with Objectives:** Evaluate how successfully the educational program aligns with the organization's goals and objectives. Ensure that the content, delivery methods, and learning outcomes directly relate to the learners' needs.

- **Learner Engagement:** Assess learners' level of engagement and participation throughout the program. Analyze attendance, online course completion rates, and participant feedback regarding the content's relevance and interest level.

- **Learning Outcomes:** Measure the educational program's knowledge and skill acquisition effectiveness. Conduct pre- and post-assessments to determine learning gains and evaluate the application of newly acquired knowledge and skills.

- **Quality of Delivery:** Evaluate the instruction and delivery methods used in the educational program. Assess the effectiveness of trainers or facilitators, the suitability of instructional materials, and the accessibility of resources for learners.

- **Feedback and Improvement:** Gather feedback from learners, trainers, and other stakeholders to spot the areas for improvement. Conduct surveys or one-on-one interviews to gather qualitative data on the strengths and weaknesses of the program.

- **Impact on Performance:** Assess the impact of the educational program on individual and organizational performance. Measure changes in key performance indicators (KPIs), such as productivity, quality of work, customer satisfaction, or learner retention rates.

- **Return on Investment (ROI):** Calculate the return on investment for the learning program by comparing development and delivery costs to the tangible benefits achieved. Quantify cost savings, revenue increases, or other financial outcomes attributable to the program.

- **Continuous Improvement:** Use evaluation findings to inform ongoing improvement efforts. Adjusting the content, delivery methods, or evaluation strategies based on feedback and data analysis results ensures that the learning program remains effective and relevant.

Evolution of Program Evaluation

The evolution of program evaluations has transitioned from a simplistic focus on assessing the achievement of stated objectives to a more comprehensive approach considering both intended and unintended outcomes. Initially, educational evaluators, influenced by Tyler's model from the 1940s, primarily concentrated on determining if programs met their predefined goals. However, the societal changes of the 1960s and 1970s, coupled with increased government spending on health, education, and welfare, necessitated a more rigorous evaluation framework.

Researchers began developing new evaluation models in response to these demands for greater accountability. Scriven (1972) was among the pioneers who advocated for a broader assessment that included the stated goals of a program and its broader impacts. This shift emphasized the importance of understanding the full range of outcomes that a program might produce, whether anticipated or unforeseen.

Similarly, Suchman (1967) emphasized the need to delve into the underlying processes that lead to program outcomes. This perspective highlighted the complexity of program implementation and the importance of understanding how various factors interact to produce results.

Weiss (1972) introduced theory-based evaluation (PTE), which revolutionized the field by combining a theoretical understanding of how a program should work with an evaluation process guided by that theory. This approach, also known as theory-driven or program theory evaluation, encourages evaluators to develop explicit models or theories of change that articulate the pathways through which a program is expected to achieve its goals.

One of the critical contributions of PTE is its emphasis on moving beyond a superficial assessment of program outcomes and instead focusing on understanding the underlying mechanisms driving those outcomes. This encourages evaluators to adopt a more nuanced and holistic approach to evaluation, often involving qualitative methods to explore the complex dynamics at play.

Purpose of Evaluation

Evaluations provide a comprehensive analysis of the strengths and weaknesses of the educational program, shedding light on its influence on learners. You gain valuable perspective by examining individual, team, departmental, and organizational insights. With this knowledge, you can refine strategies, rectify errors, and make informed decisions. The evaluation aims to assess diverse aspects of the program, but in general, they look at the following things:

- Did the program achieve its intended goals and objectives?

- What did the learners gain from the program regarding learning outcomes?

- Did the learners effectively utilize the acquired knowledge and skills in their professional and personal roles?

- What influence did the program have on the overall performance and productivity of the organization?

- Was the expenditure on the program justified by the benefits accrued to the organization?

- Did the program deliver value in proportion to the resources invested?

Types of Evaluation Models

The evaluations offer diverse insights into the effectiveness of educational programs and learning methodologies within the organization. You will gather different perspectives when conducting evaluations during the program or at the end. The following are some types of learning evaluations that offer different perspectives about the program:

- **Process Evaluation:** Assessing the implementation and delivery of the educational program to ensure alignment with organization strategies. It identifies the factors that contributed to the success and failure of the program.

- **Summative Evaluation:** Assessing the overall success and effectiveness of the program by finding the answers to questions like, did the program meet its objectives? It helps draw a conclusion about success and make informed future decisions.

- **Formative Evaluation:** Evaluating educational programs during the development or implementation phase to enhance effectiveness and address any issues by identifying the strengths, weaknesses, and areas of improvement.

- **Impact Evaluation:** Analyzing the program's long-term effects on learners' performance, retention, and productivity. Various methods are used, such as quantitative data analysis and qualitative insights.

- **Outcome Evaluation:** Examining how learning initiatives have influenced learner attitudes, behaviors, knowledge, and skills.

Numerous evaluation models have evolved to capture insights about program effectiveness comprehensively. Each model depicts the efficacy of educational programs in distinct dimensions. Various key performance indicators (KPIs) specific to the program can measure these dimensions. Table 9-5 describes the KPIs corresponding to each evaluation model.

Table 9-5. *Evaluation Models and Their KPIs*

KPIs	Model
Comprehensive analysis from initial reaction to ultimate success	Kirkpatrick model
Success and failure analysis	Brinkerhoff's success case method
Financial analysis	Phillips' ROI model
Various context analysis	Context, Input, Process, and Product (CIPP) model

Kirkpatrick Evaluation Model

Dr. Donald Kirkpatrick developed this enduring model back in the 1950s. Widely adopted by organizations, it offers valuable insights despite its limitations. The model encompasses four levels of evaluation, each crucial for identifying and addressing challenges effectively (Figure 9-2). Neglecting any level compromises the analysis, potentially skewing learning models and ROI calculations.

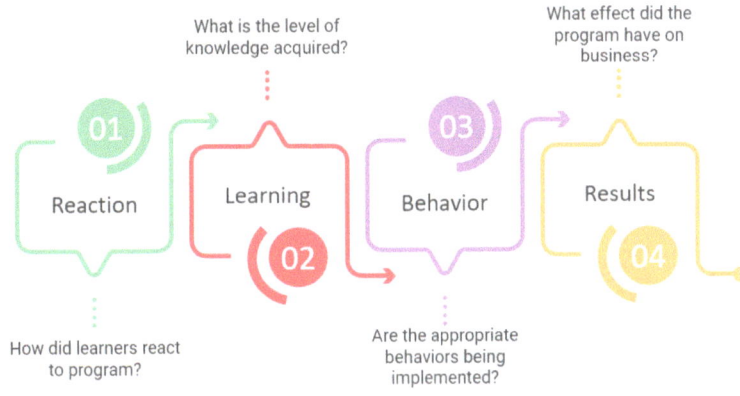

Figure 9-2. *Kirkpatrick Evaluation Model*

To optimize the benefits of the Kirkpatrick model while addressing its challenges, start with the organization's long-term goals and work backward. Identify the desired results, relevant areas, and critical benefits. Collaborate with SMEs and stakeholders to define target behaviors and measurable learning outcomes. Assess learners' existing knowledge and skills to pinpoint areas for improvement. Evaluate, test, and refine the educational program to ensure accessibility, relevance, and engagement, fostering a positive learning experience and facilitating comprehension.

Brinkerhoff's Success Case Method

Evaluating training extends beyond acknowledging successes; failures are
equally significant. Robert O. Brinkerhoff develops it. With the success case
method (SCM), Brinkerhoff's method thoroughly examines successes and
failures within each educational program (Figure 9-3). The SCM scrutinizes
the most and least successful learning initiatives in organizations, assessing
success in cost-effectiveness and returns. While SCM can be a one-time
evaluation method, it should complement other approaches.

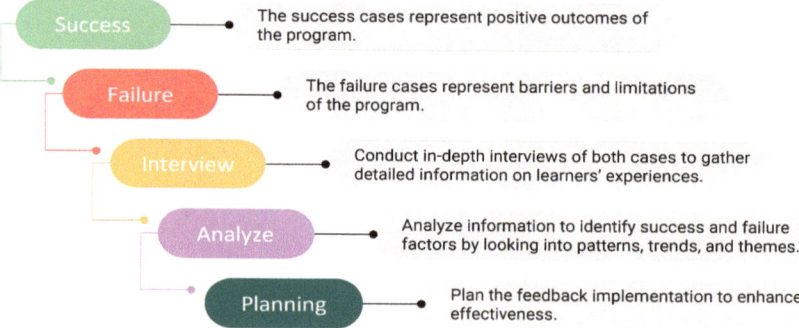

Figure 9-3. *Brinkerhoff's Success Case Method*

Phillips' ROI Model

Phillips' ROI model builds upon the Kirkpatrick model by introducing
a fifth level (Figure 9-4), providing organizations with a comprehensive
framework for assessing the ROI of their educational program initiatives.
Reaction focuses on learners' responses to the program through short
surveys, giving insight into the cohesion of the organization's L&D strategy.
Learning measures learning through pre- and post-training quizzes,
akin to the Kirkpatrick model. Unique to the Phillips model, behavior
delves deeper by examining why educational programs succeeded or
failed, offering qualitative feedback for program improvement. Results
broaden the scope of impact assessment beyond business results to
analyze multidimensional impacts, including various external factors

influencing the learner's decision and results, offering a comprehensive view of program effectiveness. Finally, ROI employs cost benefit analysis to quantify the ROI of educational program initiatives, facilitating communication of benefits and value to stakeholders.

Figure 9-4. *Phillips' ROI Model*

CIPP Model

The CIPP model, developed by Daniel Stufflebeam in 1971, presents a departure from its era's dominant experimental design model, aiming to enhance program improvement rather than prove program efficacy. This model, documented to be highly effective in various educational and noneducational settings, consists of four interrelated evaluation components: Context, Inputs, Process, and Products, allowing for a comprehensive examination of program dimensions (Figure 9-5). Each component addresses different program development and evaluation phases, with Context focusing on program planning and adaptation. Various data collection and analysis methods are suitable for conducting a Context study. Depending on the specific requirements of the situation, the evaluator could choose from a range of options, including reviewing documents, analyzing demographic data, conducting interviews, administering surveys, analyzing records such as test results or learner performance data, and facilitating focus groups.

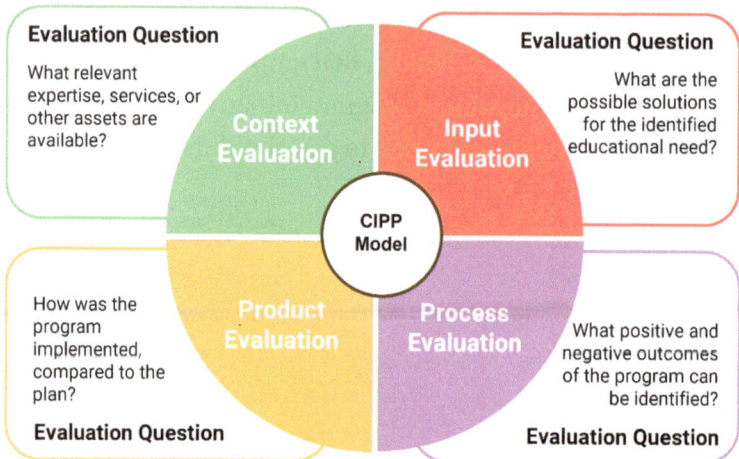

Figure 9-5. CIPP Model

Inputs assessing resource allocation feasibility. Various methods to explore and evaluate potential approaches to address an educational need in an Input assessment are conducting a literature review, visiting exemplary programs, seeking expert advice, and requesting proposals from individuals interested in addressing the identified needs.

The Process evaluates the program implementation. When designing a CIPP process evaluation study, the evaluator typically opts for minimally intrusive methods while the program is ongoing. Standard methods include observing program activities, reviewing relevant documents, and conducting participant interviews.

Products examining program outcomes involve assessing the effectiveness and impact of a program by systematically evaluating the results it produces. To conduct a comprehensive product evaluation study, the evaluator can utilize a variety of methods and data sources, including stakeholders' assessments of the project or program, comparative analyses of outcomes with similar projects or programs, evaluation of program objectives achievement, group interviews covering a wide range of program outcomes, case studies detailing selected participants' experiences, surveys gathering feedback from stakeholders, and participant reports detailing the effects of the project.

The flexibility of the CIPP model accommodates the evolving nature of educational programs and educators' need for continuous improvement data, providing a systematic framework for formative and summative evaluation studies.

Characteristics of an Evaluation Model

After reviewing various evaluation models, each offering distinct perspectives on program efficacy, it becomes evident that specific characteristics should be adhered to to ensure comprehensive assessment. These characteristics are outlined as follows:

- Structured and impartial evaluation minimizes personal biases, ensures objectivity, and delivers a precise and unbiased understanding of the subject under scrutiny.

- Goal-oriented evaluation aims to achieve specific objectives. Whether assessing program effectiveness, enhancing decision-making, or gauging performance, the evaluation's purpose drives its entire process.

- Contextually tailored evaluation considers the unique context in which it occurs, incorporating the environment's characteristics, needs, and goals. Adaptation to the specific context ensures relevance and effectiveness, acknowledging that evaluation criteria may vary across domains such as education, healthcare, or organizational settings.

- Comprehensive and diverse evaluation captures the holistic perspective of the program as it encompasses many dimensions, spanning from assessing processes and outcomes to capturing stakeholder perspectives.

Differences Between Assessment and Evaluation

Distinguishing between assessment and evaluation is vital in understanding their respective roles in educational or organizational contexts. While assessment focuses on measuring learning progress and understanding, evaluation aims to assess the effectiveness of programs or processes. Clear comprehension of these differences (Table 9-6) ensures informed decision-making and targeted improvements.

Table 9-6. *Differences Between Assessment and Evaluation*

Aspect	Assessments	Evaluations
Nature	Identifies learning progress and achievement.	Makes judgments and decisions based on collected data.
Purpose	Primarily formative, providing ongoing feedback.	Primarily summative, delivering an overall judgment.
Feedback	Focuses on individual improvement and learning.	Provides feedback to inform policy decisions and program development.
Focus	Individual focused, aiming to improve learning and performance.	System focused, aiming to improve educational programs and policies.
Methods	Uses tests, quizzes, and structured approaches.	Employs multiple methods such as observations, surveys, and interviews.
Influence	Informs instructional decisions and supports individualized learning.	Informs policy decisions and influences program development.
Emphasis	Emphasizes the process of learning and growth.	Emphasizes the outcomes and results achieved.

(continued)

Table 9-6. (*continued*)

Aspect	Assessments	Evaluations
Conducted by	Teachers or instructors typically conduct them.	Conducted by external evaluators or a team of experts.
Scope	Focuses on specific learning objectives or competencies.	Examines the overall impact and effectiveness of an educational system.
Utilization	Used to inform teaching and learning strategies.	Used to devise strategic planning and resource allocation.
Analysis	Identifies strengths and weaknesses of individuals or groups.	Determines the effectiveness and quality of programs or interventions.

Summary

- The cyclic continuous improvement approach involves four key elements: assessments, feedback, evaluation, and iterative improvement.

- Assessment goes beyond grading, measuring learner learning progress, and identifying strengths and weaknesses.

- Effective assessment aligns with learning objectives, providing insights into learning proficiency and areas of difficulty.

- Assessments serve three primary purposes: assessment of learning, assessment for learning, and assessment as learning, each contributing to learner progress and instructional strategies.

- Assessment is crucial for measuring learner learning beyond grading, providing insights into progress, strengths, and weaknesses.

- Effective assessment methods influence learner motivation and study behaviors, aligning with desired learning outcomes and integrating grading, objectives, and motivation.

- Assessments serve diverse purposes, including evaluating learning, providing real-time insights, and engaging learners actively, with various types such as diagnostic, formative, summative, ipsative, norm-referenced, and criterion-referenced assessments serving different functions.

- Feedback is vital in identifying and rectifying mistakes, benefiting learners, educators, and the teaching-learning process.

- Feedback aims to recognize strengths, provide guidance for improvement, motivate learners, and foster self-regulation skills.

- Various types of feedback, such as formative vs. summative, individual vs. generic, automatic vs. manual, and oral vs. written, cater to different needs and preferences, influencing learners' growth and development.

- Effective feedback characteristics, summarized by the mnemonic MICROSENSOR, ensure that feedback is meaningful, insightful, constructive, relevant, open, specific, educational, neutral, supported, observational, and reflective.

- Evaluating educational programs involves assessing alignment with objectives, learner engagement, learning outcomes, quality of delivery, feedback and improvement, impact on performance, return on investment (ROI), and continuous improvement.

- The purpose of evaluation includes assessing goal achievement, learning outcomes, utilization of knowledge and skills, organizational performance, cost-effectiveness, and overall value. Various evaluation models like process, summative, formative, impact, and outcome evaluation offer diverse perspectives on program effectiveness.

Let's Brainstorm

These mini-scenarios will make you scratch your head and scribble on your pad.

Mini-scenario 1

In a higher education setting, imagine you're a university professor teaching a course on psychology and you have a learner named Michael who struggles with traditional written exams but excels in practical applications. During an observation, you notice Michael actively participates in group discussions, offering insightful analyses and practical solutions to case studies. Which assessment strategy will you employ for Michael that allows him to demonstrate his knowledge?

Mini-scenario 2

As an educator working with diverse learners, you assess their needs and progress throughout the academic year. How would you design a series of pre-tests for a diagnostic assessment to identify each learner's prior knowledge and potential learning gaps in mathematics?

Mini-scenario 3

You're a technology integration specialist at a school implementing a new online learning platform. As part of the platform's features, it offers automatic feedback on quizzes and assignments. During a training session, you notice that some teachers need to be more open about relying solely on automatic feedback, expressing concerns about its effectiveness compared to manual feedback. Additionally, learners have varying preferences for feedback delivery, with some preferring written comments while others prefer oral feedback. How would you address these concerns and preferences to ensure that the feedback caters to different needs and preferences, fostering optimal growth and development among learners?

Mini-scenario 4

You're a learning and development specialist tasked with evaluating the effectiveness of a newly implemented sales training program in a retail company. Utilizing the Kirkpatrick evaluation model, describe how you would approach each phase of evaluation, starting from gathering feedback on participants' reactions to the training content, assessing the increase in their knowledge and skills, observing changes in their on-the-job behaviors, and, finally, measuring the impact of the training on key performance indicators such as sales returns and customer satisfaction.

Mini-scenario 5

You're a project manager leading a training initiative to improve customer service skills within a large telecommunications company. You must evaluate the training program's return on investment (ROI) using the Phillips ROI evaluation model as part of your responsibility. Describe how you would apply each step of the Phillips model, beginning with identifying the tangible and intangible benefits expected from the training, collecting data on the costs associated with the program, confining the effects of the training from other factors, converting data into monetary values, and finally calculating the ROI to determine the program's overall effectiveness and value to the organization.

CHAPTER 10

Publish the Learning Journey

This is a crucial chapter where we go from the theoretical foundations of course design to the real-world application. After exploring the complex terrain of instructional design's scientific and artistic aspects in previous chapters, we are entering the critical stage of creating the desired educational experience.

In Chapters 2 through 5, the scientific complexities of instructional design were examined, along with the fundamentals of human learning at different levels, the importance of various learning theories and modalities, the investigation of pedagogical approaches, and the necessity of creating accessible content.

Building on this solid foundation, Chapters 6 through 9 explained the artistic aspects of instructional design, outlining the problematic process of instructional design (ID) models, defining SMART objectives, creating intriguing storyboards, and establishing reliable systems for assessment, feedback, and evaluation. This foundational understanding serves as our compass as we navigate the terrain of instructional design.

Now, we are in the two-fold implementation phase, where the first aspect is about selecting an authoring tool and the second aspect is about publishing the developed course on a learning management system (LMS). However, to ensure learners' seamless accessibility and interoperability, the published course on an LMS must comply with

© Ankita Jiyani Mangtani 2024
A. J. Mangtani, *Instructional Design Unleashed*, Design Thinking,
https://doi.org/10.1007/979-8-8688-0416-8_10

specific eLearning standards. The effectiveness and efficiency of the course development process are significantly impacted by every choice made during this phase.

eLearning Course Development Overview

After completing the storyboarding phase, your next step as an L&D expert is to create the course using the chosen authoring tool (Figure 10-1), as determined during the project's analysis phase. Once the course is developed within the software, the next task is to prepare the course package for upload onto the LMS.

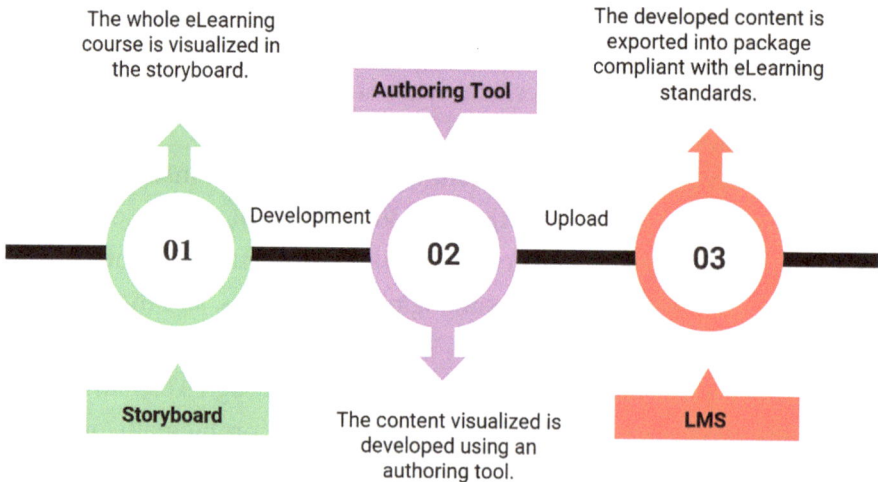

Figure 10-1. *eLearning Course Development Overview*

The course package must adhere to specific eLearning standards to ensure compatibility with the LMS. These standards are guidelines for formatting and organizing course materials to deliver and track them effectively within the learning management system. Compliance with eLearning standards ensures that the course functions smoothly and consistently across different LMS platforms, allowing for seamless deployment and user experience.

What Is an Authoring Tool?

An authoring tool is a software program or application used to create digital content, particularly for eLearning purposes. It allows users to develop interactive and multimedia-rich educational materials such as online courses, presentations, quizzes, and simulations without extensive programming knowledge. Authoring tools typically provide templates, drag-and-drop interfaces, and pre-built components to streamline content creation. They allow for the production of instructional content that is compliant with eLearning standards, such as AICC, SCORM, xAPI (Tin Can), cmi5, and LTI. Learning content created with authoring tools is usually executed via a learning management system (LMS) or in a browser (in HTML5 format). These tools play a crucial role in instructional design, enabling educators and instructional designers to create engaging and interactive learning experiences for learners.

Features of an Authoring Tool

The comprehensive features of an authoring tool, which serves as the backbone for creating engaging and interactive eLearning courses are as follows:

Content Authoring: The core feature encompasses creating lessons by adding and editing slides, images, text, video, audio, or other on-screen elements. These functionalities bring courses to life, representing what learners will see, hear, and do. Key capabilities include the following:

- Adding or importing images
- Editing images
- Importing and exporting audio files
- Recording audio for narration
- Editing audio files
- Webcam and screen recording for videos

- Importing, exporting, and editing videos

- Animating on-screen images or text

- Text and style editing

- Responsive design

- Importing PowerPoint lessons

Interactivity: Engaging learners through interactive features enhances engagement and retention. Major features include the following:

- Built-in activities like drag and drop, matching, and hotspots

- Custom activities or games

- Interactive scenarios with on-screen objects using layers, triggers, buttons, and more

Templates and Themes: Establishing a consistent design is crucial for learner focus. Template and theme features include the following:

- Custom themes for the entire course

- Background themes

- Built-in template shells

- Creation of custom templates

Content Management: Organizing assets, media, and other content saves time and keeps workflows organized. Key features include the following:

- Libraries for courses, lessons, quizzes, and resources

- Quiz and question banks

- Template libraries

- Media libraries for video, audio, and images.

Collaboration: Facilitating teamwork and stakeholder involvement streamlines course development. Features include the following:

- Review systems with feedback
- Co-authoring capabilities
- Shared resources and files
- User roles and permissions

Assessment: Assessing learner progress and understanding is vital. Features include the following:

- Quizzes with various question types and feedback options
- Course assessments with question banks and scoring
- Analytics for tracking and reporting

Accessibility: Ensuring courses are accessible to all learners is essential. Features include the following:

- Compliance with accessibility guidelines like WCAG 2.0
- Closed captions, keyboard-accessible navigation, and customizable elements for different sensory needs

Publishing: Meeting organizational requirements for course deployment is crucial. Publishing formats include the following:

- HTML5, Flash, iOS, Android, CD, and cloud-based options
- Publishing to various file formats and compliance standards like SCORM and xAPI

Administration: Customizing preferences and workflows enhances productivity. Features include the following:

- Language settings

- Flexible workspace setups

- Roles and permissions

- Single sign-on for security

Support and Training: Access to help and resources aids users at all levels. Support features include the following:

- Online community forums

- Live chat, email, or phone support

- Tutorials, guides, and video resources

- Webinars, demos, and consultancy services

- Regular updates and upgrades for continued improvement

Types of Authoring Tools

Regarding eLearning authoring tools, they fall into two main categories: desktop and cloud based. Each type has its characteristics that affect course creation differently. Let's explore both types.

Desktop Tools: Desktop authoring tools have been around for a while and are still widely used for creating digital learning content. They usually have robust features like branching, customization, and simulated experiences. Many top authoring tools today consist of a core desktop tool paired with an online collaboration tool that operates in the cloud.

Desktop tools can be used on different computers but may need multiple licenses. Users store all their data on their machines and must back up their training materials regularly.

Cloud-Based Tools: Cloud-based software allows content creators to access their eLearning projects through a web browser, collaborate in real time, and seamlessly change the content. These tools typically have a simple web interface similar to apps and offer easy content creation using templates and multimedia.

Using cloud-based eLearning authoring tools requires an Internet connection, but they offer greater flexibility in accessing content. This is beneficial for team collaboration on corporate eLearning content and for sharing training courses with instructors or learners. All data is stored on the provider's cloud hosting, with minimal security risks.

How to Choose an Authoring Tool?

It's essential to align the qualities of the authoring tool with the project's specific needs. Every authoring tool has specific features designed for instructional situations. To streamline your decision-making process, consider the following questions:

- What is the skill level of the users utilizing the software?
  - Novice users without programming experience
  - Experienced users, proficient in programming
- What type of eLearning course are you aiming to develop?
  - Simple, template-based courses featuring text and images
  - Video-centric eLearning modules
  - Animation-driven eLearning content
  - Tailored scenario-based eLearning experiences
  - Customized eLearning projects with intricate interactions and advanced graphic design elements

- What types of output standards are required for your project?

  - SCORM (Sharable Content Object Reference Model)

  - Tin Can API (Experience API)

  - Other specific output formats

eLearning Standards

The video formats and media players should be compatible with each other to provide a smooth viewing experience. Similarly, compatibility between the content and learning management systems (LMSs) in eLearning is crucial for effective delivery and engagement. Let's explore the parallels between video formats and eLearning standards to understand how adherence to standards ensures seamless functionality.

Video formats like MP4, MPEG, WAV, MOV, AVI, WMV, and FMV have their specifications and characteristics. These formats must adhere to specific standards to ensure they can be played smoothly on various video players like PotPlayer, VLC, and Windows Media Player installed on multiple devices. Without these standards, compatibility issues may arise, hindering the viewing experience for users.

eLearning standards, such as AICC, SCORM, xAPI, cmi5, and LTI, have specific functionality. They help the content to interact with LMS platforms installed on various devices. These standards ensure seamless integration and functionality within LMS platforms.

eLearning standards provide guidelines and protocols for developing and delivering content. For example, SCORM dictates how content packages should be organized, while xAPI enables tracking of learning experiences beyond traditional LMS boundaries. These standards ensure that eLearning content can be accessed, interacted with, and tracked consistently across different LMS platforms.

Benefits of Using eLearning Standards

Embracing eLearning standards offers a multitude of advantages for both learners and course creators. By adhering to these standards, courses become seamlessly compatible with various learning management systems, facilitating effortless access and completion for learners.

Moreover, these standards enable sophisticated tracking and reporting mechanisms, offering invaluable insights into learner progress and experiences. This wealth of data informs course enhancement and aids in comprehensive evaluation.

Furthermore, using standards streamlines the process for course creators, resulting in significant time and resource savings. Content can be developed once and effortlessly disseminated across diverse platforms and systems, fostering efficiency and scalability. Here are some benefits of using eLearning standards:

- **Interoperability:** eLearning standards facilitate seamless communication among various eLearning systems and tools, fostering integration and ensuring a cohesive learning journey for learners.

- **Compatibility:** These standards guarantee that eLearning content is accessible across different platforms and devices, enabling learners to engage with the material anytime, anywhere conveniently.

- **Reusability:** By adhering to standards, eLearning content becomes easily shareable and reusable across diverse learning environments, streamlining content creation efforts and maintaining consistency in learning outcomes.

- **Scalability:** Standards empower eLearning content to adapt effortlessly to varying audience sizes and learning contexts, simplifying training delivery to large groups of learners.

- **Data tracking:** With standards in place, tracking learner data and performance across multiple learning environments becomes feasible, furnishing valuable insights for enhancing learning experiences and pinpointing areas for refinement.

- **Personalization:** Leveraging standards enables the delivery of tailored learning experiences by leveraging learner data to customize content and delivery methods according to individual preferences and needs.

Implementing eLearning Standards

Navigating the implementation of eLearning standards may initially appear daunting, yet the process can be streamlined with careful consideration and strategic steps. Begin by selecting the most suitable standard for your course, a decision contingent upon your specific objectives and preferences.

The widely adopted SCORM standard proves a reliable choice for courses necessitating straightforward tracking and reporting functionalities. Conversely, if your training demands more intricate tracking capabilities, xAPI offers greater adaptability to capture complex learning interactions and performance data. Organizations seeking a middle ground between structure and flexibility may find cmi5 to be a compelling option.

Additionally, factor in elements such as your existing learning management system's compatibility, technical capabilities, and desired data interoperability levels with other systems. By comprehensively assessing these factors, you can decide to serve your training program's unique needs best, fostering a more engaging and effective learning environment.

After selecting a standard, develop course content using tools and software compatible with the chosen standard. Rigorous testing is imperative to ensure seamless functionality and compliance with the standard's requirements. Finally, upload your course onto a compatible learning management system, granting learners convenient access to the enriching educational experience you've meticulously crafted.

Timeline of eLearning Standards

Over time, several eLearning standards have emerged (Figure 10-2) to ensure eLearning modules/solutions compliance with different learning management systems (LMSs). The following section of the chapter delves into each eLearning standard depicted in the timeline, providing detailed explanations and insights.

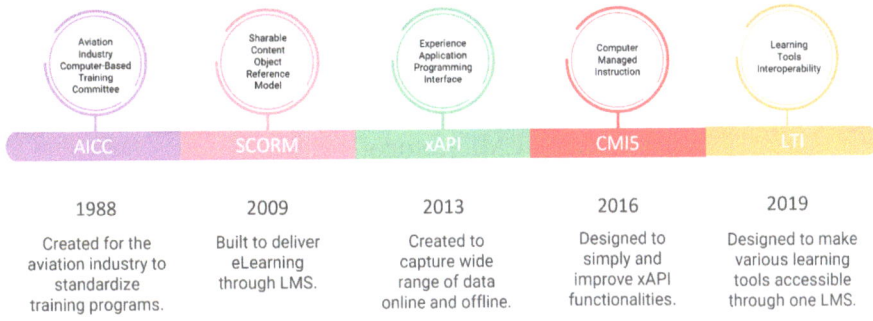

Figure 10-2. *Timeline of eLearning Standards*

AICC

Full Form: AICC stands for Aviation Industry Computer-Based Training Committee.

History: In the late 1980s, the aviation industry embraced digital methods for training its workers, leading to the development of the AICC learning technology standard in the 1990s. Spearheaded by aircraft manufacturers like Boeing, Airbus, and McDonnell Douglas under the Aviation Industry CBT Committee (AICC) banner, the initiative addressed concerns about escalating computing costs associated with multimedia training solutions. The committee formulated guidelines known as Computer Managed Instruction (CMI001) to standardize training materials and technologies used by airline workers. Through AICC, the industry achieved interoperability and compatibility among computer-based training systems, facilitating the hosting and delivery of online learning content across various compatible platforms.

Evolution of AICC

- AICC established the first technical specification for LMSs in 1993, primarily for CD-ROM and LAN-based training (AGR-006).

- With the advent of the Internet, AICC updated its CMI with a new chapter, "Runtime," detailing online communication between content and LMSs, termed "HTTP-based AICC/CMI Protocol" (HACP).

- Adding HACP to the specification in 1998 (AGR-010) facilitated AICC's expansion beyond the aviation industry.

- In 1999, AICC enhanced its CMI001 specification by introducing a JavaScript API runtime interface as an alternative to HACP, which was later adopted with minor modifications in SCORM.

How It Works: The AICC standard employs the HTTP AICC Communication Protocol (HACP) to facilitate communication between course content and a learning management system (LMS). This protocol utilizes HTML forms and simple text strings to transmit information back and forth between the LMS and the course content (Figure 10-3). The methodology of HACP is straightforward, providing an efficient means to exchange data between the system's two components.

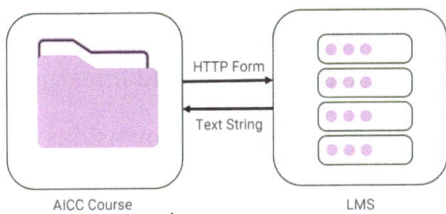

Figure 10-3.   *Working of AICC*

Complying with AICC: To achieve AICC compliance for an online learning course, the content must adhere to at least one of the nine AICC Guidelines and Recommendations (AGRs), which outline the criteria for compliance. These AGRs were introduced at various stages to accommodate the continuous technological advancements, reflecting the significant evolution witnessed during the committee's tenure.

Refer to this link to read about AICC documentation: `https://github.com/ADL-AICC/AICC-Document-Archive/`.

Integration with LMS: AICC facilitates integration with learning management systems (LMSs), allowing administrators, instructors, mentors, and trainers to track individual learner progress and performance outcomes effectively across all training programs offered within their organization's LMS platform.

Advantages: Despite its age, AICC maintains certain advantages and performs strongly in specific areas.

- **Flexible Deployments:** With AICC, content and digital learning platforms can be hosted on separate servers or domains, allowing for more adaptable deployments.

- **Enhanced Security**: AICC supports secure HTTPS data transfers, ensuring the integrity and confidentiality of training data and making it a relevant addition for those prioritizing data security and integrity.

Disadvantages: The prevalence of its downsides outweighs its advantages, contributing to the decline in the popularity of AICC.

- **Complex Data Structure:** AICC's intricate data structure necessitates a multistep process for uploading course content to LMSs, posing challenges for users who need programming backgrounds or coding skills.

- **Decreasing Support:** AICC has been losing support since the disbandment of the AICC committee in 2014, resulting in the absence of updates and diminishing backing from authoring tools and learning management systems.

- **Limited Tracking Capabilities:** AICC lacks comprehensive metrics and tracking abilities, such as the inability to monitor course progress or completion rates, limiting its effectiveness in assessing course efficacy compared to other standards.

- **Compliance Challenges:** AICC compliance relies on meeting nine guidelines or recommendations, potentially leading to systems being technically compliant but lacking essential features, necessitating manual coding efforts, and compromising overall functionality. Similarly, compliant content may not fully meet requirements due to deficiencies in other AGRs.

SCORM

Full Form: Sharable Content Object Reference Model

History: In February 1998, President Bill Clinton and the US Department of Defense joined forces to establish Advanced Distributed Learning (ADL), a pioneering initiative to advance eLearning standardization and enhance online education's cost-effectiveness.

Rather than embarking on a novel endeavor, ADL opted to draw upon existing specifications pioneered by initiatives like AICC. This approach led to the "Sharable Content Object Reference Model" (SCORM), which amalgamated the best practices of the time rather than introducing a new standard.

Evolution of SCORM: The SCORM project underwent several iterations over the years, culminating in its final release in 2009 with "SCORM 2004 4th Edition." This marked the apex of SCORM's development, as subsequent evolution beyond this point ceased. Let's have a look at the different versions of SCORM.

SCORM 1.0, released in January 2000, served as an initial blueprint for the SCORM framework, offering a preview of forthcoming developments rather than a fully implementable specification. It outlined the fundamental components of SCORM, including content packaging, runtime communication, and metadata description.

SCORM 1.1, launched in January 2001, marked the first tangible and executable iteration of the SCORM framework. It transformed the conceptual outline of SCORM 1.0 into a practical specification, prompting commercial vendors to embrace it. Early adoptions underscored the SCORM concept's validity while highlighting the need for further refinement to ensure robustness for widespread implementation.

SCORM 1.2, launched in October 2001, represented a significant milestone for SCORM adoption. Building upon the insights gleaned from early implementations of SCORM 1.1, it culminated in a resilient and actionable specification. Adopting vendors witnessed substantial cost savings attributable to enhanced content interoperability.

SCORM 2004, initially known as SCORM 1.3, addressed the shortcomings of SCORM 1.2, notably ambiguities and the absence of sequencing and navigation specifications. It introduced a mature framework with refined content packaging, runtime, and metadata standards. A new addition, "Sequencing and Navigation," allowed content creators to define rules for user progression between content objects. Despite its original designation as "1st Edition," SCORM 2004 encompassed various editions, serving as a comprehensive standard for eLearning content interoperability.

SCORM 2004 2nd Edition, launched in July 2004, swiftly followed the initial adoption of SCORM 2004, addressing identified defects in the framework. ADL's prompt response led to the widespread adoption of this updated specification, resulting in the emergence of various implementations.

SCORM 2004 3rd Edition, launched in October 2006, focused primarily on refining the sequencing and navigation specification to address complexities and enhance interoperability. It introduced user interface requirements for learning management systems (LMSs), ensuring consistent functionality across different systems.

SCORM 2004 4th Edition, released in March 2009, continues to refine the sequencing specification while introducing additional features to enhance the options available to content authors. These new features simplify the creation of sequenced content, further streamlining the content development process.

How It Works: SCORM operates on two fundamental principles: packaging content and facilitating data exchange during runtime (Figure 10-4).

- Packaging content, or the content aggregation model (CAM), outlincs thc methodology for tangibly delivering content. Central to SCORM packaging is the "imsmanifest" document, encompassing all necessary information for the learning management system (LMS) to import and initiate content seamlessly. This XML-based file delineates the course structure from both a learner's perspective and a file system standpoint, addressing document launches and content identification queries.

- Runtime communication, or data exchange, governs how content interacts with the LMS while in use. This facet encompasses delivery and tracking mechanisms. Communication entails two primary elements: content locating the LMS and subsequent interaction through "get" and "set" calls, utilizing a designated vocabulary. These interactions enable various rich and interactive experiences to be conveyed to the LMS, such as requesting learner information or reporting assessment scores.

SCORM is the conduit for communication and data modeling, facilitating the seamless collaboration between eLearning content and the LMS. To deploy e-courses via an LMS, SCORM relies on three key components:

- **Content Packaging:** This process culminates in creating a ZIP file containing the entire course hierarchy.

- **Runtime Environment:** This component launches the course within a web browser, enabling learners to access and interact with the content.

- **Sequencing:** Sequencing ensures a structured navigation experience throughout the course by dictating the progression path for learners.

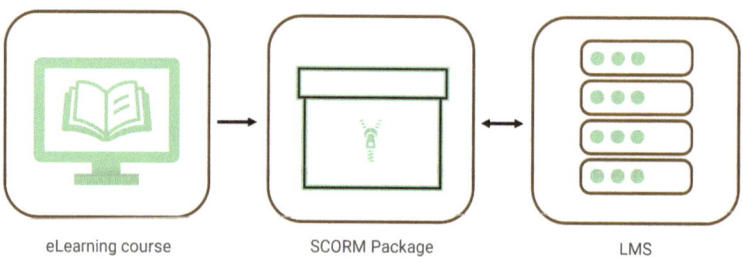

eLearning course   SCORM Package   LMS

Figure 10-4. *Working of SCORM*

Integration with LMS: By incorporating SCORM, learners gain effortless access to learning materials directly within a web browser interface. Moreover, SCORM facilitates the collection of data from diverse learning activities, enabling the tracking of learner progress and the provision of tailored educational feedback.

Advantages: Various factors contribute to SCORM's status as the industry's favored and widely adopted standard.

- **Streamlined Content Organization:** SCORM simplifies creating course catalogs by defining essential metadata standards. This facilitates the construction of catalogs irrespective of content origin, eliminating the necessity of being bound to only single content provider or authoring tool.

- **Enhanced Learning Sequencing:** SCORM enables the establishment of rules dictating the duration learners must spend on specific pages or sections before advancing. Additionally, it allows for the regulation of overall course completion time, fostering a structured learning experience.

- **Flexible Content Integration:** SCORM empowers educators to seamlessly blend content from diverse sources within a course. For instance, they can integrate eLearning modules crafted using different authoring tools, enhancing the richness and diversity of course materials.

- **Simplified Content Creation:** Utilizing SCORM, educators can leverage user-friendly authoring tools to develop interactive eLearning content without requiring advanced coding expertise. This accessibility democratizes content creation, enabling educators of varying technical proficiencies to produce engaging learning materials.

- **Wide-Ranging Authoring Tool Compatibility:** The vast majority of authoring tools extend support to both versions of SCORM, ensuring compatibility and interoperability across various content creation platforms.

- **Extensive LMS Integration:** Virtually all learning management system (LMS) providers accommodate SCORM content, facilitating the seamless migration of courses between different systems. This compatibility simplifies the transition from legacy LMS platforms to newer solutions.

- **Standardized Content Archiving:** SCORM facilitates the archival of outdated content in a standardized ZIP format, ensuring documentation integrity and facilitating efficient storage management.

Disadvantages: While SCORM remains a preferred eLearning standard, it has certain drawbacks.

- **Absence of Offline Learning Support:** SCORM courses rely on continuous Internet connectivity and a web browser. This means learners cannot access courses offline, potentially leading to disruptions and discontinuations if Internet connections are unstable.

- **Exclusive Dependency on LMS Deployment:** SCORM tracking is exclusively facilitated through learning management systems (LMSs). Deploying content directly via websites or mobile applications is not feasible with SCORM integration.

- **Limited Native HTML5 Compatibility:** Initially rooted in Flash technology, SCORM content faces challenges with the transition to HTML5. Although modern authoring tools can convert SCORM content into HTML5 format, the quality may not match native HTML5 content, particularly when incorporating multimedia elements such as videos.

- **Constrained Tracking Metrics:** SCORM restricts the range of metrics that can be tracked, typically encompassing course completions, overall time spent on courses, and assessment performance scores.

- **Constrained Reporting Capabilities:** Reporting capabilities derived from SCORM data are restricted, primarily focusing on the tracking metrics as mentioned previously. This limitation hinders the depth and breadth of insights gleaned from course analytics.

xAPI (Tin Can)

Full Form: Experience Application Programming Interface

History: In 2008, the Learning Education Training Systems Interoperability (LETSI) organization voiced concerns about the limitations of SCORM in meeting the demands of modern eLearning. They responded by unveiling a series of critical issues with the existing standard:

- Challenges in data transmission when the learning management system (LMS) and learning course are hosted on different domains

- Dependency on continuous Internet connectivity for recording learning statistics

- Identified security vulnerabilities

- Restricted scope of parameters available for capturing statistics

Acknowledging these shortcomings, the Advanced Distributed Learning (ADL) organization collaborated with Rustici Software to devise an enhanced standard aligning with contemporary requirements, dubbed "Project Tin Can."

In 2013, the initial iteration of Tin Can, labeled Tin Can 1.0, was introduced, garnering adoption from pioneering entities. Concurrently, ADL gave the project its official nomenclature: the "Experience API" (xAPI).

Continual enhancements were made to the standard until 2016, culminating in the release of version 1.03, regarded as a finalized iteration with no immediate plans for further updates.

How It Works: Let's understand the concept of an API through a real-life "application." Imagine you want to book a flight. You visit a travel app or website and enter your desired departure and arrival locations and travel dates. Behind the scenes, this website or app uses APIs to communicate with various airlines' booking systems.

When you submit your flight search, the travel website's API sends your request to the APIs of different airlines. Each airline's API checks its database for available flights matching your criteria. Once it finds suitable options, it sends back the flight details, including prices and schedules, to the travel website's API.

The travel website's API then compiles all the flight options from different airlines and presents them to you in a user-friendly format. You select your preferred flight, and the travel website's API communicates your booking details back to the respective airline's API to confirm your reservation.

In this scenario, the travel website's API bridges you (the user) and the airlines' booking systems, enabling seamless communication and interaction. Just like you don't directly interact with each airline's booking system to find and book a flight, you would need to use APIs to facilitate communication in a website or app to access different services or databases.

Differences Between API and xAPI: API-based solutions function like a Q&A session within the eLearning landscape. When the learning management system (LMS) seeks information, it poses a query to the API, like asking for the latest course list. The API then relays this inquiry to a third-party website, retrieves the information, and furnishes it back to the LMS as a response.

However, while xAPI shares similarities with traditional APIs, it operates uniquely as a set of data standards within the broader LMS framework (Figure 10-5). It regulates how questions are asked and answered, simplifying the process for eLearning solution designers and developers. This team needs to determine the specific interactions to monitor, referred to as statements. These statements are written in JSON (JavaScript Object Notation), which shares similarities with XML.

For instance, consider Mike, an employee undergoing a mobile tutorial about using company software. xAPI generates statements in a structured format detailing Mike's interactions with eLearning content. These statements, comprising a noun (Mike), a verb (action), and an object (content), offer insights into Mike's learning journey, including areas of proficiency and struggle.

xAPI empowers developers to craft richer experiential learning interactions by capturing detailed data through a specialized Learning Record Store (LRS) database. This LRS captures granular information about each learning interaction, surpassing the capabilities of standards like SCORM.

In practice, xAPI facilitates real-time tracking of diverse learning activities across various devices, including educational games, videos, and software usage. By extracting detailed navigation data and post-training software interactions, xAPI enriches the learning experience and provides developers with insights into informal learning activities beyond the LMS environment.

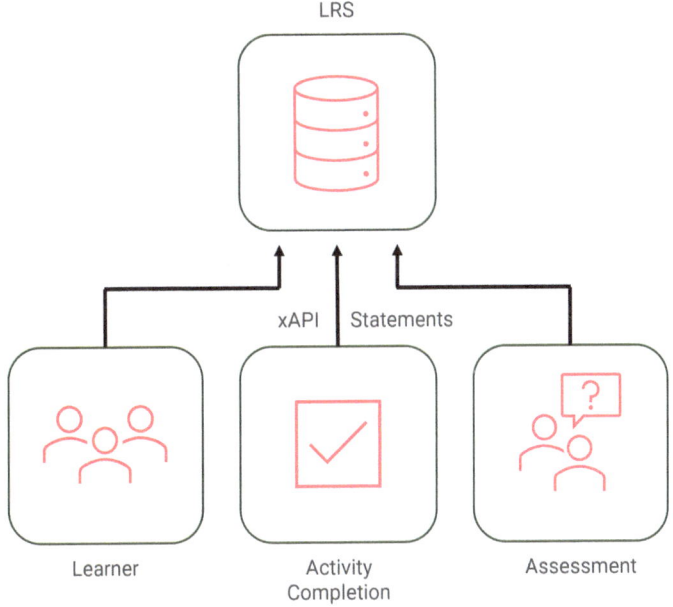

Figure 10-5. *Working of xAPI*

Integration with LMS: Integration with learning management systems (LMSs) of xAPI enables the comprehensive tracking and collection of diverse data about each employee's learning journey and advancement. This includes insights into offline and interactive learning activities, real-world performance, eBooks and white papers downloads, quiz scores, and more. Every interaction, from clicks to comments, is automatically logged, offering a holistic view of the learner's progress across various learning modalities and contexts. xAPI's ability to convert this data into quantifiable actionable insights enhances the effectiveness of the learning experience, transforming raw information into practical knowledge.

Advantages: xAPI builds upon the foundation of SCORM, offering additional benefits and functionalities.

- **Comprehensive Activity Recording:** The versatile structure of xAPI statements accommodates nouns, verbs, and objects, enabling the recording of a wide array of learning activities, thereby ensuring comprehensive tracking capabilities.

- **Expanded Tracking Capabilities:** With xAPI and Learning Record Stores (LRS), tracking learning activities extends beyond the confines of an LMS. LRS, situated anywhere on the Internet, can capture learning data from various systems, enhancing the scope of tracked eLearning scenarios.

- **Cross-Device Compatibility:** Tracking initiates seamlessly across devices, ensuring continuity regardless of the learner's choice of technology. xAPI statements can originate from diverse sources such as mobile phones, gaming consoles, medical equipment, hardware simulations, and more.

- **Enhanced Data Security:** Learners can securely store their learning data in designated "personal data lockers," facilitating secure storage and transfer of information between organizations when necessary.

Disadvantages: While xAPI boasts numerous tracking capabilities, it also has certain limitations.

- **Limited Impact Assessment:** While xAPI captures primary data indicating learner activity, its reporting often needs more depth to draw qualitative or quantitative conclusions regarding the overall impact of the activity.

- **Complex Metrics Configuration:** Due to its expansive data tracking capabilities spanning various learning activities, comprehensive performance analytics with xAPI demands significant time and effort to derive actionable insights effectively.

cmi5

Full Form: Computer Managed Instruction

History: Following the dissolution of AICC in 2014, a cohort of its experts embarked on a fresh initiative. Recognizing the broad applicability of xAPI and its potential to accommodate diverse implementations, they aimed to streamline the adoption of xAPI within standard LMS-centric learning environments.

In 2016, they introduced what they termed a "companion standard for xAPI," christened "cmi5," drawing inspiration from AICC's inaugural interoperability specification, "CMI001."

Evolution: cmi5 combines the flexibility of xAPI for tracking various learning activities with the structured framework of SCORM, which has been a cornerstone in learning technologies and systems.

How It Works: cmi5 introduces a framework comprising four essential components (Figure 10-6):

- **Assignable Units (AUs):** These represent individual learning activities within a course. Each AU is designed to capture comprehensive learner data using xAPI (Experience API), enabling a deeper understanding of learner progress and performance.

- **Course Package:** This encompasses the overall structure of the course, delineating AUs and their accompanying metadata. It serves as a blueprint for organizing learning content and facilitating seamless navigation for learners.

- **Learning Management System (LMS):** The LMS serves as the central hub for managing the learning process. It initiates the launch of AUs, prepares launch data, and meticulously records learner interactions and progress. Additionally, the LMS orchestrates data transfer to the Learning Record Store (LRS) for comprehensive tracking and analysis.

- **Learning Record Store (LRS):** Acting as the repository for learner data, the LRS receives, stores, and retrieves xAPI statements generated by AUs. By accumulating learner activities, scores, mastery levels, and completion statuses, the LRS offers valuable insights into learner behavior and performance.

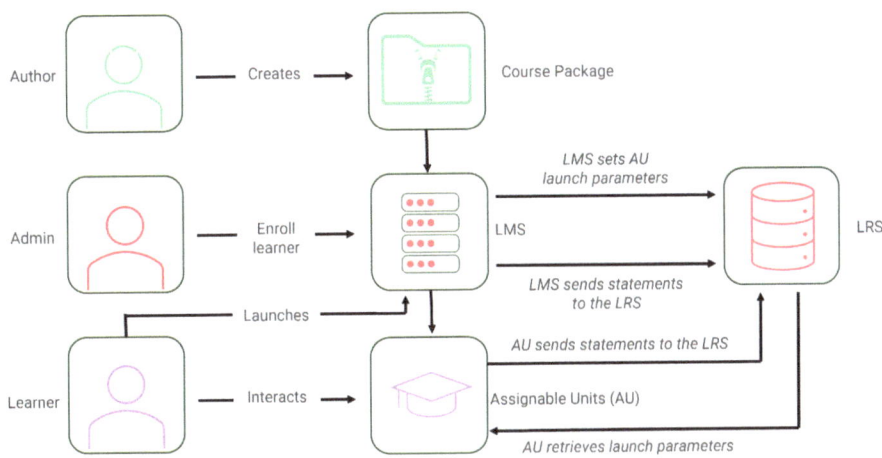

Figure 10-6. *Working of cmi5*

In practice, content authors craft AUs with embedded xAPI capabilities to capture nuanced learner data. These AUs are then integrated into the course structure, accompanied by instructions and satisfaction rules. Upon initiating the launch of an AU, learners are seamlessly redirected by the LMS to engage with the designated learning content. As learners interact with AUs, their activities, progress, and achievements are meticulously logged and relayed to the LRS for comprehensive tracking and analysis, fostering a dynamic and informed learning environment. Read more about cmi5 at `https://github.com/AICC/CMI-5_Spec_Current/blob/quartz/cmi5_spec.md`.

Integration with LMS: Integrating cmi5 with an LMS enhances instructors' control over pacing and sequencing during course delivery. It also offers detailed tracking capabilities, enabling administrators, instructors, mentors, and trainers to monitor learner performance and assess mastery levels effectively.

Advantages: Since it utilizes xAPI communication, it provides some advantages in tracking details.

- **Enhanced Statement Precision:** Leveraging cmi5 as a specific implementation of xAPI offers the advantages inherent to xAPI, with the added benefit of precise statement categorization. Assignable Units facilitate capturing session details and core eLearning metrics such as pass/fail, content completion, duration, and score, allowing for structured data grouping.

Disadvantages: Despite providing a flexible solution combining SCORM and xAPI aspects, cmi5 has some limitations.

- **Rigorous Constraints:** While cmi5 provides specific benefits, its implementation imposes stricter limitations than the more versatile xAPI. These constraints may hinder the seamless handling of large volumes of data typically managed by xAPI.

LTI

Full Form: Learning Tools Interoperability

History: The LTI standard, developed by the IMS Global Learning Consortium, revolutionizes how learning tools integrate with LMS. Originally designed to streamline the integration of various educational tools into LMS platforms, LTI has evolved into a vital component of modern educational technology.

For instance, consider a university LMS that hosts online courses. The university wants to incorporate a language learning app for foreign language learners. Without LTI, integrating this app into the LMS would require custom development work to align it with the LMS infrastructure.

However, with LTI, the language learning app seamlessly plugs into the LMS. Learners can access it directly through the LMS interface, eliminating the need for separate logins. Their progress and performance data are seamlessly synchronized with the LMS, simplifying administrative tasks and enhancing the learning experience.

LTI facilitates the effortless data exchange between the LMS and external learning tools, enabling educators to enrich their courses with diverse resources while maintaining a cohesive user experience for learners.

Evolution: LTI has undergone several versions, each addressing evolving needs in educational technology integration:

- **LTI 1.0 (2010):** Introduced the specification for seamlessly integrating external learning tools into LMS platforms, enabling users to launch into external tools from within the LMS interface.

- **LTI 1.1 (2012):** Enhanced communication between LMS and tools by allowing the passing of additional information and facilitating actions like grade passback.

- **LTI 1.2 and 2.0 (2014):** Developed to accommodate the increasing complexity of learning tools, with LTI 2.0 introducing REST web services for deeper integrations. However, due to complexities and low adoption rates, LTI 2.0 saw limited industry acceptance.

- **LTI 1.3:** LTI 1.3, built on the simplicity of LTI 1.1, adopts modern security standards like OAuth 2.0 and JSON Web Tokens.

- **LTI Advantage:** It is an asset of services that offers services like Names and Roles Provisioning, Assignment and Grade Services, and Deep Linking on the LTI 1.3 security model.

How It Works: The functionality of LTI enhances the educational experience by seamlessly integrating external tools into the LMS (Figure 10-7). This integration allows learners to access diverse content without needing to navigate away from the familiar LMS interface. As a result, learner engagement is heightened, and teaching and learning outcomes are optimized. LTI workflow operates as follows:

- **Setup by Instructor/LMS Administrator:** Initially, the instructor or LMS administrator accesses an external tool and obtains a URL, a key, and a "secret" necessary for connecting the tool to the LMS.

- **Integration into Course Structure:** The instructor incorporates a link to the external tool into the LMS's course structure. The LMS administrator may add this link manually or automatically. The URL, key, and secret are also inputted as metadata to enable seamless connectivity.

- **Learner Interaction:** When a learner clicks on the provided link within the LMS, a request is generated and sent to the external tool. This request contains the learner's identity and other pertinent information, such as their role and course details, as well as the key and a signature. This information is transmitted through an HTTP request utilizing the OAuth standard for secure authentication.

- **Seamless Transition to External Tool:** The external tool seamlessly redirects the learner upon receiving the request. This transition may occur within the same browser window or in a new one. From the learner's perspective, they haven't navigated away from the LMS interface, ensuring a smooth and uninterrupted user experience.

516

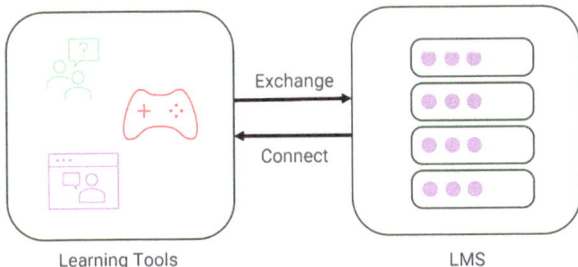

Figure 10-7. *Working of LTI*

Integration with LMS: Integrating LTI into LMSs involves adopting a technical standard, not a product, to seamlessly integrate learning tools within the institution's learning environment. It provides a secure foundation for building the institution's digital ecosystem. When people mention "LTIs," they typically refer to "LTI Tools" or applications that integrate into an LMS using the LTI standard. This integration ensures secure launches, user interface integration, single sign-on functionality, and the exchange of essential context information between the learning environment and external tools. Additionally, LTI facilitates secure communication of data such as scores and grade books from the learning application back to the learning environment, enhancing the overall functionality and usability of the LMS.

Advantages: Here are some benefits of integrating LTI into the LMS.

- **Streamlined Integration**: LTI ensures seamless integration of learning tools, enabling smooth navigation between connected tools.

- **Efficient Authentication:** LTI supports single sign-on (SSO), providing fast and secure authentication for users.

- **Expansion of Sales Channels:** Tool providers benefit from LTI integration as it allows them to plug additional tools into their software, expanding their user base.

- **Cost-Effective Enhancements:** Consumers can enrich their platforms with third-party tools without significant expenditure on eLearning development.

Disadvantages: While LTI can integrate various applications into a single LMS, it also has several drawbacks.

- **Limited LMS Support:** Not all LMSs support LTI, restricting its accessibility. For instance, popular platforms like Google Classroom and Microsoft Teams do not support LTI.

- **Incomplete Data Access:** LTI integrations may lack access to certain LMS features, such as the learner's agenda, necessitating additional solutions like xAPI for comprehensive data communication.

- **Mediocre User Experience:** Integrated tools often operate within an iframe, limiting the user experience of the LMS or application.

Comparing eLearning Standards

When it comes to eLearning initiatives, choosing the proper standard can make all the difference in achieving your organization's goals effectively. Table 10-1 lists the comparison between discussed eLearning standards. The ideal scenarios for utilizing these eLearning standards are as follows:

- **AICC Standard for Legacy Systems:** When dealing with legacy systems built around the AICC standard and where compatibility, data integrity, regulatory compliance, and cost considerations are paramount, it is ideal to continue using AICC for eLearning initiatives.

- **SCORM Standard for Corporate Compliance Training:** For corporate compliance training initiatives that require structured data collection, seamless integration with existing LMS platforms, and centralized learning analytics capabilities, choosing the SCORM standard is the ideal solution.

- **xAPI Standard for Adaptive Learning Platforms:** For adaptive learning platforms that aim to gather rich and varied learning data, integrate seamlessly with LXPs, and enhance adaptive learning algorithms through data-driven insights, choosing the xAPI standard is the ideal solution.

- **cmi5 Standard for Collaborative Training Initiatives:** A standardized data tracking and interoperability approach is essential for collaborative training initiatives involving multiple providers with disparate LMS platforms. Adopting the cmi5 standard enables seamless cross-provider collaboration, ensures compliance with industry regulations, and enhances the depth of learner data insights.

- **LTI Standard for Integrated Learning Tools**: For educational institutions seeking to integrate diverse learning tools and resources into their LMS to enhance the learning experience and streamline administrative processes, choosing the LTI standard is the ideal solution. LTI enables seamless integration, single sign-on authentication, data exchange, and customization, empowering instructors to create engaging and personalized learning environments for their learners.

Table 10-1. *Comparison of eLearning Standards*

Parameters	AICC	SCORM	xAPI	CMI5	LTI
Course sequence	Limited support for defining a sequence of learning activities.	Strong support for sequencing with prerequisites and dependencies.	Flexible sequencing capabilities.	Flexible sequencing similar to xAPI.	Limited sequencing support often relies on external tools or standards for this functionality.
Progress tracking	Basic progress tracking capabilities.	Robust progress tracking features, including completion status and score.	Extensive tracking capabilities capture a wide range of learning activities.	Comprehensive progress tracking is similar to xAPI.	Progress tracking varies depending on the LTI tool provider but can support basic tracking.
Third-party platforms integration	No integration with third-party applications.	Limited integration capabilities.	Extensive support for integrating external applications.	Extensive integration capabilities similar to xAPI.	Strong support for integrating external tools and applications within learning environments.

Advanced tracking	Basic tracking of course completion and fundamental interactions.	Detailed tracking of learner interactions, completion status, and assessment scores.	Highly flexible tracking capturing various learning experiences and interactions beyond traditional eLearning courses.	Advanced tracking capabilities similar to xAPI.	Tracking capabilities vary based on the LTI tool provider but generally support basic tracking.
Mobile support	No support for mobile devices.	It's not mobile-compatible.	Fully mobile-friendly.	Fully mobile-friendly.	Mobile-friendly, but depends on the tools and platforms used.
Popularity	Less popular and less widely adopted compared to other standards.	Widely adopted in the eLearning industry.	Increasing popularity due to its flexibility and advanced tracking capabilities.	Less widely adopted compared to SCORM and xAPI.	Widely adopted for integrating learning tools and platforms.

(continued)

521

Table 10-1. (*continued*)

Parameters	AICC	SCORM	xAPI	CMI5	LTI
Implementation	It is relatively straightforward to implement.	Requires adherence to specific packaging and data model standards.	More flexible implementation due to its data model.	More complex to implement compared to SCORM and xAPI.	Implementation complexity varies depending on the LTI tool provider and integration requirements.
Security	High, due to its implementation of tighter control mechanisms over data transmission and authentication processes, providing a more robust framework for securing eLearning content and user data.	Lower, it has limited encryption and authentication mechanisms.	High, supports advanced security measures such as authentication and encryption.	High, similar security features to xAPI.	High, security features depend on the LTI tool provider but generally support secure data transmission.

Summary

- Authoring tools are software applications used to create and develop interactive eLearning content without requiring extensive programming knowledge, enabling users to design, edit, and publish digital learning materials efficiently.

- eLearning standards provide guidelines for content interoperability, ensuring compatibility across different platforms and facilitating seamless delivery, tracking, and assessment of online learning materials.

- AICC is a standard for computer-based training in the aviation industry, providing guidelines for interoperable online learning content and systems.

- SCORM is a standardized framework facilitating the packaging, delivery, and tracking of eLearning content within learning management systems (LMSs), ensuring interoperability and ease of content management.

- xAPI, or Experience API, enables comprehensive tracking and analysis of learning experiences by capturing a wide range of data across various learning activities and environments, transcending the limitations of traditional tracking methods.

- cmi5 is a profile of xAPI that facilitates communication between learning management systems (LMS) and learning activities, offering enhanced tracking capabilities and flexibility for course delivery.

- LTI facilitates seamless integration of learning tools within an institution's LMS, providing a secure foundation for building a digital ecosystem and enabling efficient exchange of data and context information between the learning environment and external tools.

Let's Brainstorm

These mini-scenarios will make you scratch your head and scribble on your pad.

Mini-scenario 1

You're a training manager tasked with selecting an authoring tool for your team's eLearning projects. The team comprises both seasoned instructional designers and newcomers to eLearning development. The courses you plan to develop vary from interactive simulations to straightforward compliance training modules. Additionally, your organization requires compatibility with SCORM and xAPI standards for seamless integration with the existing learning management system. Considering the diverse skill levels of your team members and the varied nature of your eLearning projects, how would you prioritize the selection of an authoring tool? What factors would you consider regarding usability, complexity, and compatibility with output standards such as SCORM and xAPI?

Mini-scenario 2

You're a training coordinator tasked with implementing your organization's new eLearning course for employee onboarding. The course curriculum is structured with prerequisite modules and dependencies, requiring learners to complete certain sections before advancing to others. Progress tracking is critical for HR in monitoring employee training completion rates and performance. However, due to budget constraints, your organization's IT infrastructure limits the integration capabilities of external applications. Which eLearning standards would you choose in this situation and why?

Mini-scenario 3

You're an instructional designer tasked with developing a series of compliance training modules for a multinational corporation. The training content must be dynamic and interactive, allowing learners to engage with simulations and scenarios tailored to their specific roles and responsibilities. Additionally, the organization aims to track learner interactions beyond traditional course completion, including real-world performance metrics and on-the-job application of knowledge and skills. Which eLearning standard would you suggest to the development team and why?

Mini-scenario 4

As a learning technology specialist in a healthcare organization, you're tasked with implementing a new training program for medical staff to comply with updated regulatory requirements. The training program consists of interactive simulations and assessments tailored to different organizational job roles. Real-time tracking of learner interactions and performance is crucial to ensure compliance and identify areas for improvement. However, the organization's existing learning management system (LMS) lacks native support for cmi5, and there are concerns about the complexity and compatibility of integrating this new standard. Given the complex nature of the training program and the need for comprehensive tracking capabilities, how would you approach the implementation of cmi5 in the organization's learning ecosystem? What strategies would you use to address concerns regarding integration complexity, compatibility with the existing LMS, and ensuring seamless tracking of learner interactions and performance data in compliance with cmi5 standards?

Mini-scenario 5

You're an educational technology consultant working with a consortium of universities aiming to create an innovative online learning platform that fosters collaborative learning experiences across multiple institutions. The platform aims to seamlessly integrate various educational tools and resources, allowing instructors to customize learning paths and assessments for diverse learner populations. However, each university has its unique learning management system (LMS) and administrative protocols, posing a significant challenge in achieving interoperability and consistency across the platform. Additionally, ensuring data security and privacy while facilitating seamless data exchange between different LMSs presents a complex technical and logistical hurdle. Which eLearning standard would you suggest the consortium implement and why?

CHAPTER 11

AI in Instructional Design: Foundation, Innovations, and Ethical Considerations

In the dynamic landscape of education, the symbiosis between the science and art of instructional design has long been a catalyst for practical learning experiences. As instructional designers strive to create engaging and impactful courses, they navigate the intersection of pedagogy, technology, and creativity. However, generative artificial intelligence (AI) has ushered in a new era, propelling instructional design into uncharted territories.

In this ever-evolving educational ecosystem, staying abreast of current trends and embracing paradigm shifts are imperative for instructional designers to craft courses that resonate with learners. Combining traditional methodologies with cutting-edge AI technologies introduces unprecedented possibilities for creating instructional content. This transformation accelerates the development process and opens avenues for previously inconceivable innovation.

© Ankita Jiyani Mangtani 2024
A. J. Mangtani, *Instructional Design Unleashed*, Design Thinking,
https://doi.org/10.1007/979-8-8688-0416-8_11

This exploration into "AI in Instructional Design: Foundation, Innovations, and Ethical Considerations" goes beyond the boundaries of conventional wisdom. As we delve into generative AI, we witness the acceleration of the design process, the augmentation of creativity, and the potential for personalized learning experiences. The traditional approaches that once formed the bedrock of instructional design are now complemented and enhanced by the capabilities of AI algorithms.

Exploring Top Generative AI Platforms

This chapter comprehensively analyzes the top three tried and tested platforms across distinct categories in the ever-evolving landscape of generative AI. Delving into content generation, storyboarding, and multimedia creation tools, such as video, static image, avatar video, sound generation, and synthetic voices, we present a nuanced examination to aid instructional designers in making informed decisions.

The first category under scrutiny is content generation, a pivotal aspect of instructional design. We have meticulously evaluated three leading platforms in this domain, dissecting their capabilities and performance. Each tool's proficiency in generating diverse instructional content, from text-based content to dynamic interactive modules, is scrutinized. Our comparison focuses on the current versions of these tools, ensuring relevance and accuracy in our assessment.

Storyboarding is an art form that bridges creativity with structure in instructional design. This section scrutinizes the top three generative AI platforms for their prowess in storyboarding. We assess their ability to translate ideas into visual narratives, fostering a seamless transition from concept to course content. Our analysis explores the tools' features, ease of use, and adaptability, providing valuable insights for instructional designers seeking efficient storyboarding solutions.

The third and equally crucial category encompasses multimedia creation tools. We examine the capabilities of the chosen platforms in video creation, static image generation, avatar video production, sound generation, and synthetic voice integration. The evaluation considers the tools' efficiency, versatility, and suitability for different use cases within instructional design. Our comparisons are conducted meticulously, offering a comprehensive overview of each platform's strengths and limitations.

Each comparison within this chapter is conducted precisely within the education and learning domain, ensuring a thorough examination of the selected generative AI platforms. Parameters such as ease of use, speed of content creation, adaptability to various instructional contexts, and the overall user experience are considered. The insights provided are rooted in the generative AI tools' capabilities during this assessment, emphasizing the criticality of remaining updated in this swiftly evolving field for various instructional use cases.

The dynamism of the generative AI landscape means that the efficacy of these platforms may evolve. Readers are encouraged to try the tools themselves after this comparison, as the technological landscape is dynamic, and newer versions or capabilities may have emerged after our analysis. This chapter aims to provide a snapshot of the tools' capabilities at a specific time, offering a foundation for informed decision-making in generative AI for instructional design.

Parameters for Comparative Analysis

When comparing the quality of output generated by different generative AI tools for various use cases, the following parameters (Figure 11-1) are used to analyze the output and are categorized under output quality, user experience, and performance and efficiency. Significantly, specific parameters pertinent to the tools category are inherently clear and will be elaborated upon in subsequent sections of this chapter. Additionally, it is essential to note that not all tools were

compared across every subparameter, as depicted in the image. Please
note that all the tools used for various purposes in this comparative study
are paid and licensed versions.

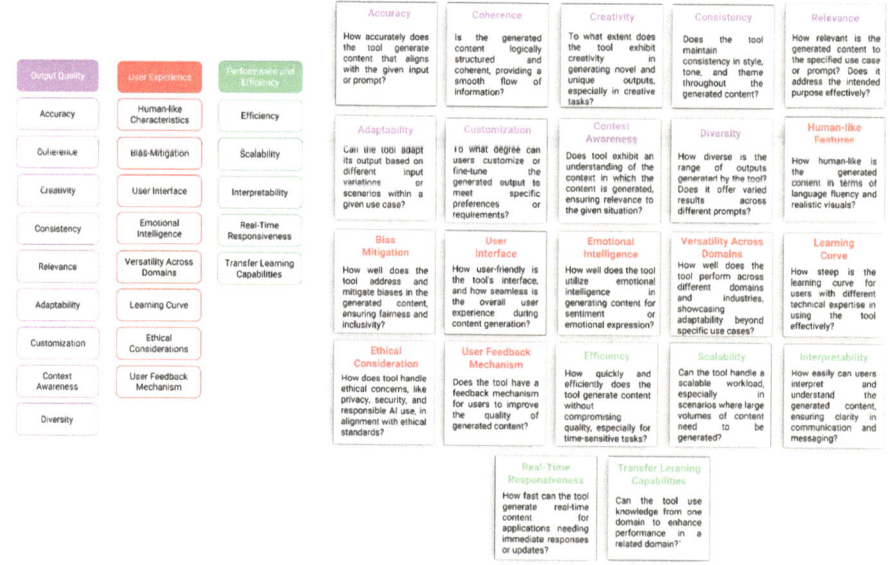

Figure 11-1. *Parameters for Comparative Analysis*

Content Generation Platforms

The content generation tool under consideration proves its versatility
across a spectrum of scenarios, showcasing its adaptability and efficiency.
From crafting concise articles to analyzing audio scripts and generating
diverse content types, this tool is a dynamic solution for various content
creation needs. Table 11-1 presents the differences between the three
tools: ChatGPT-4, Gemini, and Perplexity. The following are key
applications taken into consideration for this comparison:

- Crafting concise articles on specific topics (under
  500 words)

- Analyzing and extracting content from audio scripts
  (YouTube videos or podcasts)

- Generating a list of the top ten current affairs

- Crafting poems on specified themes following the
  writing style of the selected poet

- Writing product reviews based on provided information

- Creating engaging social media posts for
  specific themes

- Generating learning materials for educational courses

- Summarizing lengthy research papers or documents

- Translating the content into regional and foreign languages

- Creating captivating speeches or presentations based
  on key points and themes

Table 11-1. *Analysis of Content Generation Platforms*

Parameters	ChatGPT-4	Google Bard/ Gemini	Perplexity
Accuracy	Moderate, close to human accuracy and style with natural and engaging conversation. It may only produce reliable results for factual information sometimes.	Low, not reliable for factual information.	High, ensures precise and reliable answers

(*continued*)

531

Table 11-1. (*continued*)

Parameters	ChatGPT-4	Google Bard/ Gemini	Perplexity
Creativity	High, capable of providing creative and varied writing styles.	Moderate, may depend on the specific design and purpose.	Variable, depends on the fine-tuning and model customization.
Consistency	Moderate, generally consistent in maintaining the context and tone throughout a conversation.	Low to moderate, strive for consistency in maintaining poetic style and thematic coherence.	Moderate, the consistency of output depends on what kind of input was given.
Relevance	High, focuses on generating relevant responses based on input, but occasional lapses may occur.	Low, emphasizes relevance in producing poetic content aligned with given themes.	Moderate, produce relevant content and share the additional links for more information.
Integration with Services	Limited, primarily a stand-alone model.	Robust and seamless integration with Google services.	Variable, depends on how it's implemented and integrated.
Customization	Limited fine-grained control; output may be less customizable.	Moderate, provides control within Google's ecosystem.	High, users can have more control over parameters and customization.

(*continued*)

Table 11-1. (*continued*)

Parameters	ChatGPT-4	Google Bard/ Gemini	Perplexity
Context Awareness	High, demonstrates a good understanding of the context within conversation only.	Moderate to high, maintain the context awareness within a domain only.	Moderate to high, maintain a good understanding of the context in various domains.
Human-Like Features	High, strives to emulate human-like conversation, exhibiting empathy and understanding.	High, emphasizes human-like expression in poetic output, aiming for emotional resonance.	Low, focuses on gathering useful information and may lack empathy, understanding, and emotional resonance.
Bias Mitigation	Low, ongoing efforts to mitigate biases, but not perfect.	Low, limited information available, but likely focuses more on creative expression than bias mitigation.	Low, likely to focus on precise information and additional links to support the generated output.
User Interface	Easy to use and generates required output.	Easy to use and generates required output.	Moderate to use, it may take time to get used to the platform.

(*continued*)

Table 11-1. (*continued*)

Parameters	ChatGPT-4	Google Bard/ Gemini	Perplexity
Learning Curve	Low, user-friendly for nontechnical users.	Moderate, familiarity with Google's ecosystem is beneficial.	High, requires a deeper understanding of machine learning concepts for customization.
Efficiency	High, generally efficient in generating responses quickly.	Moderate, efficiency may vary depending on the complexity of poetic generation algorithms.	Moderate, efficient in generating precise and factual information.
Real-Time Responsiveness	High, capable of providing real-time responses in conversations.	Moderate to high, real-time responsiveness may depend on the implementation and computational resources available.	Moderate to high, responsiveness may depend on the amount, relevance, and context of information needed.

(*continued*)

Table 11-1. (*continued*)

Parameters	ChatGPT-4	Google Bard/ Gemini	Perplexity
Performed Extraordinarily In	All except for what was mentioned for perplexity AI.	Crafting poems on specified themes following the writing style of the selected poet. Creating engaging social media posts for specific themes. Creating captivating speeches or presentations based on key points and themes	Generate a list of the top 10 current affairs Summarizing lengthy research papers or documents.

Storyboard Generation Platforms

In the dynamic landscape of eLearning, AI storyboarding tools have
emerged as invaluable assets, revolutionizing the instructional design
process. These tools create many use cases that streamline course
development and elevate the learning experience. Table 11-2 presents
the differences between the three tools: Boords, Krock, and Storyboard
Hero. The following are key applications taken into consideration for this
comparison:

- Educational videos

- Scenario-based learning

- Microlearning modules

- Digital storytelling

- Animation projects

- Advertising and marketing campaigns

Table 11-2. *Analysis of Storyboard Generation Platforms*

Parameters	Boords	Krock	Storyboard Hero
Accuracy	Moderate to high, translating scripts into visual storyboards aligning with the initial idea.	Moderate to high, translating AI-generated scripts into visual storyboards, ensuring the final output aligns with the original idea.	Lower accuracy compared to Boords and Krock

(*continued*)

Table 11-2. (*continued*)

Parameters	Boords	Krock	Storyboard Hero
Creativity	Low to moderate, limited options to be creative with the visuals of storyboards.	High, using the advanced tools, one can explore stunning visual elements to bring into storyboards.	Moderate, gives few options or tools to be creative in visual styles.
Relevance	Low, the output generated needs to be perfectly aligned with the prompt's script.	Moderate and generates relevant output by giving the prompt and CSV files.	Low to moderate, creates customized output as needed.
User Interface and Learning Curve	Easy, beginners with minimum experience can get started with the platform's intuitive interface.	Easy to moderate, user-friendly interface, making storyboard creation intuitive and efficient, catering to users with varying experience levels.	The user interface could be more intuitive, making it easier for users to create and customize their storyboards.
Efficiency	High, allows users to create storyboards quickly but sometimes misses aligning the visuals with the script.	Moderate, generates scripts and visuals quickly and integrates the elements in the storyboard.	Moderate, generates concepts and scripts in seconds, seamlessly organized into storyboards.

(*continued*)

Table 11-2. (*continued*)

Parameters	Boords	Krock	Storyboard Hero
Visual Assets	Diverse range of storyboard styles and creative tools.	It may need to match the level of stylistic variety and creative toolsets available in Boords.	Does not have a diverse range of visual styles and creative toolsets.
AI Script Generation	Available, generates the script with voice-over and detailed notes quickly with the given prompt about the learners, content, and the depth of the content.	Available, the script generates the scene with a detailed description and on-screen text only. Doesn't generate voice-overs.	Available, generates long storyboards with voice-overs and detailed notes quickly with the given prompts with some additional parameters such as video type and video style.
Text to Image Generation	Available, generates images for one frame at a time by giving the manual prompt for each frame. It has various options to customize, such as camera angle.	Available, it creates images per frame without giving a manual prompt; it will take the details from the script and notes section for each storyboard frame.	Available, it employs AI to generate images for whole storyboard scenes in one go, with an option for hand-drawn images.

(*continued*)

Table 11-2. (*continued*)

Parameters	Boords	Krock	Storyboard Hero
Collaborative Features	Collaborative, allowing teams to work together in real time.	Collaborative with features for team collaboration and feedback.	Offers collaboration features for team-based storyboarding.
AI Integration	AI-driven storyboard, voice-over, and notes for each slide/frame.	AI-driven scene suggestions and templates for efficient story creation.	AI-powered tools for automated scene generation and recommendations.
Storyboard Customization	Provides tools for customizing storyboards with scenes, characters, and text.	Offers customization options for scenes, characters, and various elements.	Allows users to customize storyboards, characters, scenes, and AI-generated images.
Performed Extraordinarily In	• Educational videos • Microlearning modules • Scenario-based learning	• Digital storytelling • Animation projects	• Advertising and marketing campaign

Static Image Generation Platforms

In the dynamic world of visual content creation, static image generation tools are versatile assets with numerous applications. From crafting logos to manipulating typography fonts and creating engaging infographics, these tools offer a range of possibilities. Table 11-3 presents the differences

between the three tools: DALL-e, Midjourney, and Adobe Firefly.
The following are key applications taken into consideration for this
comparison:

- Logo creation

- Text generation for menu cards or content on
  blackboards

- Typography, fonts manipulation

- Generation of geometrical shapes

- Utilization of stock images

- Crafting abstract and visionary images

Table 11-3. *Analysis of Static Image Generation Platforms*

Parameters	DALL-E	Midjourney	Adobe Firefly
Accuracy	Moderate, sometimes, the characters' facial features in the generated image deteriorate.	High, almost matches the expected output.	Moderate to high, provides expected output in many cases.
Creativity	High-quality, focuses on generating diverse and creative images.	Variable, depends on the design and customization options.	High-quality, emphasis on creative image synthesis.

(*continued*)

Table 11-3. (*continued*)

Parameters	DALL-E	Midjourney	Adobe Firefly
Customization	Limited, may offer some degree of customization in image generation.	Extensive, it provides a range of customization options for artwork creation.	Extensive, it offers diverse customization options for graphic design.
Context Awareness	Exceptional nuance and context understanding. It rarely misses a line in the prompt.	Generally acceptable with low-context prompts but struggles with high-context.	Shows some mistakes, even with low-context prompts.
Output Quality	High-quality, known for producing visually appealing and novel images.	Variable, output quality may depend on user proficiency and design choices.	High-quality, emphasizes professional graphic design outputs.
Ease of Use	Moderate, requires an understanding of the tool's capabilities for optimal use.	Moderate, operates through discord server.	Moderate, may require familiarity with Adobe's design ecosystem.
Versatility	Versatile; can generate a wide array of images based on textual prompts.	Specialized for concept art; may be less versatile for nonartistic outputs.	Versatile; suitable for a range of graphic design projects and concepts.

(*continued*)

Table 11-3. (*continued*)

Parameters	DALL-E	Midjourney	Adobe Firefly
Learning Curve	Moderate to high; understanding the model's capabilities is crucial.	Moderate; user-friendly for artists with digital art experience.	Moderate; familiarity with Adobe's design tools may be beneficial.
Integration with Platforms	Stand-alone model, may not be directly integrated into design platforms.	May integrate with various digital art platforms for seamless use.	Integrates with Adobe Creative Cloud, enhancing compatibility.
Real-Time Responsiveness	On average, it takes a few minutes per generation to generate diverse outputs based on a prompt.	It offers a range of customization, but the process may take some time and effort for the desired output and is almost 1.5x slower compared to the other two.	The fastest AI image generator.
Performed Extraordinarily In	Image synthesis based on textual prompts and creative artwork.	Concept art creation, digital illustration, artistic design.	Graphic design projects, creative concepts, visual storytelling.

Video Generation Platforms

In the dynamic realm of video creation, cutting-edge AI technologies
have revolutionized the landscape, paving the way for innovative
possibilities. Table 11-4 presents the differences between the three tools:
Steve AI, Rizzle, and Pictory. The following are key applications taken into
consideration for this comparison:

- Automatic video content creation

- Multilingual video creation

- Context-aware scene transitions

- Explainer videos

- Virtual videos

Table 11-4. *Analysis of Video Generation Platforms*

Parameters	Steve AI	Rizzle	Pictory
Relevance	High relevance by offering features and templates tailored to the needs of users.	Low to medium, depends on the type of script user input. A detailed script shows medium relevance, and a short script shows very low relevance.	Medium to high, ensuring that the generated videos suit the intended purposes of users.
Customization	Offers a wide range of customization options to fine-tune every aspect of their videos, including text, images, transitions, and effects.	Offers extensive customization options. Users can edit individual components of your videos, including text, media, metadata, and the cover image.	Provides various customization options to enhance the video generated by the tool.

(*continued*)

Table 11-4. (*continued*)

Parameters	Steve AI	Rizzle	Pictory
Context Awareness	Demonstrates high context awareness, ensuring the generated videos maintain coherence and relevance to the input text.	Provides medium context awareness.	Medium to high, depending on the option selected on the tool's home page. For example, the video's script has heightened context awareness, but the URL to the video has low context awareness.
User Interface	User-friendly interface with intuitive design for quick adoption.	Intuitive platform designed for easy navigation and usability.	User-friendly, with a focus on simplicity for efficient use.
Video Editing Capabilities	Basic to advanced video editing features, depending on the version.	Limited video editing tools after the video is generated.	Provides a range of video editing capabilities for customization.
Templates and Themes	Offers a library of templates and themes for diverse video styles.	It doesn't allow users to choose any template or theme for video creation.	Includes templates and themes for quick and stylish video creation.

(*continued*)

Table 11-4. (*continued*)

Parameters	Steve AI	Rizzle	Pictory
Media Library Access	Access to stock media libraries for incorporating images, videos, and music.	Integrates with stock media and music libraries to enhance video content.	May include access to media libraries for diverse visual elements.
Learning Curve	Low to moderate; designed for simplicity and quick adaptation.	Moderate; familiarity with video editing tools may be beneficial.	Low to moderate; user-friendly design for easy learning.
Performed Extraordinarily In	• Automatic video content creation • Multilingual video creation in six languages • Explainer videos • Context-aware scene transitions • Virtual video with animated avatar	• Automatic video content creation	• Automatic video content creation • Multilingual video creation in 29 languages • Explainer videos • Context-aware scene transitions • Virtual video with animated avatar

Avatar Video Generation Platforms

In the innovative landscape of video creation, avatar-based technologies have revolutionized communication and information delivery. Avatar video creation tools mark a significant advancement, providing dynamic avenues for storytelling, education, and engagement. Table 11-5 presents the differences between the three tools: Synthesia, DeepBrain, and Yepic Studio. The following are key applications taken into consideration for this comparison:

- Realistic avatar-based videos for interactive eLearning experiences

- Language learning tutorials

- Artificial tutor/narrator/news anchor

- Interactive storytelling

- Historical figure presentation

Table 11-5. *Analysis of Avatar Video Generation Platforms*

Parameters	Synthesia	DeepBrain	Yepic Studio
Avatar Realism	High, focuses on creating realistic and lifelike avatars as it is trained on real actors' datasets.	Variable, depends on the design and customization options.	Variable, the realism may vary based on design features.
Customization	Extensive, it offers various customization options for 160+ avatars with different facial expressions, styles, and ethnicities.	Extensive, it offers customization options for 100+ avatars with different ethnicities and professions.	Limited customization of 50+ actors or avatars of different ethnicities.
Gestures	Minimum gestures can be added to avatars, like nodding the head or raising eyebrows.	Claims to add a variety of gestures while recording the video for creating a custom avatar.	Simple front-faced avatars with no gestures.

(*continued*)

Table 11-5. (*continued*)

Parameters	Synthesia	DeepBrain	Yepic Studio
Learning Curve	Moderate to high, requires familiarity with the platform for optimal use.	Easy, simple, and intuitive platform.	Easy, simple, and intuitive platform.
Output Quality	High-quality, focuses on creating visually compelling avatar videos.	Variable, output quality may depend on the customization options and design.	Variable, the output quality may vary based on the chosen features and type of avatar.
Voice Synchronization	High, emphasizes synchronized lip movements with realistic avatars.	Variable, the level of synchronization may depend on customization.	Low to moderate, the synchronization sometimes feels out of place.
Voice Cloning	Offers a feature of cloning users' voices in enterprise plans.	Provide an option to clone users' voices.	Do not provide this feature.
Digital Twin	Provides the option to create a custom user avatar by uploading their video.	Capable of creating digital twins. These custom AI avatars accurately replicate the person's voice, mannerisms, and appearance.	Capable of creating digital twins and taking photos of the user.

(*continued*)

Table 11-5. (*continued*)

Parameters	Synthesia	DeepBrain	Yepic Studio
Multilingual Video Creation	Provides 120+ languages in 400+ voices.	Provides 80+ languages in 100+ voices	Provides 65+ languages in 450+ voices.
Multiple Avatars on the Screen	Provides the option to use multiple avatars on a single screen.	It doesn't have the option to present multiple avatars on a single screen.	It doesn't have the option to present multiple avatars on a single screen.
Video Rendering Speed	It almost takes five minutes for a 40-second-long video.	It takes seven to eight minutes for a 40-second-long video.	It almost takes ten minutes for a 40-second-long video.
Performed Extraordinarily In	• Realistic avatar-based videos for interactive eLearning experiences • Language learning tutorials • Artificial tutor/narrator/news anchor • Historical figure presentation	Same as Synthesia, but it lacks avatar realism in some cases.	Same as Synthesia, but it sometimes lacks avatar realism, creativity, and narration.

Sound Generation Platforms

After the launch of the AI platform, sound generation platforms have revolutionized the way of creating and interacting with music and audio content. These platforms produce sounds, melodies, and compositions with remarkable precision and creativity. Table 11-6 presents the differences between the three tools: Soundraw, Soundful, and Beatoven. The following are key applications taken into consideration for this comparison:

- Background music (intros and outros)

- Soundscaping: music for relaxation and meditation

- Sound design: sound effects, audio textures, and ambient sounds

- Music education

Table 11-6. *Analysis of Sound Generation Platforms*

Parameters	Soundraw	Soundful	Beatoven
AI Algorithm	Uses an advanced in-house AI algorithm, trained on datasets crafted by music producers, ensuring original and high-quality music.	AI-driven sound synthesis uses machine learning algorithms to generate unique and original music tracks tailored to users' preferences and requirements.	Integrates an AI algorithm that uses advanced music theory and production concepts to deliver unique music.

(*continued*)

Table 11-6. (*continued*)

Parameters	Soundraw	Soundful	Beatoven
Sound Quality	High, produces creative sound, claims copyright-free, and can be used commercially.	High, produces original sound that can be monetized by the users on other platforms.	High, produces original production-ready music pieces that match industry standard mixing and mastering.
Text to Music	Not available	Not available	Available, allows the user to enter a prompt to generate music piece.
Customization	Offers customization to change genre, mood, tempo, duration, theme, and instruments to create the sound.	Users can pick from various genres, moods, and instruments to create unique sounds.	Extensive, users can customize the length, genre, mood, and instruments and fine-tune AI-generated music tracks.
User Interface and Learning Curve	Simple and intuitive interface with a smooth learning curve.	User-friendly, allows easy navigation for users without any prior experience or technical knowledge.	Intuitive platform and easy to navigate and work with.

(*continued*)

Table 11-6. (*continued*)

Parameters	Soundraw	Soundful	Beatoven
Collaboration	Available, makes it easy for users to collaborate on projects with others. Users can share projects with others and invite them to edit or comment on their work.	Collaboration features are not available.	Collaboration features are not available.
Royalty-Free Music	Yes, it gives users royalty-free music.	Yes, it gives access to royalty-free music.	Yes, users can create music without paying any royalties.
Performed Extraordinarily In	All use cases listed previously.	All use cases listed previously.	All use cases listed previously.

Synthetic Voice Generation Platforms

In the dynamic field of audio creation, introducing voice-generation tools has brought about a transformative shift in how we produce and convey spoken content. Voice generation tools represent a notable leap forward, offering dynamic possibilities for narration, virtual assistants, and interactive audio experiences. These tools enable the generation of natural and expressive voices that can serve various purposes, from narrating stories to providing information. Table 11-7 presents the differences between the three tools: WellSaid Labs, Speechify, and PlayHT. The following are key applications taken into consideration for this comparison:

551

- Podcast or digital radio jockey

- Audio books

- Audio historical guides

- Language translations

- eLearning

- Voice cloning

Table 11-7. *Analysis of Synthetic Voice Generation Platforms*

Parameters	WellSaid Labs	Speechify	PlayHT
Naturalness of Voice	High, focuses on natural and human-like voices.	High, emphasis on natural and clear voices with natural accents.	Variable, depends on the selected voices and customization.
Voice Customization	Limited, offers only three types of conversational tones for voice avatars.	Extensive, provides a list of voice variations, such as conversational, excited, angry, etc., tones to voice avatars.	Limited voice customization is listed according to use case, such as digital assistant, news, customer service, etc.
Reading Speed Change	Not available, the speed depends on the customization and voice avatar.	Available, capable of reading 4.5 times the average reading speed, about 200–300 words per minute.	Not available, the speed depends on the customization and voice avatar.

(*continued*)

Table 11-7. (*continued*)

Parameters	WellSaid Labs	Speechify	PlayHT
Adding Pauses	Can be achieved using commas, periods, hyphens, ellipses, or spaces between lines of the script.	Can be achieved using toolbar menu option called "add pause" in duration, generally in seconds.	Can be achieved using toolbar menu option called "add pause" in duration, generally in seconds.
Pronunciation Library	Available, users can add words to pronounce it differently and add cues to the script to adjust the loudness, pace, and pause to the script.	Available, users can add words to pronounce it differently than that of the AI voice avatar.	Available, user can select the text and choose the pronunciation option to add a new pronunciation of the text.
Voice Cloning	Available, may not explicitly offer voice cloning capabilities.	Available, users can upload their audio recordings to clone their voice in seconds, and the tool will maintain their accent, nuance, and style at their consent.	Available, user can upload hours of voice recording to get their voice cloned with their consent.
Text-to-Speech Quality	High-quality, emphasizes clear and expressive text-to-speech.	High-quality; prioritizes clarity and comprehension.	Variable, depends on the selected voices and languages.

(*continued*)

Table 11-7. (*continued*)

Parameters	WellSaid Labs	Speechify	PlayHT
Languages Supported	Supports only English language.	Supports 30+ languages, including non-English languages such as Arabic, Chinese, Danish, etc.	Supports 140+ languages such as German, French, Japanese, etc.
AI Voice Detector Results	It shows that up to 95–99% of the voices are of humans.	It shows that up to 90–93% of the voices are of humans.	It shows that up to 50–55% of the voices are of humans.
Performed Extraordinarily In	• Podcast or digital radio jockey • Audio books in English • Audio historical guides • eLearning	• Audio historical guides in different languages • Language translations	• Multilingual podcast • Audio historical guides and books in different languages • Language translations

Ethical Considerations of Using AI-Generated Content and Multimedia

Now, that we've explored the various AI tools available to assist instructional designers in generating ideas, handling repetitive tasks, and many more, it's essential to delve into the ethical considerations surrounding AI-generated content and multimedia. While these AI technologies offer significant benefits in terms of efficiency and innovation within the instructional design process, they also raise important ethical questions that deserve careful consideration. By examining the ethical implications of relying on AI to create content, we can better understand the potential risks and challenges associated with its use and work toward ensuring responsible and ethical practices in utilizing these powerful tools.

- **Authenticity and Transparency:** Ensure that AI-generated content is clearly labeled and disclosed to learners. Learners have the right to know whether the content they engage with is AI-generated or created by human instructors. Transparency fosters trust and prevents potential deception or misrepresentation.

- **Respect Copyrights:** AI-generated content may draw from extensive datasets, potentially including copyrighted material. Instructional designers must ensure they have the appropriate rights and permissions to use AI-generated content. Respect intellectual property rights and avoid any copyright infringement. Do not use the copywritten materials as part of your prompts without permission.

- **Responsible Usage:** Use AI-generated content responsibly, providing accurate and reliable information to learners. Avoid misleading or deceptive practices, and refrain from using AI-generated content to spread false or manipulated information, commonly known as AI hallucination. The prevalence of hallucinations in AI-generated content varies depending on the task's complexity and the AI model's sophistication.

- **Data Privacy and Security:** AI relies on data to
  generate content. Handle and protect learner data
  responsibly, adhering to data protection regulations
  and implementing robust security measures. Prevent
  any misuse or unauthorized access to learner data.

- **Address Bias and Discrimination:** Be aware that AI
  algorithms can inadvertently perpetuate biases in the
  data they are trained on. Regularly assess and fine-tune
  AI algorithms to ensure fairness, avoiding perpetuating
  stereotypes or discriminatory practices.

- **Human Oversight and Accountability:** While AI can
  automate various aspects of content creation, human
  oversight and accountability remain crucial. Have human
  checks to review and approve AI-generated content,
  ensuring that humans have the final say in decision-
  making and are accountable for the content produced.

- **Continuous Monitoring and Improvement:** AI
  algorithms evolve. Continuously monitor and improve
  AI-generated content, assessing its performance and
  impact and making adjustments as necessary. Stay
  informed about AI's latest developments and best
  practices to ensure responsible usage.

By addressing these ethical considerations in integrating AI-generated
content into learning and development materials, instructional designers
can uphold ethical standards, build trust with learners, and create positive
learning experiences. Transparency, copyright compliance, responsible
usage, data privacy and security, bias mitigation, human oversight, and
continuous improvement are essential pillars for navigating the ethical
challenges associated with AI-generated content in the learning and
development industry.

Conclusions

In conclusion, the findings of a comparative study of generative AI tools have unveiled each tool category's multi-faceted strengths, limitations, effectiveness, and efficiency across a diverse range of applications. Rigorous analysis was conducted on a sample space of 50 instructional designers working as freelancers and full-time employees across different time zones and industries. The diverse representation of participants underscores our findings' broad applicability and relevance, providing valuable insights into how generative AI tools impact their productivity. Furthermore, the graphical representations of time-saving, cost-saving, manpower reduction rate, and time-saving learning tools have demonstrated compelling evidence of the tangible benefits of these generative AI platforms.

The time-saving chart (Figure 11-2) shows the time that can be saved using the generative AI tools. The comparison shows a drastic reduction in the time required to complete the tasks using traditional ways.

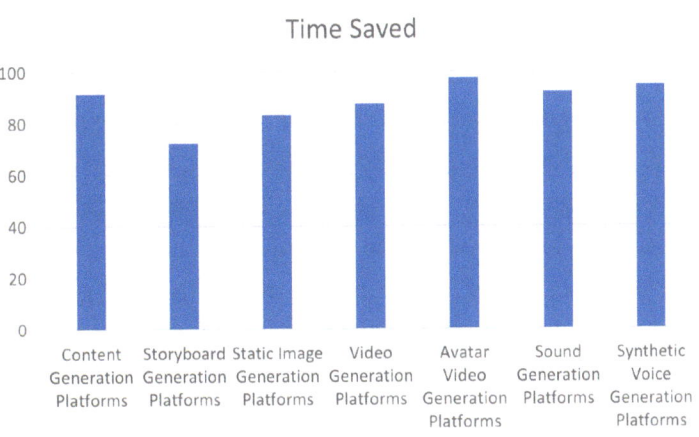

Figure 11-2. *Time-Saving Analysis*

The cost-saving chart (Figure 11-3) illustrates the monetary savings achievable by opting for the generative AI tool's yearly subscription plan instead of hiring manpower.

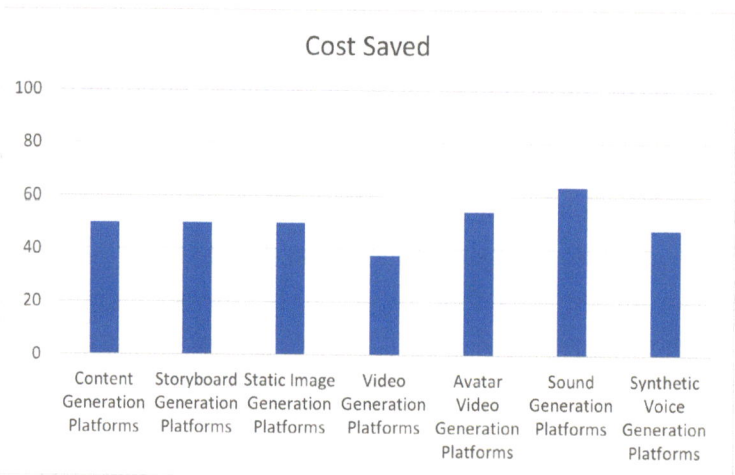

Figure 11-3. *Cost-Saving Analysis*

The manpower reduction rate chart (Figure 11-4) illustrates the decrease in the workforce needed to achieve the desired output using generative AI tools compared to the manpower required when employing individuals with the necessary skill set to perform the tasks manually. We evaluated the manpower on a scale of 10, assuming that ten individuals were initially required to complete the task. This analysis determines the number of individuals, out of the pool of 10, who can be replaced by utilizing the skills necessary for generating output through generative AI tools. It's important to note that our findings do not imply that AI will replace jobs.

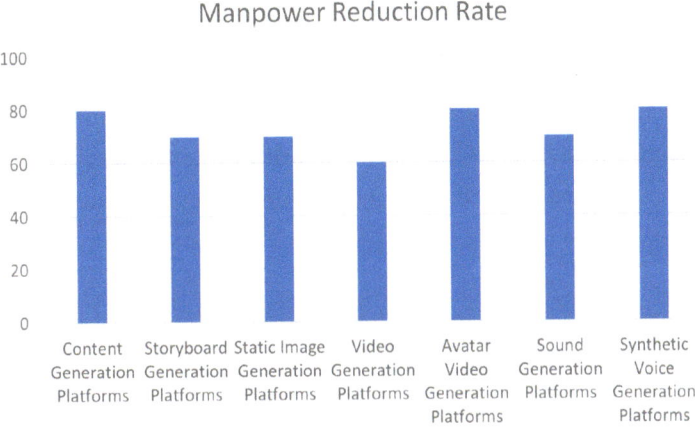

Figure 11-4. *Manpower Reduction Rate Analysis*

The time-saving learning tool (Figure 11-5) indicates the potential time saved through acquiring proficiency with the tool compared to traditional methods of training personnel to generate the desired output using conventional approaches.

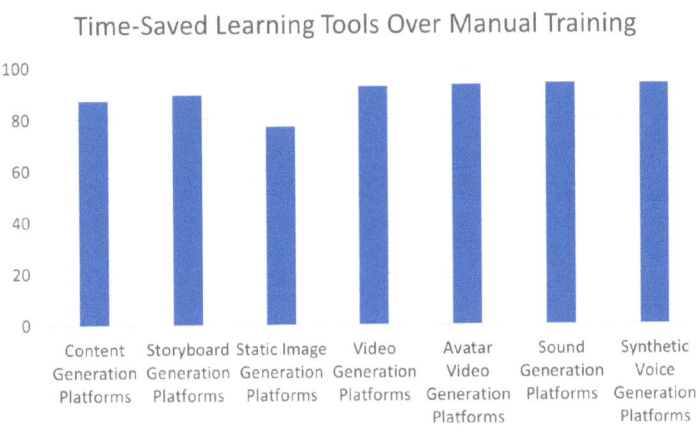

Figure 11-5. *Time-Saved Learning Tools Over Manual Training Analysis*

As we look toward the future, continued research and innovation in generative AI hold immense potential for revolutionizing workflows, driving efficiencies, and unlocking new opportunities for creativity and innovation. By responsibly and ethically leveraging AI's transformative power, we can harness its full potential to address complex challenges and propel society toward a more productive and sustainable future.

Questionnaires for Different Learning Modalities

VARK Questionnaire

1. You are planning a trip to a new city. How do you prefer to navigate and find your way around?

    a. Using a map or GPS for visual guidance

    b. Reading street signs and written directions

    c. Asking locals for directions or recommendations

    d. Exploring the city and relying on your instincts

2. You want to learn a new skill or hobby. How do you prefer to learn it?

    a. Watching tutorial videos or demonstrations

    b. Listening to podcasts or audio recordings about the skill

© Ankita Jiyani Mangtani 2024
A. J. Mangtani, *Instructional Design Unleashed*, Design Thinking,
https://doi.org/10.1007/979-8-8688-0416-8

    c.   Reading books or written instructions on the subject

    d.   Attending workshops or classes with hands-on instruction

3.   You are studying for an important exam. How do you prefer to review and retain the material?

    a.   Creating visual diagrams, charts, or flashcards

    b.   Discussing and explaining the concepts with a study group

    c.   Taking detailed notes and reading textbooks or study guides

    d.   Engaging in practice tests or hands-on problem-solving

4.   You want to learn about a historical event. How do you prefer to absorb information?

    a.   Viewing photographs, videos, or documentaries related to the event

    b.   Listening to lectures or speeches from experts on the subject

    c.   Reading books, articles, or primary sources about the event

    d.   Participating in reenactments or visiting historical sites

5. You need to understand a complex scientific concept. How do you prefer to grasp the information?

   a. Viewing visual models, diagrams, or animations that illustrate the concept

   b. Explaining and discussing the concept with a knowledgeable person

   c. Reading textbooks, research papers, or scientific articles on the topic

   d. Conducting experiments or engaging in hands-on activities related to the concept

6. You want to improve your communication skills. How do you prefer to learn and practice?

   a. Role-playing or participating in group discussions to practice communication skills

   b. Engaging in conversations and dialogues with others

   c. Reading books or articles on effective communication techniques

   d. Observing and analyzing nonverbal cues and body language

7. You want to learn about a cultural tradition or heritage. How do you prefer to explore it?

   a. Viewing visual displays, photographs, or artworks related to tradition

   b. Listening to stories, oral histories, or music from the culture

    c. Reading books, folktales, or historical accounts about tradition

    d. Participating in cultural events, festivals, or hands-on workshops

8. You need to improve your time management skills. How do you prefer to develop this skill?

    a. Creating visual schedules, calendars, or time-blocking systems

    b. Listening to podcasts or audio recordings about time management techniques

    c. Reading books or articles on time management strategies

    d. Practicing time management techniques through real-life scenarios and activities

9. You want to learn about a specific culture's cuisine. How do you prefer to learn and explore it?

    a. Viewing visual demonstrations or cooking videos showcasing traditional recipes

    b. Listening to culinary experts or chefs discussing cooking techniques

    c. Reading cookbooks, recipes, or food blogs with detailed instructions

    d. Engaging in hands-on cooking experiences, trying out recipes, and experimenting

10. You are interested in understanding different philosophical perspectives. How do you prefer to delve into this subject?

   a. Examining visual representations or infographics that outline philosophical concepts

   b. Listening to philosophical lectures or podcasts discussing different perspectives

   c. Reading philosophical texts, essays, or treatises

   d. Engaging in philosophical discussions or debates to explore various viewpoints

VARK Questionnaire Scoring Chart

Use the following scoring chart to find the VARK category to which each answer corresponds. Put a checkmark against each option you have selected. Options "a," "b," "c," and "d" correspond to Visual, Audio, Read/Write, and Kinesthetic modalities, respectively.

Question	a (V)	b (A)	c (R)	d (K)
1				
2				
3				
4				
5				
6				
7				
8				
9				
10				

Calculating Scores:

Total number of a's ticked _____

Total number of b's ticked _____

Total number of c's ticked _____

Total number of d's ticked _____

Counting the number of V's, A's, R's, and K's will give you a fair estimate of the learner preferences.

Dunn and Dunn Questionnaire

Questions	Yes	No
Sound		
I study best when it is quiet.		
I can work with some noise.		
I can ignore noise while working/studying.		
Generally, I cannot concentrate when there is noise.		
Mostly, I prefer to study with background music.		
I can work or study with any kind of music.		
I like to study with rock or dance music.		
I have difficulty concentrating while working or studying with music.		
I can concentrate when people speak quietly.		
I can study while people are talking.		
I can ignore most noises when people are talking.		
It is difficult to ignore the sound from the TV when I am concentrating.		
I get disturbed by noise when I am studying.		
Music with lyrics prevents me from studying.		

(continued)

Questions	Yes	No

Light

I like to study in bright lighting.

I study best with subdued lighting.

I enjoy reading outdoors.

I can only study for a short time if the lighting is dim.

When I am about to study, I turn on all the lights in the room.

I often read or work in subdued lighting.

I usually study under a lamp with a shade while the rest of the room is quite dark.

Temperature

I can concentrate if it is warm.

I can concentrate if it is cold.

I feel colder than most people.

I feel warmer than most.

I like the summer.

When it is cold outside, I want to be indoors.

When it is hot outside, I want to be indoors.

When it is hot outside, I go out.

When it is cold outside, I go out

I find extreme heat or cold uncomfortable.

I like the winter.

(continued)

Questions	Yes	No

Seating

When I study, I like to sit on the floor.

I like to sit in a soft armchair or on a sofa when I study. I get sleepy if I do not sit on a hard chair when I study.

I find it difficult to study at school.

I do all my homework at home.

I always study for tests at home.

I do all my homework at school.

I find it difficult to concentrate on my studies at home.

I work/study best at a library.

I can study almost anywhere.

I like to study in bed.

I like to study while sitting on a mat or a blanket.

I can study while sitting on the floor, in an armchair, on the sofa, and at my desk.

I often study in the bathroom.

Motivation

I feel good when I perform well in school.

I feel good about making my mom and dad proud of me when I do well in school.

My teacher feels good when I do well in school.

The adults are satisfied when I do well and have good grades.

I like to make people proud of me.

(*continued*)

Questions	Yes	No

I feel embarrassed when my grades are poor.

It is more important for me to do well in things outside of school than in schoolwork.

I like to make my teacher proud of me.

No one really cares if I do well in school.

My mom cares about my grades.

My dad cares about my grades.

My teacher cares about my grades.

I want to have good grades myself.

I am happy when I do well in school.

I feel bad and work less when my grades are poor.

I feel happy and proud when my grades are good.

There are many things I prefer to do more than going to school.

I love learning new things.

A good education will help me get a good job.

Task Persistence

I try to finish what I have started.

I usually finish what I have started.

Sometimes I lose interest in things.

I rarely finish what I have started.

I usually remember to finish my homework.

I often need to be reminded to do my homework.

I often forget to do or finish my homework.

(*continued*)

Questions	Yes	No
I often get tired of doing things and want to start something new.		
I usually want to finish things I have started.		
My teacher always tells me to finish what I am supposed to do.		
My parents remind me to finish tasks they have asked me to do.		
Other adults tell me to finish things I have started.		
Someone always reminds me to do something.		
I often get tired of doing things.		
I often want help to finish things.		
I enjoy getting things done.		
I like getting things done so that I can start something new.		
I remember to get things done on my own.		

Responsibility

I believe I am a responsible person.		
People say that I am responsible.		
I always do what I have promised to do.		
People say I always do what I have said I will do.		
I mostly keep my promises.		
I have to be repeatedly reminded to do the things I've been told to do.		
If my teachers tell me to do something, I try to do it.		
I consistently forget to do things I have been told to do.		
I usually remember to do what I am told.		
People always remind me to do things.		
I enjoy doing what I am supposed to do.		

(continued)

Questions	Yes	No

I believe in keeping my promises.

I need to be reminded frequently to do something.

Structured

I like being told exactly what to do.

I like being able to do things in my own way.

I like having different options in how I can do things.

I like figuring things out on my own.

I want other people to tell me how to do things.

I perform better when I know someone is checking my work.

I do my best whether the teacher checks my work or not.

I dislike working hard on something that isn't checked by the teacher.

I prefer clear and specific guidelines when starting new projects.

Social Needs

I want to work alone.

I want to work with my friend.

I want to work with some of my classmates.

I want to work in a group of five or six classmates.

I want to work with an adult.

I want to work with a peer but have an adult nearby.

I want to work with some classmates but have an adult nearby.

I want adults nearby when I work alone or with a peer.

I want the adults to stay away until my classmates and I have finished our work.

(continued)

Questions	Yes	No

What I like to do best, I do it:

   a) Alone

   b) With a friend

   c) With a few friends

   d) With a group of friends

   e) With an adult

   f) With several adults

   g) With friends and adults

Perceptual: Sensory Abilities

When I learn something new, I prefer to learn it by:

   a) Reading it

   b) Listening to a recording

   c) Listening to a tape

   d) Viewing a slide show (without sound)

   e) Seeing and hearing a film

   f) Viewing pictures with someone explaining them

   g) Hearing my teacher explain it to me

   h) Playing games

   i) Going somewhere and seeing for myself

   j) Having someone show me.

The things that I remember best are things that:

   a) My teacher told me about

   b) Someone other than my teacher told me about

(continued)

Questions	Yes	No
c) Someone showed me		
d) I learned about on trips		
e) I read about		
f) I heard about on records or tapes		
g) I saw on TV		
h) I read stories about		
i) I saw in a film		
j) I tried myself or worked on		
k) My friends and I talked about		

I like to:

   a) Read books, magazines, or newspapers.

   b) Watch films.

   c) Listen to records.

   d) Record tapes on a tape recorder.

   e) Draw or paint.

   f) Look at pictures.

   g) Play games.

   h) Talk to people.

   i) Listen when others talk.

   j) Listen to the radio.

   k) Watch TV.

   l) Go on outings.

   m) Learn new things I can do with my hands.

(continued)

Questions	Yes	No
n) Study with my friends.		
o) Build things.		
p) Experiment.		
q) Take photos or make movies/videos.		
r) Use computers, calculators, or other machines.		
s) Go to the library.		
t) Look for things in subjects like sand or clay.		
u) Shape things with my hands.		

Food Intake

I like to eat or drink while I study.

I dislike eating, drinking, or chewing while I study.

While studying, I like to chew gum, have snacks, or suck on candies.

I can only eat, drink, or have a snack after I finish studying.

I usually eat or drink when I am nervous or upset.

I rarely eat when I am nervous or upset.

I would study better if I could eat while reading.

I would be distracted if I ate while reading.

I often catch myself chewing on my pen while studying.

Time

I hate getting up in the morning.

I hate going to bed at night.

I could sleep all morning.

I stay awake long after I go to bed.

(*continued*)

Questions	Yes	No

I feel wide awake after 10 in the morning.

If I stay up very late at night, I get too sleepy to remember anything.

I feel sleepy after lunch.

When I have homework, I prefer to get up early in the morning to do it.

When I can, I do my homework in the afternoon.

I usually start my homework after dinner.

I could stay up all night.

I wish school could start around lunchtime.

I wish I could stay home during the day and go to school in the evening.

I like going to school in the morning.

I can remember things when I review them:

   a) In the morning

   b) At lunchtime

   c) In the afternoon

   d) Before dinner

   e) After dinner

   f) Late at night

Mobility

When I study, I often get up to do something (like drink or have a snack) and then go back to work.

When I study, I keep at it until I'm done, and then I get up.

I find it hard to sit in one place for a long time.

I often change positions while working and studying.

(*continued*)

Questions	Yes	No

I can sit in the same place for a long time.

I constantly change positions or fidget in my chair.

I work best in short periods with breaks in between.

I prefer to get my work over with quickly.

I like to work a bit, stop, go back to work, stop again, etc.

I prefer to stick with a task until it's finished.

I leave most tasks until the last minute and then have to work straight through.

I do a little at a time in most tasks or projects and eventually get everything done.

I prefer to work with things and not in a continuous manner when I know what I'm doing.

Analytical Approach

I enjoy analyzing data, statistics, or facts to draw conclusions.

I prefer studying by reading and taking detailed notes.

I find it helpful to create outlines or lists to organize my thoughts.

I feel more confident when I thoroughly understand the underlying principles of a topic.

I tend to excel in subjects that require logical reasoning and critical thinking.

Global Approach

I prefer learning through discussions, group activities, or interactive experiences.

I enjoy exploring concepts from different angles and perspectives.

(continued)

Questions	Yes	No

I often rely on intuition and gut feelings when making decisions.

I learn best when I can relate new information to real-life situations or examples.

I like experimenting and trying new approaches rather than sticking to a single method.

Reflective Approach

I prefer to review my notes and think about what I've learned after a study session.

I find it helpful to journal or write down my thoughts to process new information.

I enjoy discussing ideas with others and getting their input before forming my own opinions.

I prefer to work on tasks or assignments over a longer period rather than rushing through them.

I value moments of quiet reflection to better understand complex concepts.

Impulsive Approach

I tend to make quick decisions without dwelling on the details.

I prefer hands-on experiences and learning by doing rather than passive studying.

I enjoy brainstorming ideas and going with the flow rather than following a strict plan.

I learn best when I'm actively engaged and challenged.

I am comfortable adapting to changes and unexpected situations while learning.

Gregorc's Questionnaire

Read each combination of words and circle the two that best represent you.

1.
a. Imaginative
b. Investigative
c. Realistic
d. Analytical

2.
a. Organized
b. Adaptable
c. Critical
d. Inquisitive

3.
a. Debating
b. Getting to the point
c. Creating
d. Relating

4.
a. Personal
b. Practical
c. Academic
d. Adventurous

5.
a. Precise
b. Flexible
c. Systematic
d. Inventive

9.
a. Reader
b. People person
c. Problem solver
d. Planner

10.
a. Memorize
b. Associate
c. Think-through
d. Originate

11.
a. Changer
b. Judger
c. Spontaneous
d. Wants direction

12.
a. Communicating
b. Discovering
c. Cautious
d. Reasoning

13.
a. Challenging
b. Practicing
c. Caring
d. Examining

(continued)

6.

a. Sharing

b. Orderly

c. Sensible

d. Independent

7.

a. Competitive

b. Perfectionist

c. Cooperative

d. Logical

8.

a. Intellectual

b. Sensitive

c. Hardworking

d. Risk-taking

14.

a. Completing work

b. Seeing possibilities

c. Gaining ideas

d. Interpreting

15.

a. Doing

b. Feeling

c. Thinking

d. Experimenting

Circle the letters of the words you chose for each number in the following columns. Total the totals in columns I, II, III, and IV. Divide the sum of each column by four. The box with the highest number describes how you most frequently processes information.

	I (CS)	II (AS)	III (AR)	IV (CR)
1	C	D	A	B
2	A	C	B	D
3	B	A	D	C
4	B	C	A	D
5	A	C	B	D
6	B	C	A	D

(continued)

	I (CS)	II (AS)	III (AR)	IV (CR)
7	B	D	C	A
8	C	A	B	D
9	D	A	B	C
10	A	C	B	D
11	D	B	C	A
12	C	D	A	B
13	B	D	C	A
14	A	C	D	B
15	A	C	B	D

Graph your results.

To graph your preferred thinking style, place a dot on the number corresponding to your score in each of the classifications and link the dots shown in the miniature diagram.

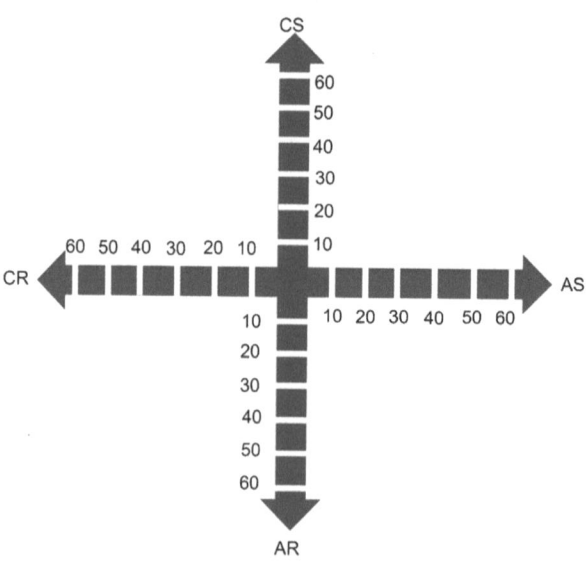

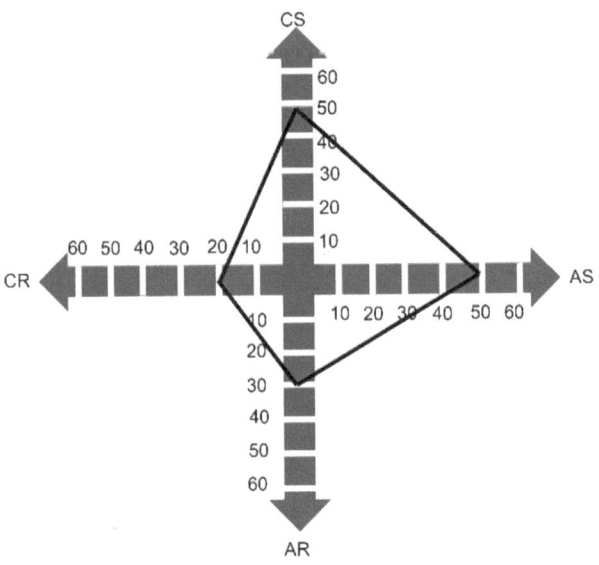

Jackson Learning Style Profiler Questionnaire

Participants will assess their agreement with each statement on a spectrum of 1 to 5, where 1 means strongly disagree, 2 means disagree, 3 means neutral, 4 means agree, and 5 means strongly agree.

Category	Question	Score
High Sensation Seeker	I am drawn to new and challenging learning activities.	
	I actively seek out novel and exciting learning opportunities.	
	I enjoy taking risks to explore unfamiliar subjects or topics.	
	I find it motivating to consciously work toward challenging learning goals.	
	I thrive in environments that provide excitement and variety.	

(continued)

581

Category	Question	Score
Goal-Oriented Achiever	I set specific and ambitious learning goals for myself.	
	I am confident in my ability to achieve the learning objectives I set.	
	I take control of my learning direction and adjust my goals as needed.	
	I am driven and motivated to excel in my learning endeavors.	
	I prefer structured and goal-oriented learning environments.	
Emotionally Intelligent Achiever	I learn from my errors and use them to help me improve.	
	I approach learning with a scientific and detached reasoning.	
	I am autonomous in my learning approach, seeking understanding beyond surface-level information.	
	I am open to feedback and use it to enhance my problem-solving skills.	
	I value learning from experiences and continuously improving my learning strategies.	

(continued)

Category	Question	Score
Conscientious Achiever	I am a responsible learner who employs complex strategies to grasp difficult concepts.	
	I use higher-level learning strategies to comprehend and synthesize information.	
	I approach learning with diligence and focus to achieve positive outcomes.	
	I am motivated to excel in academically challenging environments.	
	I approach my studies methodically to ensure a thorough comprehension of the material.	
Deep Learning Achiever	I approach learning with enthusiasm and enjoy experimenting with new concepts.	
	I am motivated by intellectual curiosity to delve deeply into knowledge.	
	I seek to thoroughly understand concepts and ideas rather than memorize facts.	
	I am open to different perspectives and viewpoints in my learning.	
	I excel in academic settings that encourage critical thinking, analysis, and exploration.	

Observing the area where you ranked yourself with the highest number of fives, which is strongly agree, will give you a fair idea of your learning style.

Apter's Motivational Style Profile

Category	Question	Yes/No
Means-Ends	I feel motivated when I achieve a goal and anticipate future benefits.	
	I find value in activities based on the positive consequences they will bring.	
	My motivation is driven by the accomplishment of a task or objective.	
	I prioritize activities that have clear and meaningful outcomes.	
	I am more motivated when I focus on the end result or future benefits.	
Rules	I feel motivated when I conform to rules and norms that are expected of me.	
	My motivation comes from adhering to established guidelines and regulations.	
	I find it rewarding to follow rules and do what is expected.	
	My motivation is driven by a sense of belonging and conformity.	
	I am more motivated when I adhere to established rules and norms.	

(continued)

Category	Question	Yes/No
Transactions	I feel motivated when I have power, control, or strength over a situation.	
	My motivation comes from demonstrating mastery and skill.	
	I find satisfaction in being capable and strong in my actions.	
	My motivation is driven by a desire for control and dominance.	
	I am more motivated when I can exhibit power and strength.	
Relationships	I feel motivated when I focus on my own needs and personal success.	
	My motivation comes from nurturing and caring for myself.	
	I find fulfillment in addressing my own emotional and personal needs.	
	My motivation is driven by self-care and personal success.	
	I am more motivated when I prioritize my own well-being.	

(continued)

Category	Question	Yes/No
Transaction Pairs	I feel motivated when I have personal power, skills, or strength.	
	My motivation comes from feeling capable and strong in my actions.	
	I find satisfaction in demonstrating my own abilities and expertise.	
	My motivation is driven by personal mastery and achievement.	
	I am more motivated when I can exhibit personal power and strength.	
Sympathy	I feel motivated when I provide or receive personal care and emotional support.	
	My motivation comes from nurturing and caring for others.	
	I find fulfillment in offering emotional assistance and support.	
	My motivation is driven by a desire to provide empathy and care.	
	I am more motivated when I can contribute to others' emotional well-being.	

CSI Questionnaire

Analytical Style			
Questions		**Score**	
	T	**F**	**U**
I prefer to analyze data and information thoroughly before making a decision.	2	1	0
I enjoy solving complex problems that require logical thinking and systematic approaches.	2	1	0
Following a structured and organized approach to tasks is essential for my work.	2	1	0
I find satisfaction in meticulously examining details and data.	2	1	0
I believe that careful planning and analysis lead to better outcomes.	2	1	0
I tend to rely on evidence and data rather than intuition when making decisions.	2	1	0
My decision-making process involves weighing pros and cons objectively.	2	1	0
I appreciate it when others present well-reasoned arguments and explanations.	2	1	0
I prefer to clearly understand all the facts before drawing conclusions.	2	1	0
I feel comfortable following established rules and guidelines in my work.	2	1	0
I value precision and accuracy in my work and decision-making.	2	1	0
I enjoy systematically organizing information and data.	2	1	0
I feel more confident in decision-making when I have a well-defined plan.	2	1	0

(continued)

	T	F	U
I often seek logical explanations for observed patterns and phenomena.	2	1	0
My work involves thorough analysis and careful consideration of options.	2	1	0
I prefer step-by-step approaches to solve problems and achieve goals.	2	1	0
I find comfort in relying on empirical evidence and data-driven insights.	2	1	0

Intuitive Style

Questions	Score		
	T	F	U
I often trust my instincts and gut feelings when making decisions.	2	1	0
I enjoy exploring new and unconventional ideas and concepts.	2	1	0
I tend to see the bigger picture and broader implications of situations.	2	1	0
My decision-making process is influenced by my intuition and instincts.	2	1	0
I enjoy exploring creative and innovative solutions to problems.	2	1	0
I am comfortable taking risks and trying new approaches in my work.	2	1	0
I am open to exploring new possibilities and alternative perspectives.	2	1	0
I have a knack for identifying patterns and connections in information.	2	1	0

(continued)

I often rely on my intuition to guide me in uncertain situations.	2	1	0
I prefer to think in terms of possibilities rather than limitations.	2	1	0
I find inspiration in unconventional ideas and out-of-the-box thinking.	2	1	0
I enjoy tackling challenges that require creative problem-solving.	2	1	0
I often have hunches or insights that guide my decision-making.	2	1	0
I enjoy brainstorming and generating ideas in a spontaneous manner.	2	1	0
I tend to trust my intuition even when there is limited evidence.	2	1	0
I am open to exploring new perspectives and unconventional approaches.	2	1	0

Calculate the score and check in which category you fall according to the following table.

Score	Category
0–28	High Intuitive
29–38	Quasi Intuitive
39–45	Adaptive
46–52	Quasi Analytical
53–76	Highly Analytical

Kolb's Questionnaire

Concrete Experience

Questions	Agree	Disagree
I prefer being actively engaged and deeply involved in hands-on activities when learning.		
I enjoy listening to others and considering different viewpoints before forming my own opinions.		
I often rely on my emotions and gut instincts when making decisions or solving problems.		
I am open-minded and willing to explore alternative points of view and ideas, even if they differ from mine.		
I often rely on my instincts and intuition when making decisions or solving problems rather than relying solely on logic or analysis.		
I prefer learning through hands-on experiences and direct interaction with the material or subject matter.		
I prefer focusing on the current tasks and immediate goals rather than planning for the future.		
I actively seek out new and unfamiliar experiences to expand my knowledge and understanding.		
I believe firsthand experiences and practical application are essential for effective learning.		
I thrive in high-pressure situations and find intense challenges motivating me to perform at my best.		

(continued)

Reflective Observation

Questions	Agree	Disagree
I prefer cautiously approaching new information or ideas and gathering more evidence before forming conclusions.		
I strive to maintain objectivity and fairness when considering different perspectives or evaluating information.		
I prefer to observe and gather information before actively engaging in a learning task or making decisions.		
I am conscious of my surroundings and pay attention to details when learning new information or skills.		
I enjoy challenging established ideas and critically questioning information presented to me.		
I value carefully observing and analyzing situations before actively participating or forming opinions.		
I prefer to take time to reflect on my experiences and thoughts before drawing conclusions or making decisions.		
I enjoy exploring different viewpoints and considering diverse perspectives when approaching a learning task.		
I value carefully observing and analyzing situations before actively participating or forming opinions.		
I tend to be more reserved and prefer to process information internally before sharing my thoughts or ideas.		

(continued)

Abstract Conceptualization

Questions	Agree	Disagree
I have a keen eye for details and enjoy carefully distinguishing between different elements or factors.		
I naturally gravitate toward analyzing information, breaking it into its components, and seeking patterns or connections.		
I tend to rely on logical thinking and reason when approaching problems or decision-making.		
I find value in critically assessing and evaluating the quality or effectiveness of information, ideas, or solutions.		
I prefer systematic and logical approaches when processing information or solving problems.		
I enjoy exploring abstract concepts and theories to deepen my understanding of a subject.		
I tend to focus on long-term goals and plan ahead for future possibilities and outcomes.		
I regard intelligence to be more than just information, but also the capacity to use knowledge effectively.		
I find satisfaction in conceptualizing and forming abstract frameworks or models to understand complex ideas.		
I tend to make decisions based on logical reasoning and objective analysis rather than emotions or intuition.		

(continued)

Active Experimentation

Questions	Agree	Disagree
I prefer learning through practical application and hands-on experiences rather than theoretical concepts alone.		
I try to connect new knowledge or skills to real-life situations and understand their practical relevance.		
I learn best when actively engaging in tasks and taking a hands-on approach rather than passively observing.		
I am open to taking risks and trying new approaches, even if there is a possibility of failure.		
I strive to maximize my efficiency and accomplish tasks promptly and productively.		
I prefer learning by actively applying knowledge and skills in practical situations rather than passively observing.		
I value competence and strive to develop a high level of proficiency in the subjects or skills I pursue.		
I enjoy exploring new ideas and approaches through experimentation and trial and error.		
I take ownership of my learning and responsibilities and strive to fulfill my commitments and obligations.		
I am open to adapting and adjusting my learning strategies or approaches based on feedback and changing circumstances.		

Kolb Questionnaire Scoring Chart

Count the number of ticks put against the Agree column of the questionnaire.

CE	AC	RO	AE
5	3	4	6

Calculate the number of learners falling into a particular category as follows:

	Concrete Experience (Feeling)	Abstract Conceptualization (Thinking)
Reflective Observation (Watching)	Diverging 5 + 4 = 9	Assimilating 3+4 = 7
Active Experimentation (Doing)	Accommodating 5+6 = 11	Converging 3+6 = 9

Honey and Mumford Questionnaire

Activist		
Questions	**Yes**	**No**
I actively seek out new experiences and challenges.		
I often act without considering the possible consequences.		
I'm known for being straightforward and expressing my thoughts directly.		
I am keen on self-discipline and sticking to a fixed routine.		

(continued)

I actively seek to apply new ideas and approaches in practice.

I thrive on the challenge of tackling something new and different.

I am drawn to innovative, unconventional thoughts rather than practical ones.

I take pride in doing a thorough job.

I do not like disorganized things and prefer to fit things into a coherent pattern.

I enjoy fun-loving, spontaneous people.

Theorist

Questions	Yes	No

I usually solve difficulties in a step-by-step manner.

I regularly question people about their basic assumptions.

I am cautious about jumping to conclusions.

I prefer to relate my actions to a general principle, standard, or belief.

I enjoy seeing other participants in debates as they plot and scheme.

I pay careful attention to detail before concluding.

I'm drawn to tactics like flowcharts and contingency plans.

I believe in coming to the point immediately.

I prefer to have as many sources of information as possible.

I prefer to respond to events spontaneously and flexibly rather than plan things out in advance.

(continued)

Reflector

Questions	Yes	No
I frequently discover that actions based on emotions are just as sound as those based on rigorous thought and investigation.		
I like the sort of work where I have time for thorough preparation and implementation.		
I take care of how I interpret data and avoid jumping to conclusions.		
I prefer to carefully consider all options before making a decision.		
I do not like rushing work to meet tight deadlines.		
I tend to evaluate people's ideas based on their practicality.		
I like to stand back from a situation and consider all perspectives.		
I frequently see inconsistencies and flaws in other people's arguments.		
I consider other people's points of view before expressing my own.		
I like to ponder many alternatives before making up my mind.		

Pragmatist

Questions	Yes	No
I focus on what matters most, whether something works in practice.		
I think that formal procedures and policies restrict people.		
I get along best with logical, analytical folks and least well with impulsive, "irrational" people.		
I want to try things out to see if they work in practice.		
It bothers me when I have to speed through work to meet a tight deadline.		

(*continued*)

More often than not, rules are there to be broken.

I prefer to relate current actions to the longer-term bigger picture.

I do whatever is practical to get the job done.

I prefer to work with colleagues who are analytic and systematic.

I prefer to explore innovative solutions beyond established norms.

Check these following links to perform the mentioned test and know your learning preferences:

HBDI Questionnaire

www.pdffiller.com/jsfiller-desk12/?requestHash=82f49b28ede21
66b2943f6b7b8aac8c8b87a36d259786e84280f24dc9ea625aa&projectId=1
335357749&loader=tips&MEDIUM_PDFJS=true&PAGE_REARRANGE_V2_MVP=
true&isPageRearrangeV2MVP=true&jsf-page-rearrange-v2=true&jsf-
new-header=false&routeId=28c7e4d987534ab0d90beee0659fc449&mode=
force_choice#a1ea359f87e14ec8aaf317a053905480

MBTI Questionnaire

www.16personalities.com/free-personality-test

Felder-Silverman Questionnaire

www.webtools.ncsu.edu/learningstyles/

4MAT Questionnaire
https://aboutlearning.dk/minitest_UK/quiz.html

NIMAS, DAISY, and WCAG

NIMAS

The National Instructional Materials Accessibility Standard (NIMAS) is a set of guidelines and specifications developed to standardize the format and structure of instructional materials for students with print disabilities. Print disabilities can include visual impairments, physical disabilities, and certain learning disabilities that affect reading.

NIMAS was established in the United States as part of the Individuals with Disabilities Education Act (IDEA) to ensure that students with print disabilities have timely access to textbooks and other instructional materials in formats that are accessible to them. The objective is to offer students with disabilities equivalent educational opportunities as their peers who do not have disabilities.

Key features of NIMAS include the following:

XML-Based Format: NIMAS specifies an XML-based file format that serves as a standardized representation of the content in educational materials. This format allows for the creation of accessible versions of the materials.

© Ankita Jiyani Mangtani 2024
A. J. Mangtani, *Instructional Design Unleashed*, Design Thinking,
https://doi.org/10.1007/979-8-8688-0416-8

Consistent Structure: NIMAS provides guidelines for consistently structuring educational content, including headings, paragraphs, lists, and other elements. This consistency facilitates the creation of accessible versions that students with disabilities can easily navigate.

Metadata: NIMAS includes metadata elements that provide information about the content, such as the title, author, copyright information, and other relevant details. This metadata helps in cataloging and managing accessible educational materials.

Conversion to Accessible Formats: Educational publishers are encouraged to produce their materials per the NIMAS specifications. This allows for more efficient and accurate conversion of the materials into various accessible formats, such as braille, large print, audio, and digital text.

By following the guidelines outlined in NIMAS, publishers and educational institutions can actively contribute to the establishment of a more inclusive learning environment for students with print disabilities. Accessible formats produced based on NIMAS specifications help ensure that students with disabilities can independently access and participate in the educational content used in schools.

DAISY

The DAISY (Digital Accessible Information System) Consortium is an international organization that develops and promotes standards for digital talking books, also known as DAISY books. DAISY is designed to make books and other information accessible to individuals who are blind, visually impaired, or print disabled. The DAISY format offers a range of features that enhance the reading experience for people with disabilities. Here are some key aspects of DAISY:

Structured Format: DAISY books are created in a structured format that allows for easy navigation. This structure includes hierarchical organization, headings, subheadings, page numbers, and other navigational elements. Readers can easily move between sections, chapters, and pages.

Text and Audio Integration: DAISY books often include both text and synchronized audio. This means that the text is highlighted as it is being read aloud. This capability proves especially advantageous for individuals with visual impairments or learning disabilities.

Navigation and Bookmarks: DAISY books allow users to navigate through the content using various navigation levels. Readers can jump to specific sections, headings, or pages. They can also set bookmarks to mark specific points in the book for quick reference.

Flexible Playback: DAISY books support flexible playback options. Readers can control the playback speed, pause, rewind, and fast forward. This flexibility allows individuals to customize the reading experience based on their preferences.

Accessibility Features: The DAISY format incorporates accessibility features such as text-to-speech, which allows the content to be read by a synthetic voice. This is especially helpful for individuals who may have difficulty reading standard print.

International Standard: DAISY is an international standard, and the Consortium works to ensure that its guidelines and specifications are adopted globally. This helps create a consistent, interoperable system for producing and consuming accessible digital content.

The DAISY Consortium collaborates with publishers, libraries, and organizations to promote the production and distribution of DAISY books. The goal is to increase access to information for people with print disabilities and to provide them with a more inclusive reading experience.

WCAG Success Criteria
WCAG 1.0 Checkpoints

Priority 1 (A)

Checkpoints	Example
1. Provide a text equivalent for every nontext element.	\<img src="example.jpg" alt="A red apple on a white background"\>
2. Provide redundant text links for each active region of a server-side image map.	Ensure that text links are provided in addition to server-side image maps.

(*continued*)

3. Ensure that all information conveyed with color is also available without color.	If using color to convey information, also provide text or other noncolor alternatives.
4. Use markup rather than images to convey information.	Instead of using images for buttons, use HTML buttons and style them with CSS.
5. Create documents that validate to published formal grammars.	Ensure that your documents adhere to the syntax and rules defined by their respective specifications.
6. Clearly indicate alterations in the natural language used in the text of a document and in any text equivalents.	Use HTML attributes to specify the language of document sections: <div lang="en">English content</div>.
7. For data tables, identify row and column headers.	Use <th> for header cells and <td> for data cells, and associate them using the headers attribute.

Priority 2 (AA)

8. Guarantee that the content remains comprehensible and navigable even when stylesheets are turned off.	Ensure that the content is understandable and navigable even when stylesheets are disabled.
9. Until user agents allow users to freeze moving content, avoid movement in pages.	Avoid using auto-refreshing or blinking content; provide controls for users to pause or stop motion.
10. Utilize client-side image maps instead of server-side image maps, unless the regions cannot be defined using an available geometric shape.	Ensure scripts and applets are compatible with assistive technologies or provide alternative methods.

(continued)

11. Provide client-side image maps instead of server-side image maps except where the regions cannot be defined with an available geometric shape.

Use client-side image maps for accessibility.

12. Ensure that scripts are readable and usable.

Ensure that scripts are designed to be readable and usable by people with disabilities.

13. Guarantee that pages remain usable even when scripts, applets, or other programmatic objects are deactivated or not supported.

Provide alternative methods for users who cannot access scripts or applets.

14. Ensure that dynamic content is accessible or provide an alternative presentation or page.

Ensure that dynamic content is accessible and provide alternative methods to access the content.

15. Ensure that forms are accessible.

Design forms to be accessible to all users, including those with disabilities.

16. Make sure that tables include the required markup to be transformed by accessible browsers and other user agents.

Use appropriate markup for tables to ensure accessibility.

17. Ensure that pages featuring new technologies transform gracefully.

Ensure that pages work well and remain accessible when using new technologies.

18. Ensure user control of time-sensitive content changes.

Provide controls for users to pause, stop, or adjust the timing of time-sensitive content.

19. Design for device independence.

Design content that can be accessed and used on various devices.

(continued)

20. Use interim solutions.	Provide temporary solutions until user agents support explicit associations between labels and form controls.
21. Employ W3C technologies when they are both available and suitable for a given task, and utilize the latest versions when supported.	Use the latest versions of W3C technologies for tasks when supported.
22. Provide clear navigation mechanisms.	Clearly label links and provide navigation mechanisms that are easy to understand.
23. Ensure that documents are clear and simple.	Use clear and straightforward language throughout your documents.

Priority 3 (AAA)

24. Until user agents support explicit associations between labels and form controls, ensure that all form controls are explicitly associated with a label or title element.	Explicitly associate form controls with labels or titles.
25. Title each frame to facilitate frame identification and navigation.	Provide descriptive titles for frames to aid users in identifying and navigating frames.
26. Provide a site map or table of contents.	Offer a site map or table of contents to help users navigate through the content.

(continued)

27. Ensure that access to information and functionality is independent of the ability to identify specific sensory characteristics of components.	Ensure that information and functionality are accessible without relying on specific sensory characteristics.
28. Ensure that tables have headers or titles.	Use <th> for header cells and provide appropriate titles or captions for tables.
29. Ensure that data tables have row or column headers for each row and column.	Ensure that data tables have headers to enhance their accessibility.
30. Ensure that tables have an appropriate caption.	Provide captions for tables to give users context and understanding.
31. Ensure that pages are readable and usable without style sheets.	Design pages to be readable and usable even when style sheets are disabled.
32. Create a style of presentation that is consistent across pages.	Maintain a consistent style of presentation throughout your website.
33. Ensure that foreground and background color combinations provide sufficient contrast.	Ensure that text is easily readable by using color combinations with sufficient contrast.
34. Ensure that documents are clear and simple.	Use clear and straightforward language throughout your documents.
35. Provide consistent navigation mechanisms.	Keep navigation mechanisms consistent across pages.
36. Ensure that navigation mechanisms are redundant.	Provide multiple methods for users to navigate your content.

(continued)

37. Ensure that all information conveyed with color is also available without color.	Provide alternative cues or information for users who cannot perceive color.
38. Use style sheets to control layout and presentation.	Use style sheets to control the layout and presentation of your content.
39. Avoid movement in pages.	Minimize or avoid movement in pages to prevent potential issues for users.
40. Use server-side image maps when possible.	Use server-side image maps as an alternative to client-side image maps when possible.
41. Use client-side image maps if you must use image maps.	If using image maps, prefer client-side image maps over server-side image maps.
42. Ensure that image maps are accessible.	Make image maps accessible by providing alternative text and clear navigation.
43. Use correct markup to convey document structure.	Use appropriate markup to convey the structure and hierarchy of your documents.
44. Associate labels explicitly with their controls.	Use explicit associations between form labels and their corresponding controls.
45. Cluster related links, specify the group for user agents, and, until user agents inherently do so, offer a means to skip the group.	Group related links, identify them, and provide a mechanism to bypass the group.
46. Associate a table with its summary.	Use the <summary> element to provide a summary for data tables.

(continued)

47. Ensure that deprecated elements do not render text content.	Avoid using deprecated elements that may not render text content correctly.
48. Use the link element to associate metadata documents with a document.	Use the <link> element to associate metadata documents, such as style sheets, with your HTML document.
49. Associate style sheets explicitly with documents.	Explicitly associate style sheets with your HTML documents.
50. Associate table data cells and header cells with a scope.	Use the scope attribute to associate data cells and header cells in tables.
51. Clearly indicate alterations in the natural language used in the text of a document and any text equivalents.	Indicate changes in language using HTML attributes like lang or <span> elements.
52. Create a logical tab order among form fields and links.	Design forms and links with a logical tab order to improve keyboard navigation.
53. Provide a means to skip over multi-line ASCII art.	Include a skip link for users to bypass multi-line ASCII art or other nonessential content.
54. Do not use tables for layout unless the table makes sense when linearized.	Use tables for data, not for layout, to ensure proper linearization.
55. Utilize relative units instead of absolute units in markup language attribute values and style sheet property values.	Use relative units like percentages or ems instead of absolute units for better accessibility.
56. Include default, place-holding characters in edit text fields and text areas.	Provide default characters in edit text fields or text areas to guide users on expected input.

(*continued*)

608

57. Ensure that embedded user interfaces are accessible or provide an accessible alternative.

Ensure embedded user interfaces, like plug-ins, are accessible or provide alternatives.

58. Ensure that moving, blinking, scrolling, or auto-updating objects or pages may be paused or stopped.

Provide controls for users to pause or stop moving or auto-updating content.

59. Ensure that every message (e.g., prompt, alert) that is conveyed by color is also conveyed by shape or some other noncolor visual cue.

Ensure color-coded messages are also conveyed through other visual cues.

60. Ensure that documents are authored so that they can be presented according to the reader's preferences.

Author documents to allow presentation according to user preferences, such as text size or color.

61. Avoid causing the screen to flicker.

Minimize screen flickering, which can be problematic for some users, especially those with photosensitive epilepsy.

62. Design content to avoid causing seizures.

Avoid content that may cause seizures in users with epilepsy.

63. Provide a second, user-selectable style sheet that achieves the same effect.

Provide options to choose an alternative style sheet with the same visual effect.

64. Provide a mechanism for finding other pages in the same collection.

Include a mechanism, like a site map, for users to find other pages within the same collection.

WCAG 2.0 Success Criteria

Perceivable

Success Criterion	Description	Conformance Level
1.1 Text Alternatives	**Provide text alternatives for nontext content.**	**Level A**
1.1.1 Nontext Content	All nontext content has a text alternative.	Level A
Controls, Input	Nontext content functioning as a control or receiving user input should possess a name that describes its purpose.	Level A
Time-Based Media	Text alternatives for time-based media provide descriptive identification.	Level A
Test	Text alternatives for tests provide descriptive identification.	Level A
Sensory	Text alternatives provide descriptive identification for content creating specific sensory experiences.	Level A
CAPTCHA	Text alternatives for CAPTCHAs identify and describe the purpose, with alternative forms for different disabilities.	Level A
Decoration, Formatting, Invisible	Nontext content used for decoration, formatting, or not presented is implemented to be ignored by assistive technology.	Level A

(*continued*)

Success Criterion	Description	Conformance Level
1.2 Time-Based Media	**Provide alternatives for time-based media.**	**Level A, AA, AAA**
1.2.1 Audio-Only and Video-Only (Prerecorded)	Alternative for time-based media is provided for audio-only and video-only content.	Level A
1.2.2 Captions (Prerecorded)	Ensure that captions are available for all prerecorded audio content in synchronized media.	Level A
1.2.3 Audio Description or Media Alternative (Prerecorded)	Alternative for time-based media or audio description is provided for prerecorded video content.	Level A
1.2.4 Captions (Live)	Captions provided for all live audio content in synchronized media.	Level AA
1.2.5 Audio Description (Prerecorded)	Audio description provided for all prerecorded video content in synchronized media.	Level AA
1.2.6 Sign Language (Prerecorded)	Sign language interpretation provided for prerecorded audio content in synchronized media.	Level AAA
1.2.7 Extended Audio Description (Prerecorded)	Extended audio description provided for prerecorded video content in synchronized media.	Level AAA
1.2.8 Media Alternative (Prerecorded)	Alternative for time-based media provided for all prerecorded synchronized and video-only media.	Level AAA

(continued)

Success Criterion	Description	Conformance Level
1.2.9 Audio-Only (Live)	Alternative for time-based media providing equivalent information for live audio-only content.	Level AAA
1.3 Adaptable	**Develop content that maintains its information and structure when presented in various formats.**	**Level A, AA**
1.3.1 Info and Relationships	Information, structure, and relationships can be programmatically determined or available in text.	Level A
1.3.2 Meaningful Sequence	Correct reading sequence can be programmatically determined when content sequence affects meaning.	Level A
1.3.3 Sensory Characteristics	Instructions do not depend solely on sensory attributes.	Level A
1.4 Distinguishable	**Facilitate improved visibility and auditory access to the content for users.**	**Level A, AA, AAA**
1.4.1 Use of Color	Information is not exclusively conveyed through the use of color.	Level A
1.4.2 Audio Control	Audio playing automatically can be paused or stopped.	Level A
1.4.3 Contrast (Minimum)	Text and images have a minimum contrast ratio, with exceptions.	Level AA
1.4.4 Resize text	The text can be enlarged to 200% without any loss of content or functionality.	Level AA

(continued)

Success Criterion	Description	Conformance Level
1.4.5 Images of Text	Text is used instead of images of text, with exceptions.	Level AA
1.4.6 Contrast (Enhanced)	Text and images have an enhanced contrast ratio, with exceptions.	Level AAA
1.4.7 Low or No Background Audio	Prerecorded audio with primarily speech has no background sounds or controllable background.	Level AAA
1.4.8 Visual Presentation	Mechanism available for customizable foreground and background colors, width, text justification, line spacing, and resizing.	Level AAA
1.4.9 Images of Text (No Exception)	Images of text are employed solely for either decorative purposes or to convey essential information.	Level AAA

Operable

Success Criterion	Description	Conformance Level
2.1 Keyboard Accessible	**Make all functionality available from a keyboard.**	**Level A**
2.1.1 Keyboard	The keyboard interface allows for the operation of all functionalities without the need for specific timings.	Level A

(*continued*)

Success Criterion	Description	Conformance Level
Path-Dependent Input	Exceptions for path-dependent input, not input technique.	
Additional Input Methods	Mouse input or other methods can be provided in addition to keyboard operation.	
2.1.2 No Keyboard Trap	Users can shift keyboard focus away, and they are informed of the procedure for doing so.	Level A
2.1.3 Keyboard (No Exception)	All features can be operated using a keyboard interface without the need for precise timings.	Level AAA
2.2 Enough Time	**Provide users enough time to read and use content.**	**Level A, AA, AAA**
2.2.1 Timing Adjustable	Time limits can be turned off, adjusted, or extended.	Level A
Turn off	Users can turn off time limits before encountering them.	
Adjust	Users can adjust time limits over a wide range.	
Extend	Users receive a warning before time expires, and they have the option to extend it.	
Real-Time, Essential, 20 Hours	Exceptions for real-time events, essential limits, and durations longer than 20 hours.	
2.2.2 Pause, Stop, Hide	Mechanism provided to pause, stop, or hide moving, blinking, scrolling, or auto-updating information.	Level A
Moving, Blinking, Scrolling	Mechanism for users to control such information.	

(continued)

Success Criterion	Description	Conformance Level
Auto-updating	Mechanism for users to pause, stop, hide, or control update frequency.	
2.2.3 No Timing	Timing is not a crucial component of the event or activity, except for noninteractive synchronized media and real-time events.	Level AAA
2.2.4 Interruptions	Users have the ability to postpone or suppress interruptions, with the exception of emergency interruptions.	Level AAA
2.2.5 Re-authenticating	Users can continue the activity without loss of data after re-authenticating when an authenticated session expires.	Level AAA
2.3 Seizures	**Avoid creating content in a manner that is recognized to induce seizures or physical discomfort.**	**Level A, AAA**
2.3.1 Three Flashes or Below Threshold	Flashes on web pages do not occur more than three times within a one-second period.	Level A
2.3.2 Three Flashes	Web pages refrain from featuring any element that flashes more than three times within a one-second period.	Level AAA
2.4 Navigable	**Offer methods to assist users in navigating, discovering content, and establishing their location.**	**Level A, AA, AAA**
2.4.1 Bypass Blocks	Mechanism available to bypass blocks of content repeated on multiple pages.	Level A
2.4.2 Page Titles	Titles on web pages accurately describe the topic or purpose of the content.	Level A

(*continued*)

Success Criterion	Description	Conformance Level
2.4.3 Focus Order	Focusable components receive focus in an order preserving meaning and operability.	Level A
2.4.4 Link Purpose (In Context)	The purpose of each link can be discerned either from the link text alone or in conjunction with its programmatically determined link context.	Level A
2.4.5 Multiple Ways	There are multiple ways to find a web page within a set of web pages.	Level AA
2.4.6 Headings and Labels	Headings and labels describe topic or purpose.	Level AA
2.4.7 Focus Visible	A keyboard-operable user interface displays a visible indicator for keyboard focus.	Level AA
2.4.8 Location	Information regarding the user's location within a set of web pages is accessible.	Level AAA
2.4.9 Link Purpose (Link Only)	There is a mechanism in place that enables the identification of the purpose of each link based on the link text alone.	Level AAA
2.4.10 Section Headings	Content is organized using section headings.	Level AAA

Understandable

Success Criterion	Description	Conformance Level
3.1 Readable	**Make text content readable and understandable.**	**Level A, AA, AAA**
3.1.1 Language of Page	The usual human language of each web page can be determined programmatically.	Level A
3.1.2 Language of Parts	The default language of each passage or phrase can be determined programmatically, with some exceptions.	Level AA
3.1.3 Unusual Words	Mechanism available for identifying specific definitions of unusual or restricted words or phrases.	Level AAA
3.1.4 Abbreviations	Mechanism available for identifying the expanded form or meaning of abbreviations.	Level AAA
3.1.5 Reading Level	Supplemental content or a simplified version is available when text requires advanced reading ability.	Level AAA
3.1.6 Pronunciation	Mechanism available for identifying specific pronunciation of ambiguous words.	Level AAA
3.2 Predictable	**Ensure that web pages display and function in consistent and expected manners.**	**Level A, AA, AAA**
3.2.1 On Focus	Components receiving focus should not initiate a change of context.	Level A
3.2.2 On Input	Changing settings of user interface components does not automatically cause a change of context.	Level A

(*continued*)

Success Criterion	Description	Conformance Level
3.2.3 Consistent Navigation	Navigational mechanisms repeated on multiple pages occur in the same relative order unless initiated by the user.	Level AA
3.2.4 Consistent Identification	Components with identical functionality across a set of web pages are consistently identified.	Level AA
3.2.5 Change on Request	Changes of context occur only upon user request, or there is a mechanism available to disable such changes.	Level AAA
3.3 Input Assistance	**Help users avoid and correct mistakes.**	**Level A, AA, AAA**
3.3.1 Error Identification	Input errors are automatically detected, identified, and described to the user through text.	Level A
3.3.2 Labels or Instructions	Labels or instructions are supplied when user input is necessary for the content.	Level A
3.3.3 Error Suggestion	Automated detection of input errors is in place, with correction suggestions provided unless it compromises security.	Level AA
3.3.4 Error Prevention (Legal, Financial, Data)	For pages causing legal commitments, financial transactions, or data modifications, at least one is true: Reversible, Checked, Confirmed.	Level AA
3.3.5 Help	Context-sensitive help is available.	Level AAA
3.3.6 Error Prevention (All)	For pages requiring user submissions, at least one is true: Reversible, Checked, Confirmed.	Level AAA

Robust

Success Criterion	Example	Conformance Level
4.1 Compatible	**Enhance compatibility with both existing and future user agents, encompassing assistive technologies.**	**Level A**
4.1.1 Parsing	In content implemented using markup languages, elements must have complete start and end tags, nested according to specifications, no duplicate attributes, and unique IDs (except where allowed).	Level A
4.1.2 Name, Role, Value	For user interface components (e.g., form elements, links, scripted components), name and role must be programmatically determined; user-settable states, properties, and values must be programmatically set; and changes must be notifiable to user agents, including assistive technologies.	Level A

WCAG 2.1 Success Criteria

New Success Criteria Under Perceivable Principle

Success Criterion	Description	Example	Conformance Level
1.3 Adaptable			
1.3.4 Orientation	Content should avoid restricting its view and operation to a single display orientation unless it is mandatory.	Examples: A bank check, a piano application, slides for a projector or television, or virtual reality content where orientation is not necessarily restricted to landscape or portrait display orientation.	Level AA
1.3.5 Identify Input Purpose	The purpose of each input field collecting user information must be programmatically determined when serving a purpose identified in the Input Purposes for user interface components section.	Implement the content using technologies that facilitate identifying the expected meaning for form input data.	Level AA
1.3.6 Identify Purpose	In content implemented using markup languages, user interface components, icons, and regions' purposes can be programmatically ascertained.		Level AAA

(continued)

Success Criterion	Description	Example	Conformance Level
1.4 Distinguishable			
1.4.9 Images of Text (No Exception)	Images of text should only be used for pure decoration or where a specific presentation of text is essential to the conveyed information.	Example: Logotypes (text that is part of a logo or brand name) are considered essential.	Level AAA
1.4.10 Reflow	The content can be displayed without loss of information or functionality and without the need for scrolling in two dimensions under specific conditions.	These conditions include vertical scrolling content at a width equivalent to 320 CSS pixels and horizontal scrolling content at a height equivalent to 256 CSS pixels. Exceptions are allowed for parts requiring two-dimensional layout.	Level AA
1.4.11 Nontext Contrast	The visual presentation of user interface components and graphical objects must maintain a contrast ratio of at least 3:1 against adjacent colors.	Examples: User interface components and states, graphical objects. Exceptions for inactive components or when appearance is determined by the user agent and not modified by the author.	Level AA

(continued)

621

Success Criterion	Description	Example	Conformance Level
1.4.12 Text Spacing	No loss of content or functionality should occur by setting specific text style properties in content implemented using markup languages.	Properties: Line height, spacing following paragraphs, letter spacing (tracking), word spacing. Exception for languages and scripts not utilizing certain text style properties.	Level AA
1.4.13 Content on Hover or Focus	Additional content triggered by pointer hover or keyboard focus should be dismissible, hoverable, and persistent.	Criteria: Dismissible – A mechanism is available to dismiss additional content; Hoverable – If pointer hover triggers content, the pointer can be moved over it without it disappearing; Persistent – Additional content remains visible until trigger removal, dismissal, or no longer valid. Exceptions for user agent-controlled visual presentation. Example: Custom tooltips, submenus, and other nonmodal popups.	Level AA

New Success Criteria Under Operable Principle

Success Criterion	Description	Example	Conformance Level
2.1 Keyboard Accessible			
2.1.4 Character Key Shortcuts	Keyboard shortcuts using letters, punctuation, numbers, or symbols should have at least one of the following: Turn off, Remap, or Active only on focus.	Example: A website allows users to turn off or remap keyboard shortcuts.	Level A
2.2 Enough Time			
2.2.6 Timeouts	Users should be warned of inactivity duration leading to potential data loss unless data is preserved for more than 20 hours without user actions.	Note: Privacy regulations may impact the approach, and explicit user consent might be required.	Level AAA
2.3 Seizures and Physical Reactions			
2.3.3 Animation from Interactions	Disable motion animation triggered by interaction unless it is essential for functionality or conveying information.	Example: Disabling motion animation in a mobile app unless it is crucial for conveying information.	Level AAA

(continued)

Success Criterion	Description	Example	Conformance Level
2.5 Input Modalities			
Enhance operability through various inputs beyond the keyboard.			
2.5.1 Pointer Gestures	Functionality using multipoint or path-based gestures should be operable with a single pointer unless a multipoint or path-based gesture is essential.	Note: Applies to web content interpreting pointer actions.	Level A
2.5.2 Pointer Cancellation	For functionality operable with a single pointer: no down-event execution, abort/undo options, up-event reversal, or essential completion on down-event.	Note: Applies to web content interpreting pointer actions.	Level A
2.5.3 Label in Name	For UI components with labeled text or images, the name should contain the visually presented text.	Note: Best practice is to have the label text at the start of the name.	Level A

(continued)

624

Success Criterion	Description	Example	Conformance Level
2.5.4 Motion Actuation	Functionality operated by device or user motion can be disabled, except when motion is essential or used through an accessibility-supported interface.	Example: Disabling motion-based interactions on a website to prevent accidental triggers, unless motion is essential for a specific function.	Level A
2.5.5 Target Size	The target size for pointer inputs should be at least 44 by 44 CSS pixels unless equivalent, inline, user agent controlled, or essential.	Example: Ensuring buttons or interactive elements have a size of at least 44 by 44 CSS pixels for better touch interaction.	Level AAA
2.5.6 Concurrent Input Mechanisms	Web content should not restrict the use of input modalities on a platform, except when essential, required for security, or respecting user settings.		Level AAA

New Success Criteria Under Robust Principle

Success Criterion	Description	Example	Conformance Level
4.1 Compatible			
4.1.3 Status Messages	Status messages in content implemented using markup languages should be programmatically determined through role or properties, allowing presentation by assistive technologies without receiving focus.	Example: A web page with dynamically updating content provides status messages that are programmatically determined, ensuring users with assistive technologies can be informed without focus.	Level AA

WCAG 2.2 Success Criteria

New Success Criteria Under Operable Principle

Success Criterion	Description	Example	Conformance Level
2.4 Navigable			
2.4.11 Focus Not Obscured (Minimum)	When a user interface component gains keyboard focus, it should not be entirely concealed by author-created content.	Example: A drop-down menu should not be completely obscured by other content when it receives keyboard focus.	Level AA

(continued)

Success Criterion	Description	Example	Conformance Level
2.4.12 Focus Not Obscured (Enhanced)	When a user interface component gains keyboard focus, no portion of the component should be obscured by author-created content.	Example: An interactive button should not have any part obscured by overlapping content when it receives keyboard focus.	Level AAA
2.4.13 Focus Appearance	When the keyboard focus indicator is visible, a portion of the focus indicator should adhere to specific size and contrast ratio criteria.	Example: The visible area of the keyboard focus indicator must be at least as large as the area covered by a 2 CSS pixel–thick perimeter of the unfocused component, with a contrast ratio of at least 3:1 between the corresponding pixels in the focused and unfocused states.	Level AAA

2.5 Input Modalities

Success Criterion	Description	Example	Conformance Level
2.5.7 Dragging Movements	Functionality using dragging movements should be achievable by a single pointer without dragging, unless dragging is essential or determined by the user agent.	Example: Drag-and-drop functionality should have an alternative method (e.g., buttons or keyboard controls) for users who cannot perform dragging actions.	Level AA

(continued)

Success Criterion	Description	Example	Conformance Level
2.5.8 Target Size (Minimum)	The target size for pointer inputs should be a minimum of 24 by 24 CSS pixels, with exceptions permitted for spacing, equivalent controls, inline targets, user agent control, or essential presentation.	Example: Interactive elements like buttons or links should have a minimum size of 24 by 24 CSS pixels, ensuring ease of interaction for users with varying motor abilities.	Level AA

New Success Criteria Under Understandable Principle

Success Criterion	Description	Example	Conformance Level
3.2 Predictable			
3.2.6 Consistent Help	Help mechanisms (human contact details, human contact mechanism, self-help option, fully automated contact mechanism) should occur in the same order across multiple pages within a set unless a change is user initiated.	Example: On various pages of a website, the help options (contact details, self-help, automated contact) appear consistently in the same order relative to other page content, providing a predictable experience.	Level A

(continued)

Success Criterion	Description	Example	Conformance Level
3.3 Input Assistance			
3.3.7 Redundant Entry	Information entered or provided by the user that needs to be re-entered is auto-populated or available for user selection, except in specific cases such as essential re-entry or when security is a concern.	Example: When a user fills out a form and navigates to the next step, the information they previously entered is automatically populated or available for selection, reducing the need for redundant data entry.	Level A
3.3.8 Accessible Authentication (Minimum)	Cognitive function tests (e.g., password recall or puzzle-solving) are not required in authentication unless alternative methods or mechanisms to assist the user are provided.	Example: Instead of relying solely on password recall, users have alternatives like using password managers or copy-pasting passwords, minimizing the cognitive burden associated with authentication.	Level AA

(continued)

Success Criterion	Description	Example	Conformance Level
3.3.9 Accessible Authentication (Enhanced)	Similar to the minimum level, cognitive function tests are not required unless alternatives or mechanisms are provided, but this criterion is enhanced to Level AAA.	Example: In addition to alternative authentication methods, enhanced mechanisms, such as advanced support for password managers or personalized recognition of nontext content, further improve accessibility.	Level AAA

Taxonomies: Action Verbs

© Ankita Jiyani Mangtani 2024
A. J. Mangtani, *Instructional Design Unleashed*, Design Thinking,
https://doi.org/10.1007/979-8-8688-0416-8

Action Verbs: Bloom's Taxonomy

Remember	Understand	Apply	Analyze	Evaluate	Create
define	explain	solve	analyze	reframe	design
identify	describe	apply	compare	criticize	compose
describe	interpret	illustrate	classify	evaluate	create
label	paraphrase	modify	contrast	order	plan
list	summarize	use	distinguish	appraise	combine
name	classify	calculate	infer	judge	formulate
state	compare	change	separate	support	invent
match	differentiate	choose	explain	compare	hypothesize
recognize	discuss	demonstrate	select	decide	substitute
select	distinguish	discover	categorize	discriminate	write
examine	extend	experiment	connect	recommend	compile
locate	predict	relate	differentiate	summarize	construct
memorize	associate	show	divide	assess	develop
quote	contrast	sketch	order	choose	generalize
recall	convert	complete	prioritize	convince	integrate
reproduce	demonstrate	construct	survey	defend	modify
tabulate	estimate	dramatize	calculate	estimate	organize
tell	express	interpret	conclude	grade	prepare
copy	identify	manipulate	correlate	measure	produce
discover	indicate	paint	deduce	predict	rearrange
duplicate	infer	prepare	devise	rank	rewrite
enumerate	relate	teach	diagram	score	adapt
listen	restate	act	dissect	select	anticipate
observe	select	collect	estimate	test	arrange
omit	translate	compute	evaluate	argue	assemble
read	ask	explain	experiment	conclude	choose
recite	cite	list	focus	consider	collaborate
record	discover	operate	illustrate	critique	facilitate
repeat	generalize	practice	organize	debate	imagine
retell	group	simulate	outline	distinguish	intervene
visualize	illustrate	transfer	plan	editorialize	make
	judge	write	question	justify	manage
	observe		test	persuade	originate
	order			rate	propose
	report			weigh	simulate
	represent				solve
	research				support
	review				test
	rewrite				validate
	show				
	trace				

Action Verbs: Krathwohl's Taxonomy

Receive	Respond	Value	Organization	Characterization
ask	answer	complete	adhere	act
choose	assist	demonstrate	alter	discriminate
describe	aid	differentiate	arrange	display
follow	compile	explain	combine	influence
give	conform	follow	compare	listen
hold	discuss	form	complete	modify
identify	greet	initiate	defend	perform
locate	help	join	formulate	practice
name	label	justify	generalize	propose
select	perform	propose	identify	qualify
reply	practice	read	integrate	question
use	present	share	modify	revise
	read	study	order	serve
	recite	work	organize	solve
	report		prepare	verify
	select		relate	use
	tell		synthesize	
	write			

Action Verbs: Simpson's Taxonomy

Perception	Set	Guided Response	Mechanism	Complex Overt Response	Adaptation	Origination
choose	begin	assemble	It is the same list as a guided response with the skill being demonstrated independently and at a higher level of proficiency.	It is the same list as a guided response with greater skill, independence, and fluidity.	adapt	arrange
describe	display	build			alter	combine
detect	explain	calibrate			change	compose
differentiate	move	construct			rearrange	construct
distinguish	proceed	dismantle			reorganize	create
identify	react	display			revise	design
isolate	respond	dissect			vary	originate
relate	show	manipulate				
select	start	measure				
separate	volunteer	organize				

Action Verbs: Dave's Taxonomy

Imitation	Manipulation	Precision	Articulation	Naturalization
attempt	act	achieve automatically	adapt	create
copy	build	excel expertly	construct	design
imitate	execute	perform masterfully	combine	develop
mimic	perform	demonstrate skillfully	create	invent
follow	complete	calibrate perfectly	customize	manage naturally
repeat	accomplish		modify	or perfectly
duplicate	follow		formulate	
replicate	play		alter	
reproduce	produce		originate	

Action Verbs: Harrow's Taxonomy

Reflex Movement	Basic Fundamental Movements	Perceptual	Physical Activities	Skilled Movements	Non-discursive Communication
to flex	to crawl	to catch	to endure	to waltz	to gesture
to stretch	to creep	to bounce	to improve	to type	to stand
to straighten	to slide	to eat	to increase	to play the piano	to sit
to extend	to walk	to write	to stop	to plane	to express facially
to inhibit	to jump	to balance	to start	to file	to dance skillfully
to lengthen	to run	to bend	to move precisely	to skate	to perform skillfully
to shorten	to grasp	to draw from memory	to touch	to juggle	to paint skillfully
to tense	to reach	to distinguish by touching	to bend	to paint	to play skillfully
to stiffen	to tighten	to explore		to dive	
to relax.	to support			to fence	
	to handle			to golf	
				to change	

APPENDIX D

Storyboard Templates

Template 1

Slide Title	
Audio Script/Narration	**On-Screen Content (Text and Multimedia)**
Text	Text **Insert rows as needed.**
Interactivity and Development Notes	
Text **Insert a separate table for a new section.**	

© Ankita Jiyani Mangtani 2024
A. J. Mangtani, *Instructional Design Unleashed*, Design Thinking,
https://doi.org/10.1007/979-8-8688-0416-8

Template 2

Program/Course/Module Name:

Section Name:

Page No. -	Lesson/Topic No.
Page Title - Welcome	Lesson/Topic Name

On-Screen Visuals/Text

Graphics Description

Audio Script/ Voice Over

Interactivities

<A description of the Interactivities>
Notes to Developers/Programmers:
<Specific instructions to developers, if any>

Insert a separate table for a new section.

Template 3

Section Name:			
Screen #	**Voice-Over Script/ Narration**	**On-Screen Assets**	**Slide Treatment**
	Insert rows as needed. Insert a separate table for a new section.		Notes to Developers/Programmers: <Specific instructions to developers, if any>

Template 4

Lesson/Section Title:	Slide ID/#:	Programming and Interactivity Notes:
On-Screen Elements:		Insert a separate table for a new section.
Audio/Narration:		Accessibility Notes:
Reviewer Comments:		

Template 5

Slide No:		Title:		Time Duration:

Developer/Programmer Notes:	On-Screen Text:	Audio Script:

Insert a separate table for a new section.

Media Elements:

Image:

Audio:

Video:

Music/SFX:

Interaction:	Quiz:
None *[if Interaction, insert the name of **.intr** here]*	None *[if Quiz, insert the name of **.quiz** here]*

Branching:	Advance:
Next:	By User
Prev:	Automatic:

Template 6

Slide Overview	Slide Mockups	Slide Notes:
Description:		Development Notes:
Duration:		
Next Slide:		
Advanced Actions?:		Accessibility Notes:
Other:		

Multimedia Assets	Scripting

Template 7

Project Title		Module Title		Developer Notes	
Screen #		Screen Type		Screen Title	

Visuals (Video/Pics/Characters)	On-Screen Text
	Animation & Interactivity

Developer Notes

Reviewer Notes

Audio

Navigation/ Branching		Additional Notes	

Index

© Ankita Jiyani Mangtani 2024
A. J. Mangtani, *Instructional Design Unleashed*, Design Thinking,
https://doi.org/10.1007/979-8-8688-0416-8

D

Z